THE BIOCHEMISTRY
OF THE TISSUES

2nd Edition

THE BIOCHEMISTRY of the TISSUES

2nd Edition

P. Banks
*Senior Lecturer in Biochemistry,
University of Sheffield*

W. Bartley
*Professor of Biochemistry,
University of Sheffield*

L. M. Birt
*Vice Chancellor,
University of Wollongong,
N.S.W., Australia*

JOHN WILEY & SONS

Chichester · New York · Brisbane · Toronto

Copyright © 1968, 1976, by John Wiley & Sons Ltd.

Reprinted May 1978
Reprinted January 1979

All rights reserved.

No part of this book may be reproduced by any means, nor transmitted, nor translated into a machine language without the written permission of the publisher.

Library of Congress Cataloging in Publication Data:

Banks, Peter.
   The biochemistry of the tissues.

   First ed. by W. Bartley, L. M. Birt, and P. Banks.
   Includes bibliographies.
   1. Biological chemistry. 2. Metabolism.
3. Human physiology. I. Bartley, Walter, joint author.
II. Birt, Lindsay Michael, joint author. III. Title.

QP514.2.B35   1976   612'.015   75-26739
ISBN 0 471 05471 2 (Cloth)
ISBN 0 471 019232 (Pbk)

Printed in Great Britain by Spottiswoode Ballantyne Limited
Colchester and London

This book is dedicated to
Professor Sir Hans Krebs, F.R.S.
by three of his grateful students

Preface

This book is based on a course of lectures given to medical students at Sheffield in preparation for the 2nd M.B. examination. Although it is primarily intended for medical students, it will probably be of use also to students of physiology and pharmacology, as well as being a supplementary book for students of biochemistry who are interested in the more physiological aspects of the subject. It presupposes a knowledge of chemistry somewhat more than 'A' level and may require reference to a standard textbook, for example *Organic Chemistry* by Fieser and Fieser and *Textbook of Biophysical Chemistry* by West.

The book is intended as an introductory text rather than a comprehensive treatment of the subject and although we feel it may be of interest to the medical practitioner it will require supplementation by a text of clinical biochemistry to meet postgraduate needs.

Although the length of the book and the topics discussed have been limited by the needs of the 2nd M.B. course we have endeavoured in all cases to relate biochemical pathways to the functioning of some specific tissue or to the whole organism. Our experience suggests that students view biochemistry as something separate from the cells themselves. We have tried to show that the physiological and structural properties of cells and tissues follow from the metabolism which provides their energy, the materials from which they are made and the control required for their proper functioning. In this way we hope we have moved from the consideration of biochemistry as a series of arbitrary chemical reactions to an attitude in which biochemistry is another viewpoint of the total anatomy and physiology of the living organism.

W. BARTLEY
L. M. BIRT
P. BANKS

Preface to Second Edition

This book has grown somewhat since the first edition. This is partly due to the request that some basic chemistry should be added, partly due to the inclusion of the biochemistry of several organs such as the skin, lung and kidney, and partly by the sheer growth of biochemical knowledge. Reproduction and development have been dealt with much more fully and there is now a chapter on the immune response. Several of the metabolic pathways, for example those of purine and pyrimidine synthesis, have been dealt with in much greater detail. On the other hand, an attempt has been made to describe the integrated metabolism of the body in broad terms, bearing in mind the anatomical relationships of the tissues. Finally, we have attempted to look at the whole body's response to the situation of overfeeding and starvation and the peculiar metabolism of alcohol and sucrose.

P. BANKS
W. BARTLEY
L. M. BIRT

Acknowledgements

We are indebted to Dr W. Ferdinand and Dr C. W. Potter for reading parts of the manuscript and for making helpful suggestions. We would also like to thank Mrs E. M. Wilson and Mrs R. Tonks for their expert typing of the several drafts, Mr J. Kugler for supplying electron micrographs and Miss L. B. Young for drawing many of the figures.

P.B., W.B., L.M.B.

Contents

Introduction 1

SECTION 1   CHEMISTRY AND LIFE

Chapter 1	The Origins of Life.	6
Chapter 2	The Properties of Atoms and Molecules	9
Chapter 3	Bioenergetics	38
Chapter 4	Water	52
Chapter 5	Proteins	59
Chapter 6	Enzymes	74
Chapter 7	The Quantitative Man	88
	References	90

SECTION 2   MUSCLE

Chapter 8	Muscle Contraction and Relaxation	93
Chapter 9	Glycolysis	104
Chapter 10	The Citric Acid Cycle	119
Chapter 11	The Respiratory Chain and Oxidative Phosphorylation .	135
Chapter 12	Special Metabolism of Muscle Types.	148
	References	156

SECTION 3   THE INTESTINE

Chapter 13	The Digestion and Absorption of Food	159
Chapter 14	Protein Synthesis	168
Chapter 15	Purine and Pyrimidine Synthesis	185
	References	195

SECTION 4   THE LIVER

Chapter 16	Carbohydrate Metabolism	199
Chapter 17	Nitrogen Metabolism	213
Chapter 18	Nitrogen Excretion	218
Chapter 19	Detoxication	225
Chapter 20	Fat Metabolism	230
Chapter 21	Synthesis of Phospholipids and Cholesterol	236
	References	245

SECTION 5 ADIPOSE TISSUE

Chapter 22 The Biochemistry of Adipose Tissue 249
References 260

SECTION 6 SKIN AND STRUCTURAL TISSUE

Chapter 23 The Biochemistry of Skin and the Structural Tissues . 263
References 274

SECTION 7 THE BLOOD

Chapter 24 Red Blood Cells 277
Chapter 25 White Blood Cells 288
Chapter 26 Plasma 290
Chapter 27 The Metabolism of Iron 294
Chapter 28 Specific Immune Responses 297
References 303

SECTION 8 THE LUNGS

Chapter 29 The Biochemistry of the Lungs 307
References 310

SECTION 9 THE KIDNEYS

Chapter 30 The Biochemistry of the Kidneys 313
References 322

SECTION 10 THE NERVOUS SYSTEM

Chapter 31 General Metabolism of the Brain 325
Chapter 32 Membrane Potentials and Ion Transport 333
Chapter 33 Secretion of Neurotransmitters 340
Chapter 34 Functional Aspects of Neuronal Biochemistry . . . 345
References 349

SECTION 11 THE EYE

Chapter 35 The Biochemistry of the Eye 353
References 360

SECTION 12 REPRODUCTION AND DEVELOPMENT

Chapter 36 The Biochemistry of Sex 363
Chapter 37 Foetal and Perinatal Biochemistry 373

Chapter 38	The Control of Fertility	378
Chapter 39	The Metabolism of the Mammary Gland	379
Chapter 40	Genetic Defects	383
Chapter 41	Terminal Differentiation	390
Chapter 42	Senescence	393
	References	397

SECTION 13 HORMONES AND ENDOCRINE TISSUES

Chapter 43	General Biochemistry of Hormones	401
Chapter 44	The Adrenal Cortex	404
Chapter 45	The Adrenal Medulla	410
Chapter 46	The Thyroid	416
Chapter 47	Pancreatic Hormones	421
	References	427

SECTION 14 THE CONTROL OF METABOLIC PROCESSES IN CELLS

Chapter 48	The Control of Metabolic Processes in Cells	429
	References	449

SECTION 15 TISSUE INTERRELATIONSHIPS

Chapter 49	Tissue Interrelationships	453

SECTION 16 SELECTED TOPICS IN HUMAN METABOLISM

Chapter 50	Obesity and Starvation	465
Chapter 51	Alcohol and Sucrose Consumption	470
	References	473
Index		475

Introduction

Biochemistry is a subject which has now penetrated into all other biological disciplines, partly because the exploration of life often requires biochemical techniques. The use of these techniques shows that life in all its different forms tends to use the same chemical devices for achieving its goals.

Although this book is concerned with the biochemistry of the human, it must not be supposed that this is something very different from that of other species. Most of the major metabolic pathways described are common to all mammals and, a great number of them, to all life.

Biochemistry is an attempt to describe life in chemical terms, but it does not stop at the simple statement of the analysis of the living material and of the various reactions whereby the components of life are synthesized and degraded. It is also concerned with how the interlocking systems of chemical reactions are controlled, so that there is growth in infancy, maintenance in maturity, and degeneration in old age.

Besides the chemical pathways, the biochemist is much concerned with the rate at which the chemical changes take place. Probably life, as we understand it, is possible only because reactions which are inherently slow can be made fast enough to supply energy at a comparatively low temperature. The catalysis of the reactions is brought about by special proteins called enzymes. These are simply catalysts that happen to be proteins, and therefore their properties reflect their protein nature. Although enzymes are very important they require the cooperation of other molecules before they can act as controllers of the chemistry of life, and the study of the flow of chemical information to enzymes, and their response to this information, is becoming an increasing part of biochemistry.

Apart from the normal chemical pattern of life, there is the disordered biochemical pattern of disease. Ultimately all diseases can be considered as reflections of a change in the normal biochemistry of the organism, which produces sufficient dysfunction to make the organism markedly inefficient. Changes in the rates of comparatively few metabolic reactions can produce changes in the functioning of the organism, since ultimately all the body's metabolic reactions are interconnected and interdependent. Changes in the rates of metabolic reactions result from some change in function or quantity of an enzyme. Measurements of enzyme activity are now therefore important tests in medicine, supplementing the more usual tests of the gross quantities of inorganic and organic substances in blood, urine and other body fluids. Enzymes may slowly change as a result of chronic disorders, or there may be abrupt changes associated with an acute condition such as myocardial infarction. The pattern of enzymes in a particular tissue is characteristic of the tissue and so, on damage, for example to the heart muscle, the tissue will liberate its characteristic pattern of enzymes into the blood stream, thus altering the normal pattern of enzymes characteristic of the blood. It is now possible in some cases, by studying the pattern of enzymes in the blood, to distinguish which of the organs is damaged. Further, the change in the pattern of enzymes with time gives a prognosis of the course of the disease since, when the lesion heals, tissue enzymes are no longer liberated into the blood.

Biochemistry is not therefore simply a subject which must be learned on the road to medical qualification; it is a subject of increasing use and relevance in the practice of medicine. It is the aim of this book to illustrate the relevance of biochemistry to the understanding of the dynamic pattern of chemical reactions that we call 'The Human'.

SECTION 1

Chemistry and Life

The distinction between living and non-living objects is so profound that whilst we sometimes think of ourselves as animals we seldom consider that we are objects of geochemical origin. It is now widely held that life did not exist on Earth at the time of its formation from interstellar dust some 4.5×10^9 years ago. The general consensus of opinion seems to be that during the first 1.5×10^9 years of the Earth's history a succession of novel chemical systems developed and that some of them became differentiated from non-living matter. Despite this differentiation, biochemical and geochemical evolution have always been intimately connected. For example, the production of oxygen from water by the photosynthetic activity of primitive plants led to the replacement of the early reducing atmosphere containing methane, ammonia and water by one of great oxidizing potential. A geochemical consequence of this was that iron ores were deposited in the ferric form rather than in the ferrous as had been the case under the earlier reducing conditions. On the biological front, the appearance of oxygen in the atmosphere resulted in the evolution of aerobic forms of metabolism to supplement and extend the anaerobic forms that had developed under the reducing conditions of the earlier periods. Furthermore, oxygen entering the upper atmosphere was converted by irradiation into ozone which absorbed much of the ultraviolet light that formerly reached the surface of the Earth. This attenuation of radiation, which is capable of destroying important biochemicals, eventually allowed organisms to colonize the land. Connections between the living and non-living states of matter still persist on a huge scale in those processes that have been termed the carbon, nitrogen, oxygen, sulphur and phosphorus cycles. For example about 6×10^{10} tons of carbon are fixed by green plants each year. The magnitude of the interactions between bio- and geo-chemical systems in the past can be gauged by the sizes of those geological deposits such as coal, oil, limestone and atmospheric oxygen formed as a result of biological activity.

During the course of prebiotic evolution it is probable that many more varieties of molecule were produced than ultimately found a role in biological systems. In other words, a period of chemical selection preceded the emergence of truly living systems. This process was responsible for establishing the underlying biochemical unity that characterizes all known living organisms. Once this unity had been established, a phase of morphological evolution produced untold variations on the common chemical theme. These variations are evident in the species of the fossil record and in the surviving flora and fauna.

With the emergence of man, the process of morphological evolution reached a stage of complexity in which matter in the living state could speculate upon the mechanism of its own formation and organization.

CHAPTER 1

The Origins of Life

In recent years there has been an upsurge of interest in the problem of how and when living organisms first made their appearance on this planet. Although the fossil record is sparse before the Cambrian Period (about 8×10^8 years ago), there is now some fossil evidence that bacteria-like organisms existed as long as 3×10^9 years ago. This evidence is supported by the presence in the same sedimentary formations as the earliest fossils of some hydrocarbons which probably had a biological origin. Assuming that the original forms of life did not arrive from outer space, a period of not more than $1 \cdot 5 \times 10^9$ years must have been sufficient for living systems to evolve from non-living ones on Earth. In order to understand how complex organic compounds might have been formed during the early history of the Earth, it is necessary to know what sources of energy and what reactants were available. It is now generally believed that by the time the Earth was formed its atmosphere consisted of H_2O, NH_3 and CH_4. A little later CO and CO_2 were probably added from the Earth's crust and there may have been some loss of NH_3 and CH_4. For these materials to react together to form more complex molecules, there would need to be sources of energy to drive the syntheses and means for removing the products from the sites of reaction to safer regions, where they could persist long enough to undergo further reactions such as polymerization. The sources of energy for driving chemical syntheses on the primitive Earth were probably ultraviolet radiation from the Sun, radioactivity of ^{40}K, vulcanism, meteorite shock waves and electrical discharges. Solar radiation, of course, still remains the chief source of energy for syntheses in the biosphere although its action spectrum has shifted, since the prebiotic era, from the ultraviolet to the visible region. Recently experiments have been performed to discover whether energy inputs of the types listed above can induce the synthesis of complex compounds in gas mixtures similar to those believed to have existed in the primitive atmosphere. In these experimental systems, means have been provided to remove any newly formed molecules from the point of synthesis so as to mimic their removal from the atmosphere by rain or from sites of radioactivity or vulcanism by diffusion. Under such experimental conditions a remarkable range of compounds has been formed, including sugars, fatty acids, amino acids, purines and porphyrins. Furthermore, some of these substances can polymerize in aqueous solutions to form peptides, nucleosides and nucleotides in the presence of dehydrating agents such as carbodiimide which can also be formed in 'primitive atmosphere' experiments.

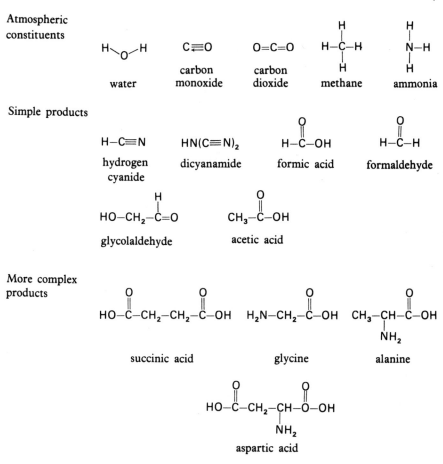

Figure 1.1 Components of the primitive atmosphere and products derived from them in 'primitive earth' experiments

How complex systems similar to those of modern cells might have developed from such simple beginnings is still obscure, but already interesting ideas and experimental data concerning the formation of polypeptides of non-random amino acid sequence are available. Moreover, the way in which such non-random sequences of amino acids may first have become encoded in the base sequences of polynucleotides is considered by some workers to be a reasonable subject for experimental study. Mechanisms whereby small droplets of the 'primitive soup' of pre-biochemicals may have become invested with membranes to form protocells can be imagined and these have already led to experiments involving lipid films and other two-phase systems.

At the present time there appears to be no reason for believing that living organisms arose other than by the slow development of more and more complex systems from the components of the primitive atmosphere and seas.

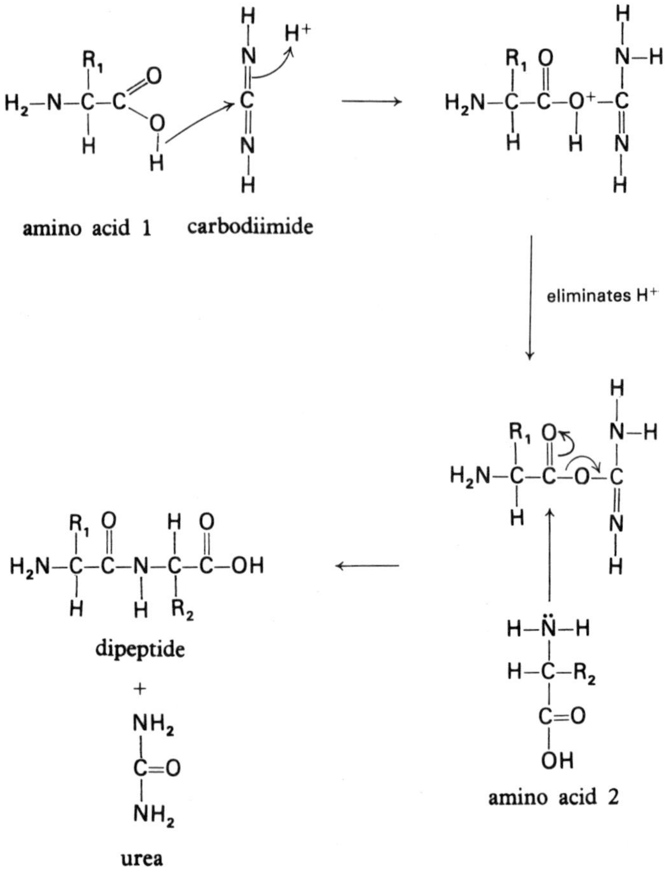

Figure 1.2 Carbodiimide facilitating the formation of peptide bonds

Living matter is characterized by the variety and complexity of its molecular components; these form a multitude of discrete polyphasic systems or organisms, all of which normally have a potentiality for growth, reproduction, short-term adaptation and evolutionary development.

In order to understand the relationship between the structures and activities of organisms (that is in order to correlate their anatomy and physiology) it is necessary to consider the properties of their component molecules. Where this can be done in detail, it appears that structure and function at the physiological and anatomical levels arise entirely from the properties and interactions of the molecules constituting the macroscopic assemblage.

CHAPTER 2

The Properties of Atoms and Molecules

In order to understand the structure of molecules and the processes involved in their formation and breakdown (behaviour central to the existence of living systems) it is necessary to know how atoms form stable linkages with one another.

Atoms are composed of a compact central nucleus surrounded by a cloud of electrons. The nucleus is formed from protons, each carrying one unit of positive charge, and a variable number of uncharged neutrons. The number of protons is equal to both the atomic number of the element and to the number of satellite electrons. Each electron carries one unit of negative charge so that the atom as a whole is uncharged. The masses of protons and neutrons are almost identical and very much greater than that of electrons, hence the mass of an atom is concentrated in its nucleus. The volume and the chemistry of atoms are, on the other hand, very largely determined by the size and properties of their electron clouds. To a first approximation the electrons can be considered to occupy discrete, well defined, regions of space around the nucleus termed orbitals. These have been given letters to designate their shape and numbers to indicate their relative extension from the nucleus. Each orbital is filled when it contains two electrons. Orbitals closest to the nucleus have lower energies than those further away since energy is required to move a negatively charged electron away from a positively charged nucleus. The periodic table of the elements is an expression of the tendency of electrons to enter the orbital having the lowest energy level. The energy levels of the orbitals closest to the nucleus are in the order of $1s < 2s < 2p < 3s < 3p < 4s < 3d < 4p \ldots$, hence the electronic configuration of the lightest elements of the periodic table is as shown in Figure 1.4. The filling of the 3d orbitals after 4s but before 4p is responsible for the existence of the closely related transitional elements of the first long period (i.e. Sc ... ... Zn.)

The inert gases are characterized by having all their outermost s and p orbitals filled with electron pairs, i.e.

$$\text{He} \qquad 1s\ \uparrow\downarrow$$

$$\text{Ne}\ \ 1s\ +\ 2s\ \uparrow\downarrow\ \ 2p\ \overset{x}{\uparrow\downarrow}\ \overset{y}{\uparrow\downarrow}\ \overset{z}{\uparrow\downarrow}$$

$$\text{A}\ \ 1s + 2s + 2p_{xyz} + 3s\ \uparrow\downarrow\ \ 3p\ \uparrow\downarrow\ \uparrow\downarrow\ \uparrow\downarrow$$

Atoms without an inert-gas configuration can obtain one and so gain stability by sharing electron pairs and thus forming bonds. For example, carbon has a total of only four electrons in its outer s and p orbitals. If it is to obtain the stable

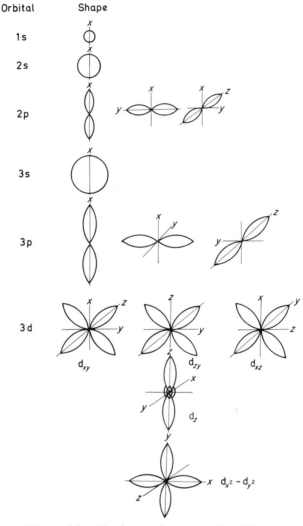

Figure 1.3 The forms of some atomic orbitals

electronic configuration of neon, it requires four more electrons. Hydrogen with one electron in its 1s orbital, requires one more to obtain the configuration of helium. In forming methane, carbon attains the stable configuration of neon.

$$\overset{\times}{\underset{\times}{\times}} C \times + 4\ H\cdot \longrightarrow H \overset{\times\,\cdot}{\underset{\cdot\,\times}{\times}} C \overset{\times}{\underset{}{\times}} H$$
$$\ H$$

Similarly in carbon tetrachloride, chlorine with 7 electrons in its outer s and p orbitals attains an argon-like structure whilst carbon obtains shares in eight electrons to achieve a neon-like configuration. The molecular orbitals formed

Hydrogen	(H)	1s ↑							
Helium	(He)	1s ↑↓							Shell 1 full
Lithium	(Li)	1s ↑↓	2s ↑						
Beryllium	(Be)	1s ↑↓	2s ↑↓		x	y	z		
Boron	(B)	1s ↑↓	2s ↑↓	2p	↑				
Carbon	(C)	1s ↑↓	2s ↑↓	2p	↑	↑			
Nitrogen	(N)	1s ↑↓	2s ↑↓	2p	↑	↑	↑		
Neon	(Ne)	1s ↑↓	2s ↑↓	2p	↑↓	↑↓	↑↓		Shell 2 full

Figure 1.4 The electronic configuration of some of the elements in the first two periods of the periodic table. The symbol ↑↓ indicates that when two electrons occupy the same orbital they must differ in spin

when atomic orbitals overlap to form bonds lie in specific directions relative to one another. This means that molecules have characteristic shapes and exhibit various forms of symmetry. Anatomy thus begins at the atomic and molecular levels.

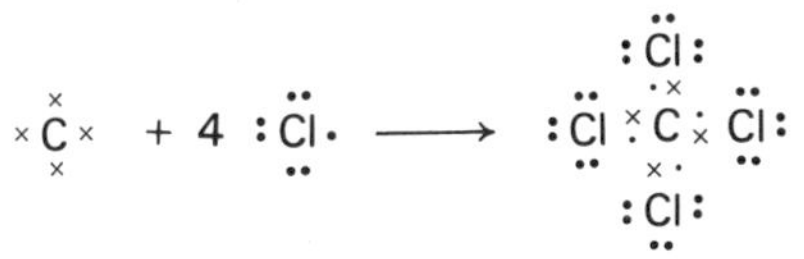

$$\overset{\times}{\underset{\times}{C}}\overset{\times}{} + 4\ :\!\ddot{C}\!l\cdot \longrightarrow \begin{array}{c} :\ddot{C}l: \\ :\ddot{C}l\overset{\times}{\underset{\times\cdot}{}}\!C\overset{\times}{\underset{\cdot}{}}\!\ddot{C}l: \\ :\ddot{C}l: \end{array}$$

Let us look again at the structure of methane. The single 2s and the three 2p orbitals of carbon can hybridize to form four identical sp^3 hybrid orbitals, each of which is directed towards the corner of a regular tetrahedron. Each hybrid orbital, containing an unpaired electron, overlaps with the 1s orbital of a hydrogen atom (also containing a single electron) to form a σ bond containing a pair of electrons. Ethane contains a carbon–carbon bond formed by the overlap of two sp^3 hybrid orbitals. The bond resulting from this union is also called a σ bond.

1s	H 1							He 2	
2s + 2p	Li 3	Be 4	B 5	C 6	N 7	O 8	F 9	Ne 10	
3s + 3p	Na 11	Mg 12	Al 13	Si 14	P 15	S 16	Cl 17	A 18	
4s + 3d + 4p ↑	K 19	Ca 20	Ga 31	Ge 32	As 33	Se 34	Br 35	Kr 36	
	Sc 21	Ti 22	V 23	Cr 24	Mn 25	Fe 26	Co 27	Ni 28	Cu 29	Zn 30

Figure 1.5 The first four periods of the periodic table

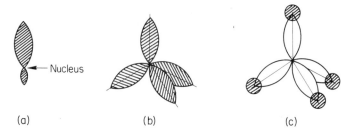

Figure 1.6 sp³ Hybrid orbitals of carbon and the structure of methane. (a) A single sp³ hybrid orbital. (b) Four sp³ orbitals distributed tetrahedrally about the nucleus of a carbon atom. (c) Methane showing overlap between sp³ hybrid orbitals of carbon and 1s orbitals of hydrogen

If a carbon atom is bonded to four different atoms or groups the resultant molecule can exist in two forms related to one another as mirror images. The chemical properties of these two forms are identical but they can be distinguished by their ability to rotate the plane of polarized light in opposite directions. Consequently these two forms of the same compound are called optical isomers and are

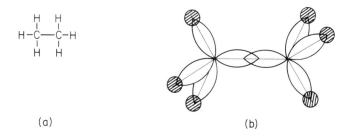

Figure 1.7 The structure of ethane. (a) Ethane. (b) Ethane showing overlap between two sp³ hybrid orbitals to form a σ carbon–carbon bond

designated by the signs (+) and (−) depending upon whether they rotate the plane of polarized light to the right or to the left. The absolute configurations of + and − glyceraldehyde are as shown below.

The carbon atom carrying the four different groups is termed an asymmetric centre. Proteins are composed of many amino acids all of which are optically active except for glycine ($CH_2NH_2 \cdot COOH$) which has no asymmetric centre. The optically active amino acids in proteins belong to what is called the L-series because they are configurationally related to (−) glyceraldehyde. Amino acids related to (+) glyceraldehyde (the D-series) are not found in proteins. The terms D and L applied to amino acids and sugars indicate that the compound in question is related configurationally to either (+) or (−) glyceraldehyde; they do *not* indicate the optical activity of the compound, that is given by the signs (+) and (−). Individual members of the L series of amino acids may be dextrorotatory (+) or laevorotatory (−). Thus the form of alanine found in proteins is L(+) alanine, that is, it has the configuration of (−) glyceraldehyde, i.e. (L) but is dextrorotatory, i.e. (+)

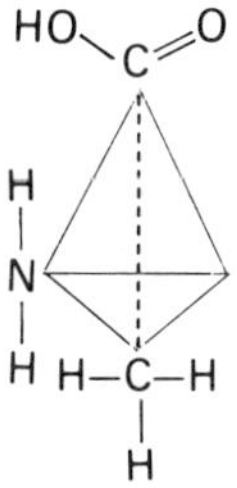

L(+) alanine

Rotation is possible about a σ bond and is important in allowing molecules to assume different conformations; for example, succinic acid can exist in the eclipsed and staggered conformations shown below.

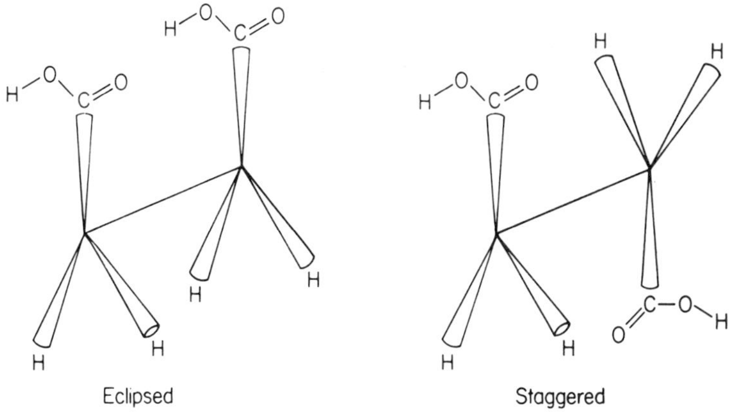

Eclipsed   Staggered

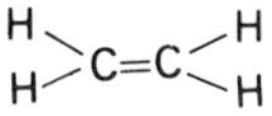

The staggered conformation is the more stable because steric interactions between the two carboxyl groups are prevented. Sometimes double or triple bonds are formed between atoms and again both partners obtain shares in eight electrons. In ethylene the 2s and two of the 2p orbitals of each carbon atom hybridize to form three coplanar, sp² hybrid orbitals and one 2p orbital remains unchanged. The bonds between the two carbon atoms are formed by the 2 sp² hybrid orbitals overlapping to give a σ bond and the two unmodified p orbitals forming a pi (π) bond. Carbon–hydrogen σ bonds are formed by the overlap of sp² hybrid orbitals of carbon and 1s orbitals of hydrogen. Free rotation about double bonds is not

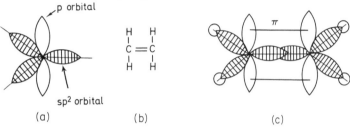

Figure 1.8 The structure of ethylene. (a) Carbon atom with 3 coplanar sp² hybrid orbitals and one p orbital. (b) Ethylene. (c) Ethylene showing overlap of two sp² orbitals and overlap of two p orbitals to form σ and π molecular orbitals respectively

possible so that maleic and fumaric acids are two quite distinct chemical compounds called geometrical isomers. Another important type of bonding is that en-

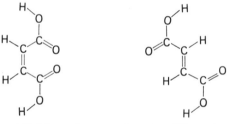

maleic acid (*cis* isomer)   fumaric acid (*trans* isomer)

countered in aromatic ring systems such as benzene. In benzene six carbon atoms, all sp² hybrids, are bonded by 6 σ and 3 π bonds. Since the six carbon

atoms of benzene are all equivalent, each π orbital is not associated with any particular pair of carbon atoms but is delocalized above and below the plane of the ring thus:

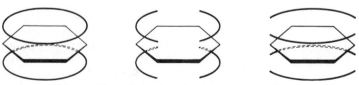

The positions of the three bonding π orbitals of benzene

As these orbitals are not easily represented benzene is often drawn thus ⌬ to indicate the delocalization of the six electrons. Electron delocalization reduces electron–electron repulsion and hence, by reducing intramolecular strain, tends to stabilize molecules. This type of stabilization is often referred to as resonance stabilization.

The shared electrons in a carbon–carbon σ bond similar to that found in ethane are located, on average, midway between the nuclei of the two carbon atoms. However, when two different atoms are bonded together, the shared electrons tend to be nearer to one of the pair than the other. Oxygen with its 6 nuclear protons has a greater affinity for electrons than carbon with its 4 nuclear protons. Hence in glyceraldehyde

$$\begin{array}{c} H\diagdown\!\!\!\diagup O \\ C \\ | \\ H-C-OH \\ | \\ H-C-OH \\ | \\ H \end{array}$$

the electrons in the carbonyl double bond will be displaced towards the oxygen thus

$$\overset{\delta+}{C} \; \overset{\delta-}{:O}$$

As a result of this tendency to attract electrons, oxygen is said to be more electronegative than carbon. Asymmetries in electron distribution are responsible for the existence of dipoles within molecules and molecules containing dipoles are said to be polar. If a molecule does not possess a permanent dipole it is termed non-polar or apolar. Permanent irregularities of charge distribution in molecules, and those induced transitorily by the presence of polar groups in nearby molecules, are important in reaction mechanisms and in the self-assembly of macromolecular structures.

Reaction Mechanisms

Now that we have discussed the way in which molecules are held together by covalent bonds, it is necessary to learn something about the way in which such bonds are made and broken. A few examples taken from reactions occurring in living organisms will suffice to illustrate the general principles involved.

Substitution reactions frequently occur in biochemical systems. In these processes single bonds are made and broken as a new or incoming group (I group) displaces the original or leaving group (L group). An example of such a reaction is the hydrolysis of an acetyl thioester, acetyl coenzyme A, by water. The biochemical role of coenzyme A will be discussed later.

The C=O group in acetyl thioesters is polarized because the electronegative oxygen atom is able to attract electrons away from the carbon atom. The latter acquires, in consequence, a partial positive charge. This electrophilic site can now be attacked by a nucleophilic reagent such as $:\overline{O}H$ having an electron pair to

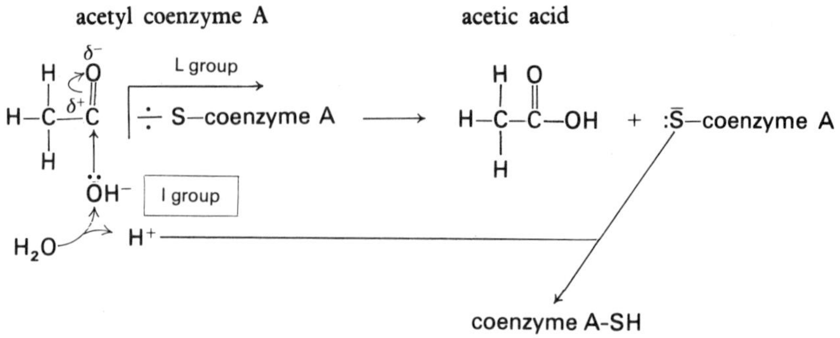

acetyl coenzyme A　　　　　　　　**acetic acid**

[reaction scheme: H-C(H)(H)-C(δ+)=O(δ−) with L group :S-coenzyme A, attacked by :ÖH⁻ (I group), H₂O → H⁺, giving H-C(H)(H)-C(=O)-OH + :S̄-coenzyme A → coenzyme A-SH]

Figure 1.9 Hydrolysis of acetyl coenzyme A. A nucleophilic group (:ŌH) attacks an electrophilic carbon displacing the leaving group :S̄-coenzyme A (see Figure 2.19 for structure of coenzyme A)

donate. As the incoming nucleophile approaches the electrophilic carbon atom, the thiolate ion (:S̄—coenzyme A) leaves. Other examples of nucleophilic reagents encountered in biochemistry are uncharged compounds with pairs of unshared electrons such as NH_3, H_2O and coenzyme A—SH; carbanions such as that derived from acetyl coenzyme A (namely

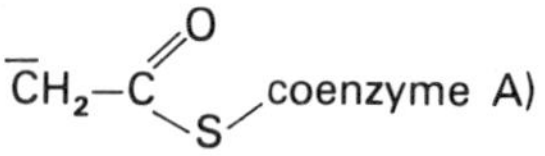

$$\bar{C}H_2-C\underset{S}{\overset{O}{\lessgtr}}\text{coenzyme A})$$

and anions such as phosphates.

Addition reactions are also frequently encountered in metabolic pathways; an example is the conversion of fumaric acid to malic acid. The approaching proton

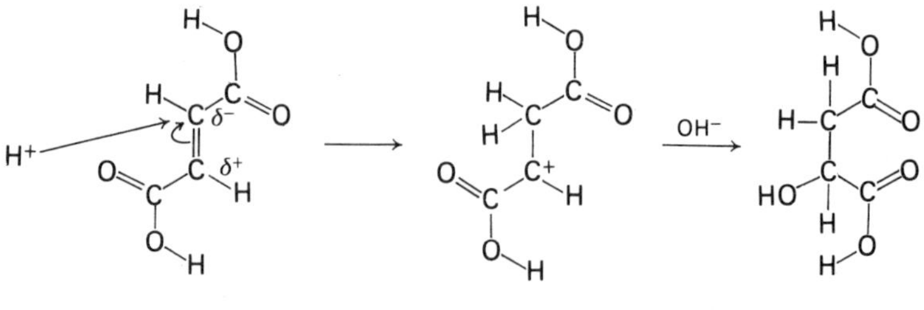

[reaction scheme showing fumaric acid → carbonium ion → malic acid, with H⁺ and OH⁻ additions]

　　　　fumaric acid　　　　　　carbonium ion　　　　　　malic acid

induces a dipole in the double bond and then binds to the induced nucleophilic site to form a positively charged carbonium ion. The latter finally reacts with an hydroxyl ion to form malic acid. Carbonium ions, deficient in electrons and carrying a positive charge on a carbon atom, are frequently intermediates in reactions although they cannot be isolated.

The last two transformations selected to illustrate reaction mechanisms are the condensation between dihydroxyacetone phosphate and glyceraldehyde-3-phosphate to form fructose-1,6-diphosphate

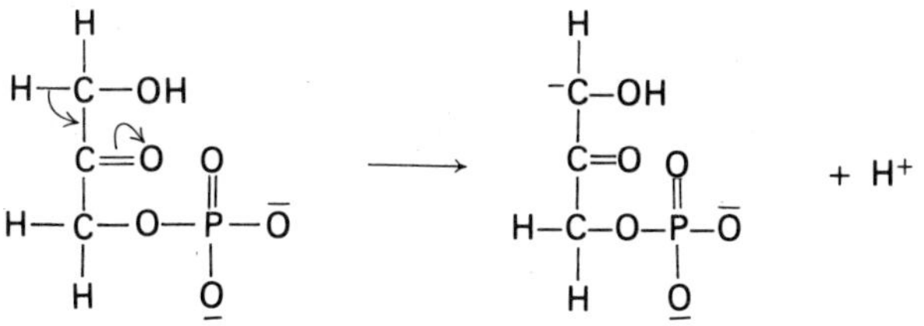

[Reaction scheme: dihydroxyacetone phosphate → carbanion (nucleophile) + H⁺]

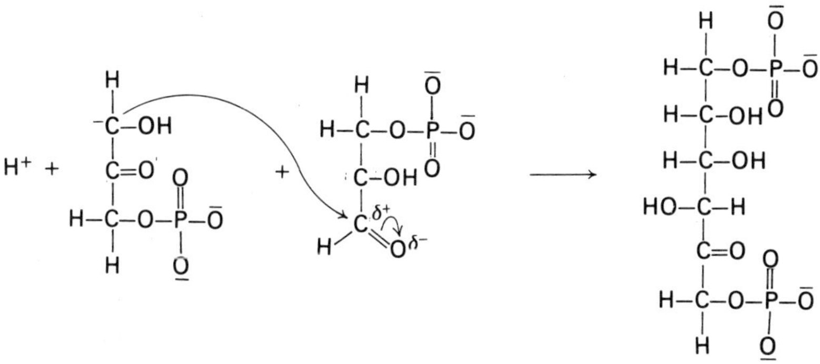

[Reaction scheme: H⁺ + carbanion + glyceraldehyde-3-phosphate (electrophile) → fructose-1,6-diphosphate]

and the decarboxylation of acetoacetic acid

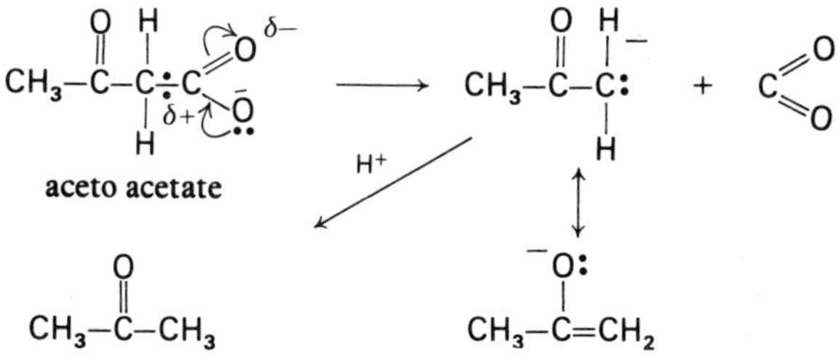

[Reaction scheme: aceto acetate → CH₃−C(=O)−C(H)(H) + CO₂; with H⁺ leading to acetone (CH₃−CO−CH₃) via enol CH₃−C(O⁻)=CH₂]

Both of these reactions involve the participation of carbanions.

Properties of Small Molecules

Cells use small molecules as sources of energy and as building blocks for the formation of polymers and multimolecular assemblages such as membranes. These versatile small molecules are compounds of carbon, so an appreciation of the chemical properties of living organisms requires some understanding of the chemistry of simple carbon compounds.

Carbon does not have the capacity to accept electrons into d orbitals, hence tetravalent carbon atoms do not accept lone electron pairs from water to become solvated. Moreover tetravalent carbon atoms do not possess lone electron pairs which can become involved in H bonds with water molecules. As a consequence of these two factors hydrocarbons are insoluble in water. Those carbon compounds that are soluble, and they are many, owe this property to polar functional groups which can be hydrated. The inability to form coordination compounds by accepting electron pairs into d orbitals is also responsible for the stability or lack of reactivity of hydrocarbons since most reactions proceed via the formation of coordination intermediates. Thus the reactivity of carbon compounds as well as their solubility is largely a property of their functional groups.

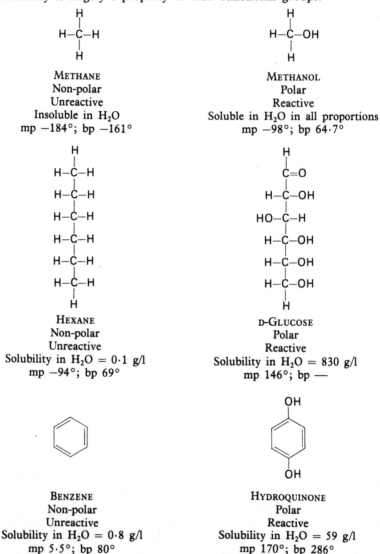

Figure 1.10 Influence of polar groups on the properties of carbon skeletons

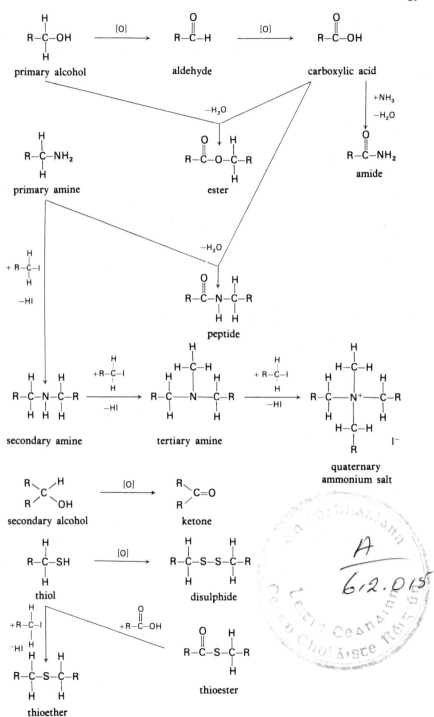

Figure 1.11 Relationships between functional groups

Carbon displays great versatility in being able to form stable covalent bonds with hydrogen, which is electropositive, and with oxygen, sulphur and nitrogen, which are electronegative. Furthermore the stability of the C—C bond and the possibility of forming multiple bonds with carbon, oxygen and nitrogen permit the existence of an almost boundless variety of carbon compounds.

Figure 1.11 indicates the chemical relationships between some of the functional groups commonly encountered in molecules of biological interest.

Carbohydrates

Carbohydrates which, as monomers, have the general formula $(CH_2O)_n$ are found in all living organisms as intermediary metabolites, food stores and structural elements. The simplest carbohydrates are the monomers or monosaccharides. These can link together to form oligosaccharides containing a few monomers and polysaccharides composed of hundreds of monosaccharide residues.

The monosaccharides can be divided into two groups, the aldoses carrying an aldehyde group and the ketoses with a keto group. Most of the monosaccharides found in biological systems are those aldoses and ketoses related configurationally to (+) glyceraldehyde. Although monosaccharides can be represented as straight-chain compounds, those containing five or six carbon atoms (pentoses and hexoses) normally exist in solution in a ring form. Thus D-glucose exists largely as a six-membered ring (II); this arrangement accounts for

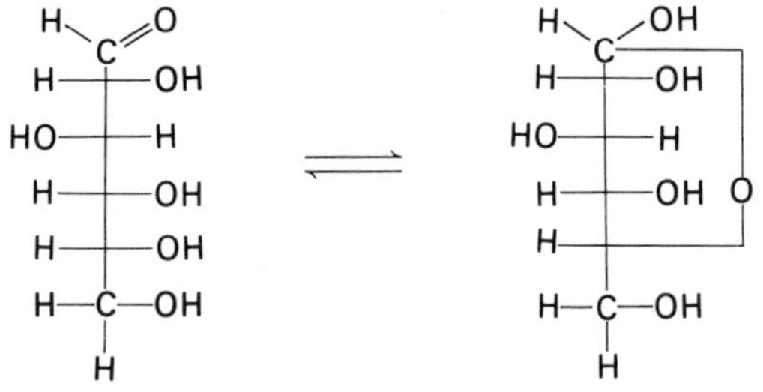

the observation that the aldehyde group of glucose is less reactive than might be expected from formulation I. The ring structure, frequently represented thus:

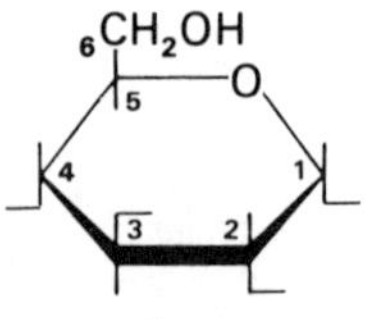

D-glucose

can exist in two forms termed α and β differing only in the configuration of the —H and —OH groups on carbon 1 relative to the plane of the ring.

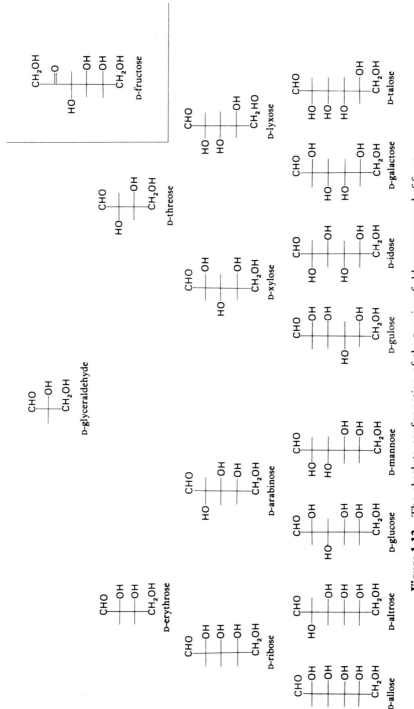

Figure 1.12 The absolute conformation of the D-series of aldose sugars and of fructose

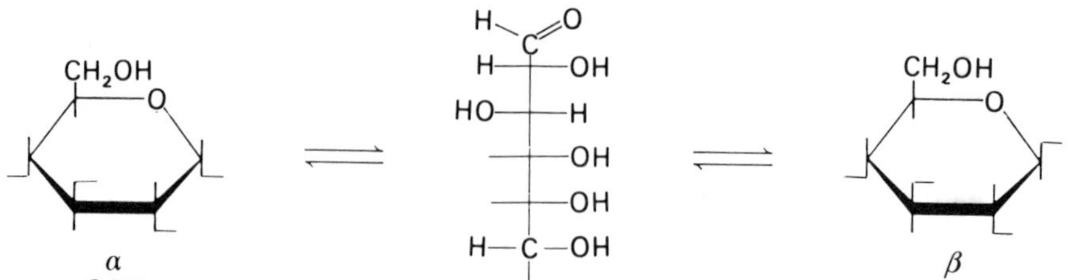

[Structures showing α (36%) ⇌ open chain ⇌ β (64%) forms of glucose]

The spontaneous interconversion of these two forms which occurs via the open configuration is termed mutarotation.

Owing to the sp³ hybridization of carbon, the six-membered ring does not lie in a single plane but takes on a puckered chair conformation of lower energy.

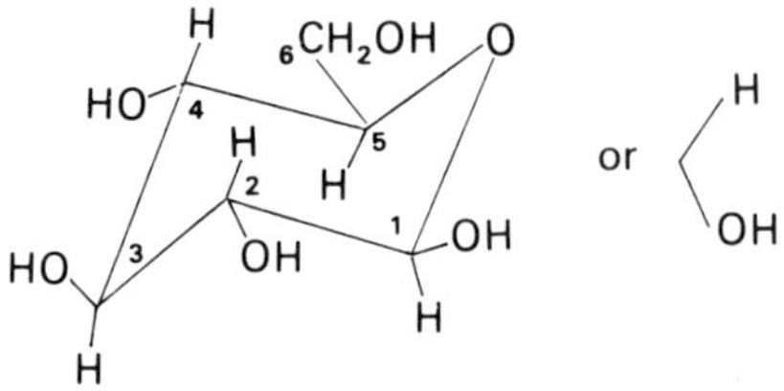

[Chair conformation of glucose]

The ring forms of some sugars frequently encountered in biochemistry are shown below.

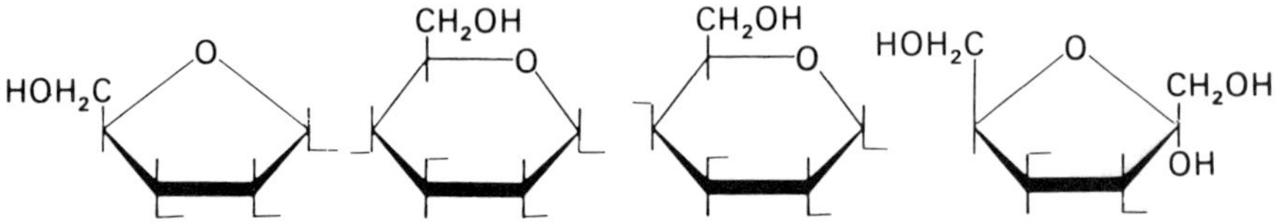

[Ring structures of D-ribose, D-glucose, D-galactose, D-fructose]

The hydroxyl groups of monosaccharides can be modified to give derivatives of the type indicated in Figure 1.13, many of which are found in biological systems.

Two monosaccharides can condense by eliminating water from two hydroxyl groups to form a disaccharide held together by a glycosidic bond. Linear and branched-chain polysaccharides can be formed in a similar manner. Some of these are composed of a single type of simple sugar, such as glucose, whilst others

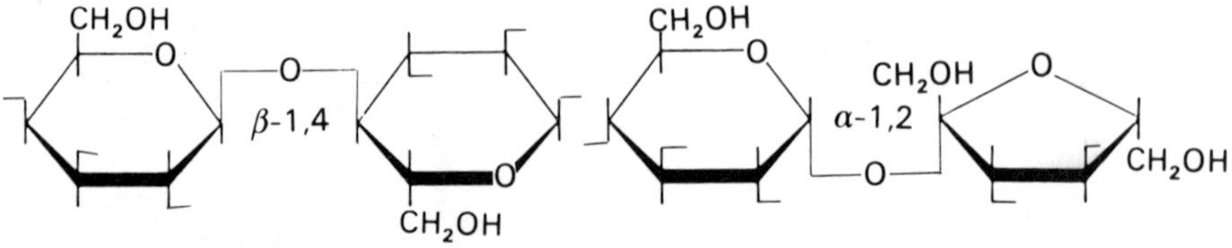

[Structures of Lactose (D-galactose residue β-1,4 D-glucose residue) and Sucrose (D-glucose residue α-1,2 D-fructose residue)]

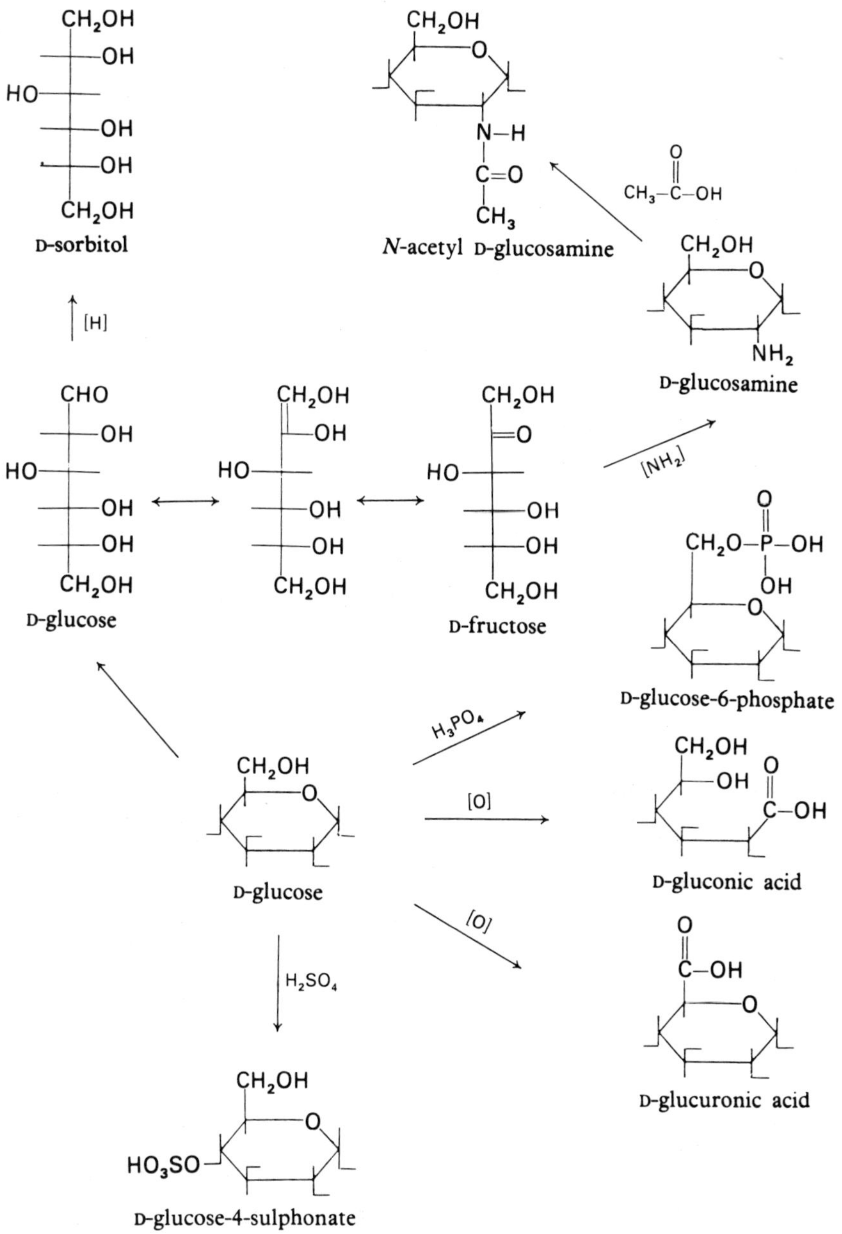

Figure 1.13 D-Glucose and related compounds

contain modified sugars such as glucuronic acid and *N*-acetyl glucosamine. Glycogen, a storage form of carbohydrate found in mammals, is composed only of glucose residues linked together to form a branched chain structure. On the other hand, chondroitin sulphate (see p. 270), a component of connective tissues,

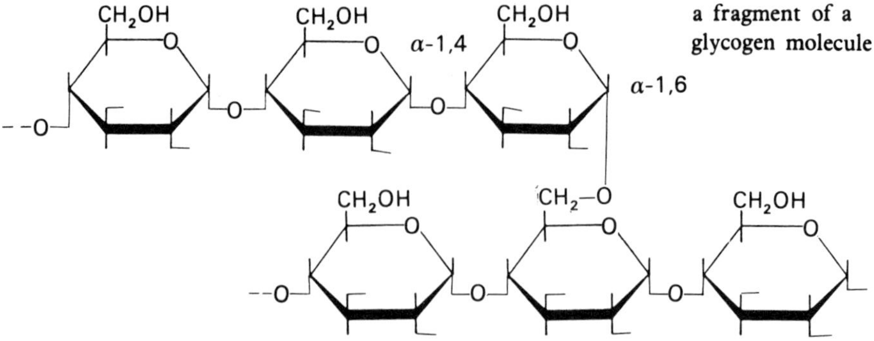

a fragment of a glycogen molecule

is an unbranched molecule having a disaccharide composed of glucuronic acid and sulphated *N*-acetyl galactosamine as its repeating unit.

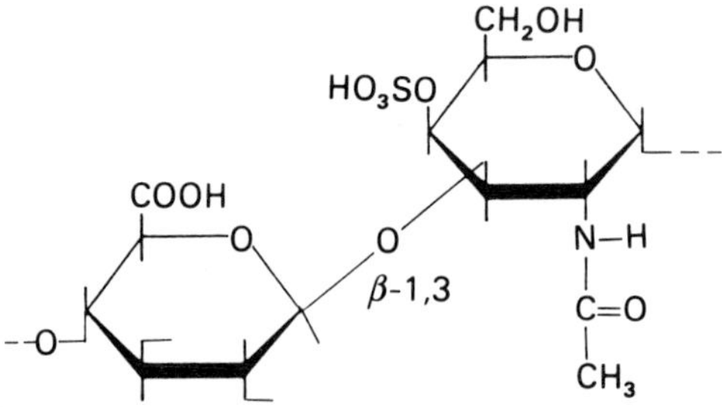

the repeating unit of chondroitin sulphate

Chondroitin sulphate tends to assume an expanded, linear form owing to the repulsive interactions between its ionized carboxyl and sulphate groups; it is thus a long, fairly rigid molecule conferring properties of rigidity and low friction upon the structures in which it occurs. The closely related molecule of hyaluronic

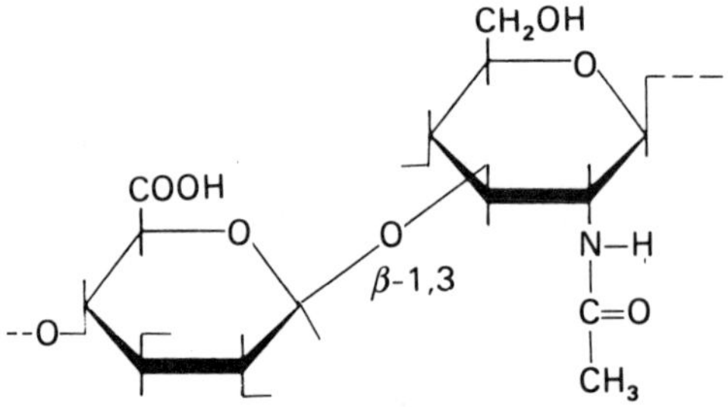

the repeating unit of hyaluronic acid

acid carries a smaller proportion of negative charges than chondroitin sulphate and so tends to be more of a random coil than a rigid rod; this behaviour confers the property of high viscosity on solutions of hyaluronic acid. Glycogen, chondroitin sulphate and hyaluronic acid are all water-soluble molecules because their polar groups can interact with water molecules. Cellulose, a polymer of glucose units joined by β-1,4 links has a linear, non-helical structure that encourages strong intra- and interpolymer hydrogen bonding (see p. 52) at the expense of solvation; in consequence cellulose forms insoluble fibres of great tensile strength.

Amino Acids

α-Amino acids are important as the monomeric units from which polypeptides, including proteins, are constructed; they have the general formula:

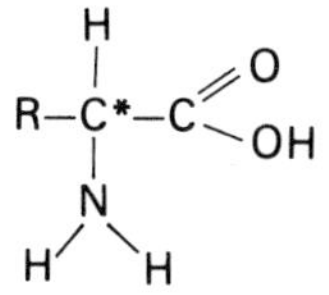

$$R-\overset{H}{\underset{\underset{H}{\overset{|}{N}}\diagdown H}{\overset{|}{C}}}*-C\overset{\diagup O}{\diagdown OH}$$

where * denotes the α carbon atom which is an asymmetric centre except when R = H. It is the presence of an α-amino group and an α-carboxyl group that enables amino acids to condense together with the elimination of water to form polypeptides.

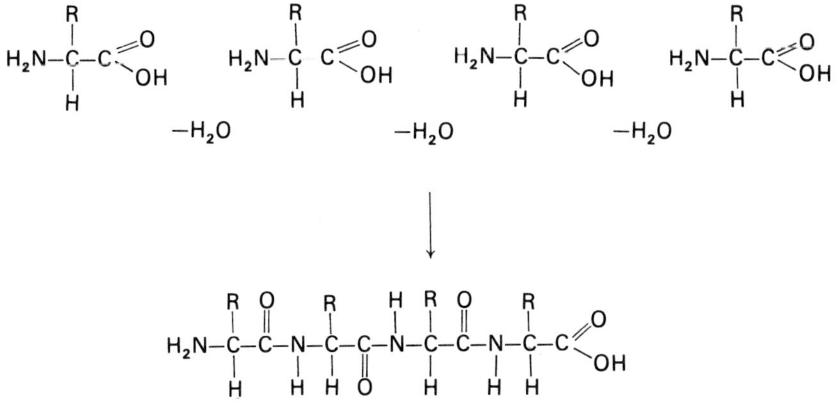

Thus all polypeptides share a common backbone of peptide bonds separated by α carbon atoms, but differ in the number, nature and sequence of side groups (R) attached to the α carbon atoms. The diversity of side chains in amino acids ensures that individual amino acids vary greatly in properties such as solubility, acidity and reactivity. Figure 1.14 shows the structural formulae of those amino acids commonly found in proteins. Amino acids such as leucine with hydrocarbon side chains are less soluble in water than those with polar groups such as glutamic acid. The side chains of non-polar amino acids in proteins tend to cluster together to form an oily centre to the molecule whilst the polar residues project into the solvent to confer water solubility. Cysteine residues in polypeptides can be oxidized so that two residues become linked by an —S—S— bridge. In this way two polypeptide chains can be linked together or a loop can be introduced into a single chain. Interactions between amino acid side chains of the

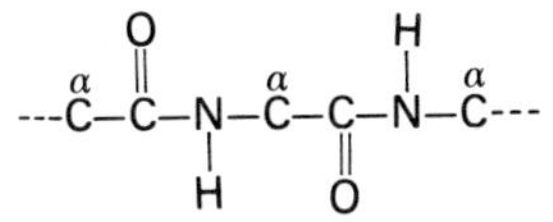

Group 1
Aliphatic

Amino acids containing one carboxyl group and one amino group

glycine $\quad NH_2-CH_2-COOH$

Glycine, the simplest amino acid, does not possess an asymmetric carbon atom.

L-alanine
$$CH_3-\underset{H}{\overset{NH_2}{C}}-COOH$$

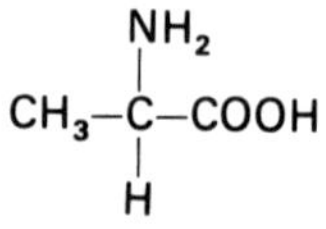

L-valine
$$\begin{matrix} CH_3 \\ CH_3 \end{matrix}\!\!\!\underset{H}{\overset{H}{C}}-\underset{H}{\overset{NH_2}{C}}-COOH$$

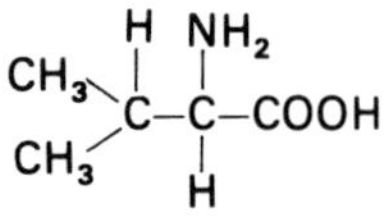

L-leucine
$$\begin{matrix} CH_3 \\ CH_3 \end{matrix}\!\!\!\underset{}{\overset{H}{C}}-CH_2-\underset{H}{\overset{NH_2}{C}}-COOH$$

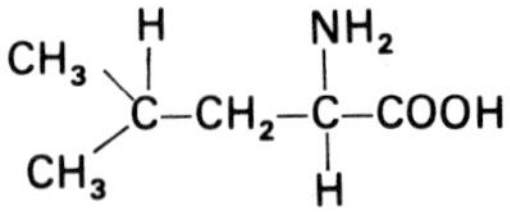

L-isoleucine
$$\begin{matrix} CH_3-CH_2 \\ CH_3 \end{matrix}\!\!\!\underset{H}{\overset{H}{C}}-\underset{H}{\overset{NH_2}{C}}-COOH$$

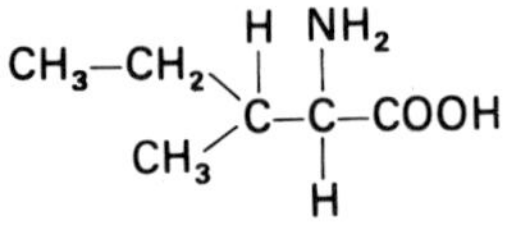

L-serine
$$HO-CH_2-\underset{H}{\overset{NH_2}{C}}-COOH$$

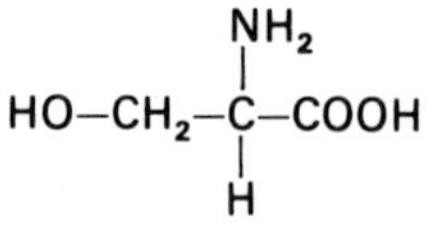

L-threonine
$$CH_3-\underset{\underset{H}{O}}{\overset{H}{C}}-\underset{H}{\overset{NH_2}{C}}-COOH$$

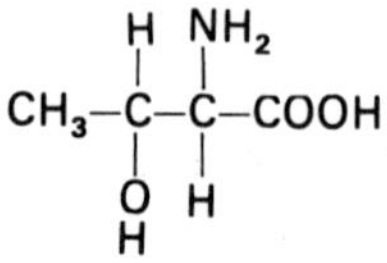

Monamino-monocarboxylic amino acids containing sulphur

L-cysteine
$$HS-CH_2-\underset{H}{\overset{NH_2}{C}}-COOH$$

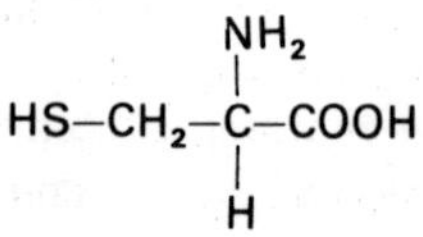

Fig. 1.14—*continued*

L-cystine
(dimer of cysteine)

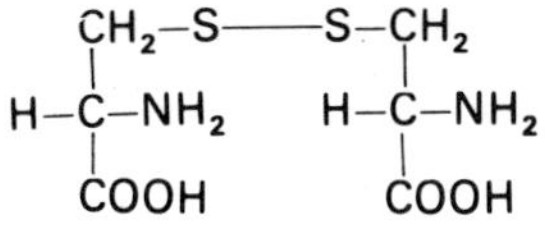

$$\begin{array}{cc} CH_2-S\!-\!\!-\!\!-\!S-CH_2 \\ H-C-NH_2 \quad H-C-NH_2 \\ COOH \quad\quad COOH \end{array}$$

L-methionine

$$CH_3-S-CH_2-CH_2-\underset{H}{\overset{NH_2}{C}}-COOH$$

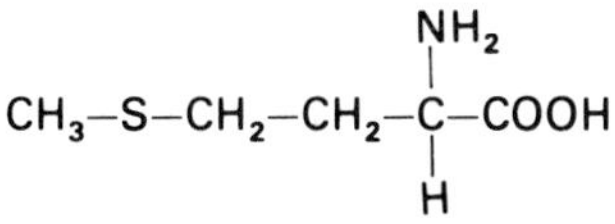

Dicarboxylic amino acids

Since these amino acids contain an additional carboxyl group, aqueous solutions of the free amino acid will be acidic.

L-glutamic acid

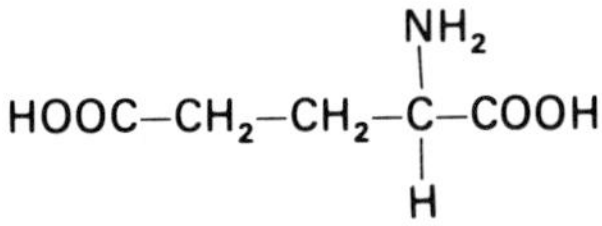

$$HOOC-CH_2-CH_2-\underset{H}{\overset{NH_2}{C}}-COOH$$

L-aspartic acid

$$HOOC-CH_2-\underset{H}{\overset{NH_2}{C}}-COOH$$

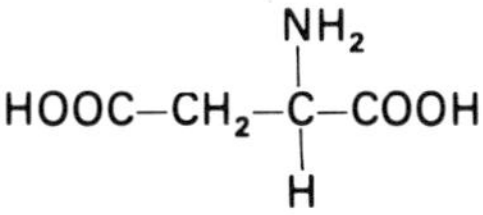

Basic amino acids

L-lysine

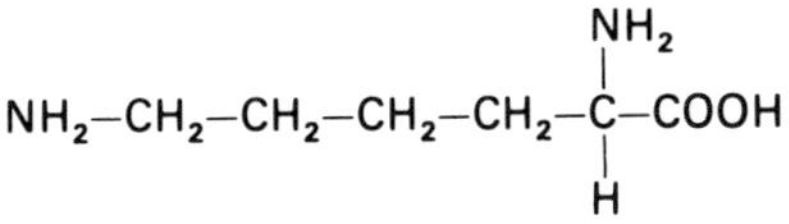

$$NH_2-CH_2-CH_2-CH_2-CH_2-\underset{H}{\overset{NH_2}{C}}-COOH$$

Aqueous solutions of lysine are basic because lysine has the additional amino group in its structure.

L-arginine

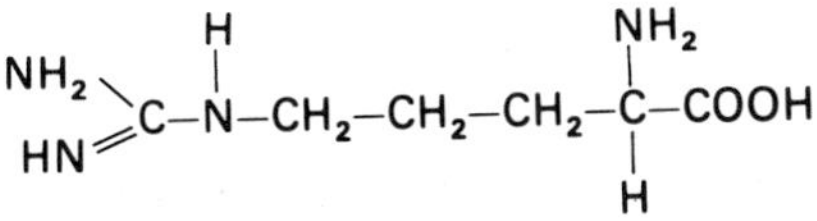

$$\underset{HN}{\overset{NH_2}{>}}\!\!C-\underset{}{\overset{H}{N}}-CH_2-CH_2-CH_2-\underset{H}{\overset{NH_2}{C}}-COOH$$

This amino acid is strongly basic because of its guanidino group.

L-histidine

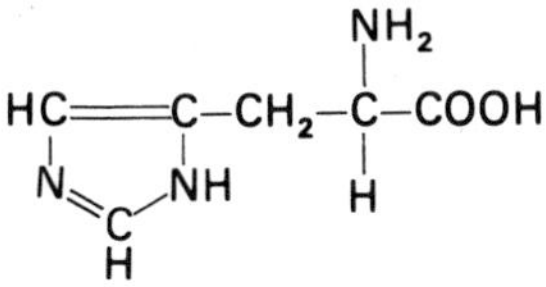

$$\begin{array}{c} HC\!=\!\!=\!C-CH_2-\underset{H}{\overset{NH_2}{C}}-COOH \\ N\diagdown\;\;\diagup NH \\ \underset{H}{C} \end{array}$$

(continued)

Fig. 1.14—*continued*

Group II   Aromatic amino acids

These amino acids contain the aromatic or benzene ring.

L-phenylalanine

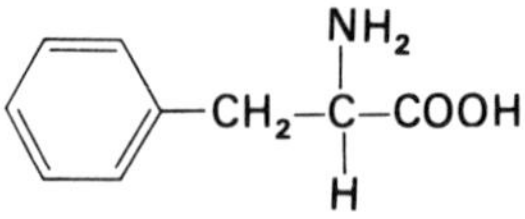

$$\text{C}_6\text{H}_5-\text{CH}_2-\underset{\underset{H}{|}}{\overset{\overset{NH_2}{|}}{C}}-\text{COOH}$$

L-tyrosine

$$\text{HO}-\text{C}_6\text{H}_4-\text{CH}_2-\underset{\underset{H}{|}}{\overset{\overset{NH_2}{|}}{C}}-\text{COOH}$$

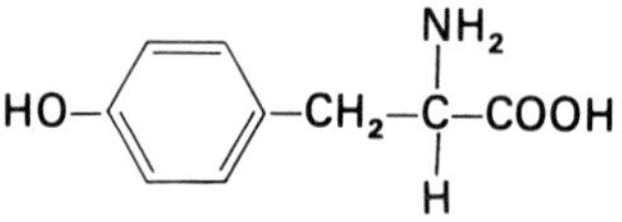

Group III   Heterocyclic amino acids

L-tryptophan

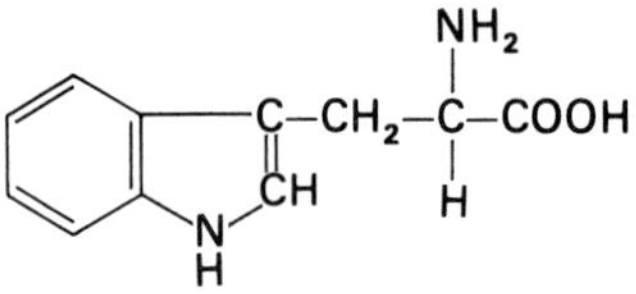

$$\text{(indole)}-\underset{\underset{CH}{|}}{C}-\text{CH}_2-\underset{\underset{H}{|}}{\overset{\overset{NH_2}{|}}{C}}-\text{COOH}$$

L-proline

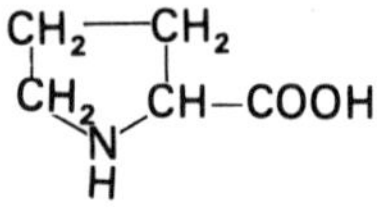

$$\begin{array}{c} \text{CH}_2-\text{CH}_2 \\ | \quad\quad | \\ \text{CH}_2 \quad \text{CH}-\text{COOH} \\ \backslash\;\;\;/ \\ \text{N} \\ | \\ \text{H} \end{array}$$

L-hydroxyproline

$$\begin{array}{c} \text{HO}-\text{CH}-\text{CH}_2 \\ | \quad\quad | \\ \text{CH}_2 \quad \text{CHCOOH} \\ \backslash\;\;\;/ \\ \text{N} \\ | \\ \text{H} \end{array}$$

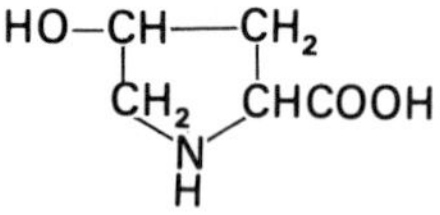

Figure 1.14   The amino acids

type outlined above largely determine the three-dimensional structure of proteins. The polar groups of amino acids also provide the nucleophilic and electrophilic atoms which form an important part of the catalytic centres or active sites of enzymes. It is the almost limitless possibilities for variation of polypeptide structure in terms of amino acid composition and sequence that makes possible the diversity of forms found in living systems.

The acidic and basic properties of amino acids are discussed on p. 57.

Lipids

The lipids are a heterogeneous collection of compounds sharing the common feature (on account of their predominantly hydrocarbon nature) of being only sparingly soluble in water. They are found chiefly in membranes and in fat droplets such as chylomicra in the blood and fat storage granules in cells.

The most common simple lipids are saturated fatty acids having the general structure

$$H-\underset{H}{\overset{H}{C}}-\left(\underset{H}{\overset{H}{C}}\right)_n -C\underset{OH}{\overset{O}{\lessgtr}}$$

where n can have any value between 0 and about 20. When n is between 0 and 5 the acids are reasonably soluble in water but when n is greater than 10 they are very insoluble because the interaction of water with the carboxyl group does not provide sufficient solvation to carry the methyl and methylene groups (CH_3- and CH_2-) into solution. The sodium salts of long-chain fatty acids are soaps able to dissolve in water with formation of micelles. These are clusters of molecules arranged so that their hydrocarbon tails are packed together and their charged carboxylate groups project into the water.

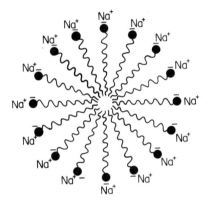

Triglycerides, the major energy store of the body, are esters of long-chain fatty acids with glycerol. The three fatty acids involved can be of different chain lengths and degrees of saturation. Triglycerides are, like their parent long-chain fatty acids, very insoluble in water; this property enables large quantities of

$$\begin{array}{c} H \\ | \\ H-C-OH \\ | \\ H-C-OH \\ | \\ H-C-OH \\ | \\ H \end{array} \quad + 3\ R \cdot C\underset{OH}{\overset{O}{\lessgtr}} \quad \longrightarrow \quad \begin{array}{c} H \\ | \quad O \\ H-C-O-C-R \\ | \quad O \\ H-C-O-C-R \\ | \quad O \\ H-C-O-C-R \\ | \\ H \end{array}$$

glycerol   fatty acid   triglyceride

energy to be stored as insoluble fat droplets without a requirement for water as a solvent.

Another important group of lipids are the phospholipids which are major components of cellular membranes. Two of the major subgroups of phospholipids are the glycerophosphatides containing glycerol and the sphingolipids containing sphingosine. As can be seen from their structures, these compounds possess both polar, hydrophilic groups that can be solvated, and non-polar, hydrophobic regions that are insoluble in water but able to aggregate with similar regions of related molecules. It is this duality of solubility which allows the molecules to form lipid bilayers, structures that are important components of cellular membranes.

The fatty acids found in triglycerides and phospholipids can be saturated or unsaturated. Unsaturated fatty acids possessing one or more *cis* double bonds

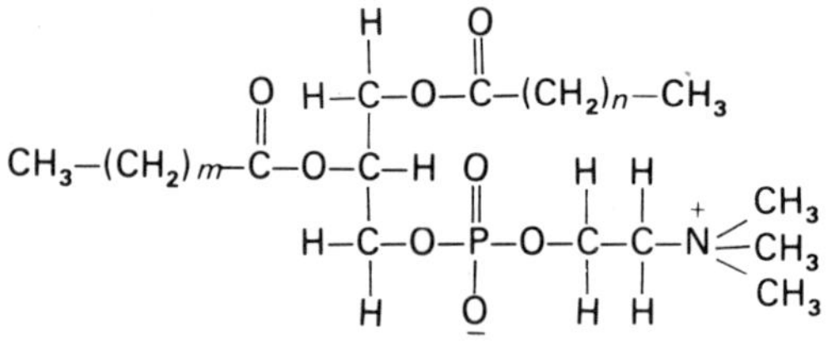

phosphatidyl choline (lecithin)

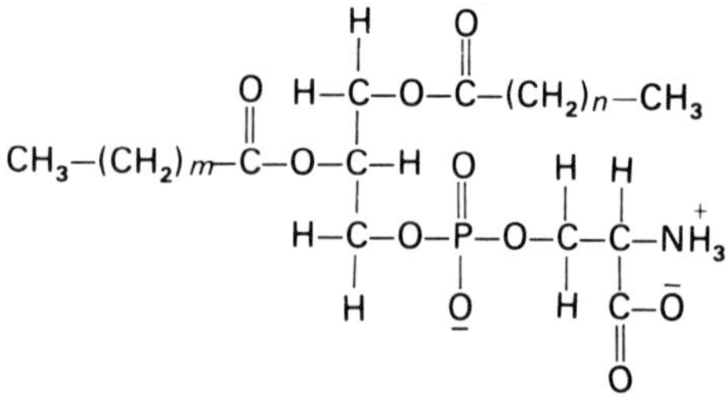

phosphatidyl serine (cephalin)

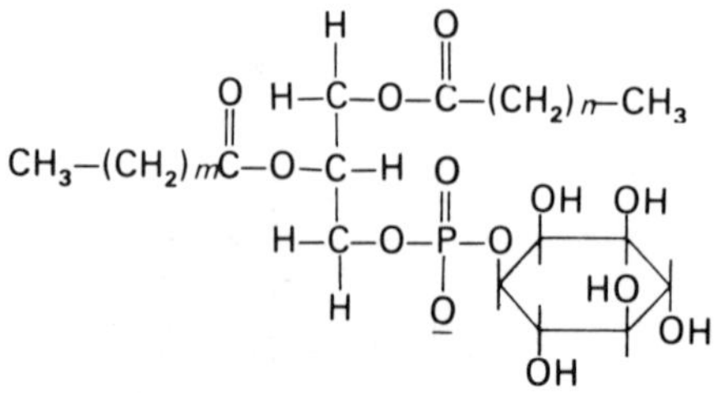

phosphoinositide

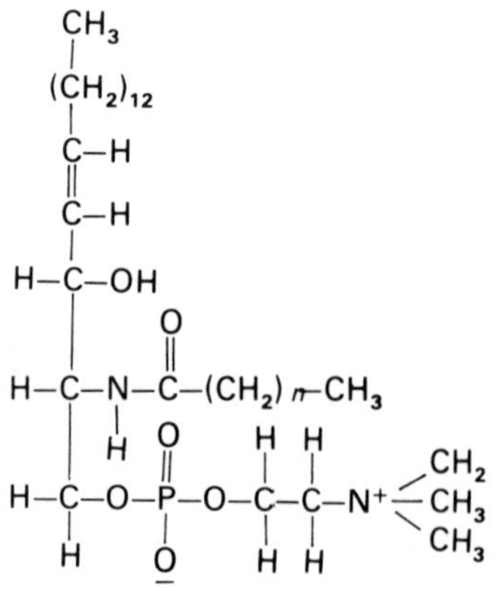

sphingomyelin

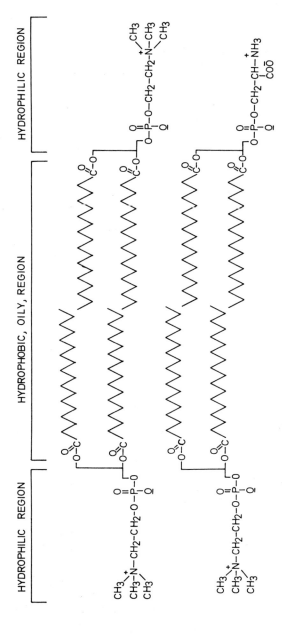

Figure 1.15 Part of a lipid bilayer formed from molecules of phosphatidylcholine and phosphatidyl serine

have kinked hydrocarbon tails which prevent them packing in an orderly manner with saturated fatty acids. As a result of this, triglycerides and phospholipids with a high proportion of unsaturated fatty acids have lower melting points than those containing only the corresponding fully saturated acids. Consequently, membranes with a high proportion of unsaturated fatty acids in their phospholipids tend to be more mobile and flexible at a given temperature than those with more saturated ones.

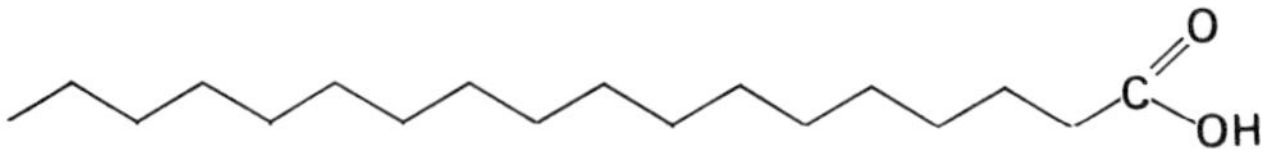

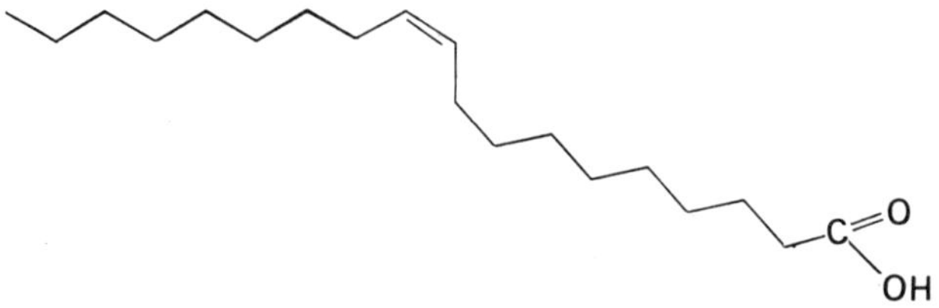

Cholesterol is another lipid found widely as a component of cellular membranes; it is a planar, insoluble, molecule able to be inserted into membranes between the hydrocarbon tails of phospholipids where its presence can affect the mobility of the membrane.

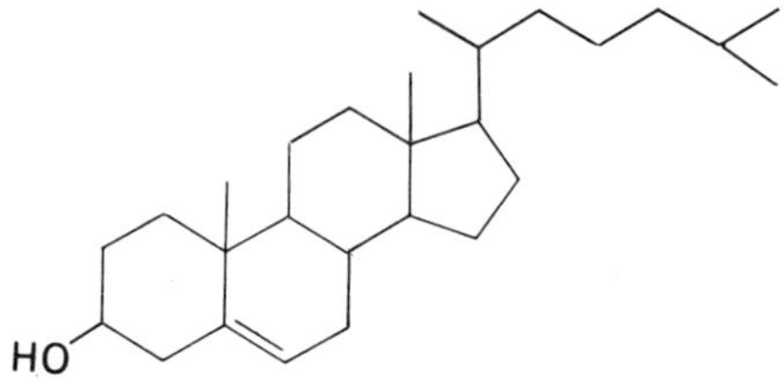

Membrane Structure

The importance of membranes in cellular functioning can hardly be overestimated; they are found in many different situations and are involved in a wide variety of processes. It is not surprising, therefore, that membranes differ greatly in composition; the relatively inert membranes comprising the myelin sheath would not be expected to have the same composition as the metabolically active inner membranes of mitochondria. However, all membranes seem to consist of a lipid bilayer of variable composition and associated globular proteins. Some of the proteins (peripheral proteins) are loosely attached to the hydrophilic surfaces of the bilayer whilst others are embedded within the hydrophobic matrix. The proteins embedded in the matrix or extending right through it are termed integral proteins and can only be removed by treatment with detergents or organic solvents. Since the integral proteins are soluble in the non-polar

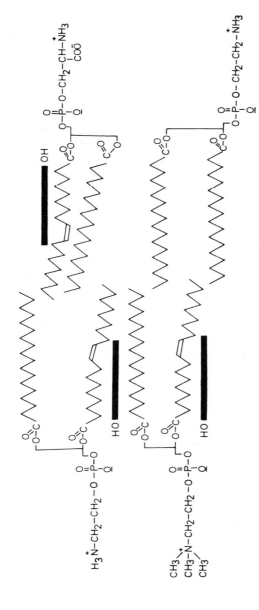

Figure 1.16 An indication of how cholesterol molecules ▬OH may fit into lipid bilayers containing phospholipids with unsaturated fatty acid residues

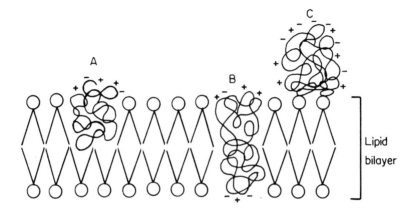

Figure 1.17 Fluid mosaic model of membrane structure. A and B are integral proteins. C is a peripheral protein, easily removed by gentle procedures such as buffer washing

matrix of membranes it follows that their surfaces within the matrix must also be non-polar. This arrangement contrasts with that found in water-soluble proteins whose surfaces are rich in polar amino acid residues (see p. 70).

Integral proteins are able to move within the plane of the membrane but not to rotate or to move out of the plane of the membrane as both these manoeuvres would expose hydrophobic parts of the molecule to water, which would be energetically unfavourable. Hence membranes must not be regarded as rigid or static structures but as fluid regions of restricted permeability within which integral proteins can diffuse in two dimensions.

Heterocyclic Compounds

There are many biologically important molecules that contain unsaturated ring systems involving atoms other than carbon; for example

tryptophan

adenosine triphosphate (ATP)

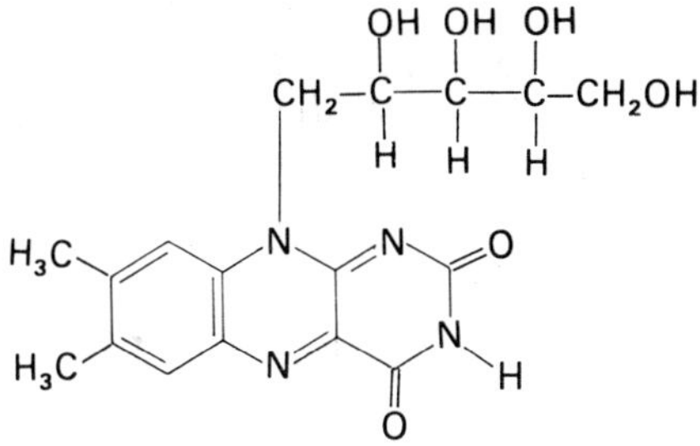

riboflavin

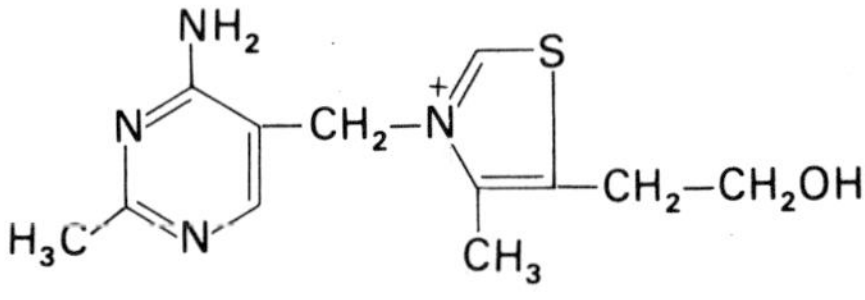

vitamin B₁ (thiamine)

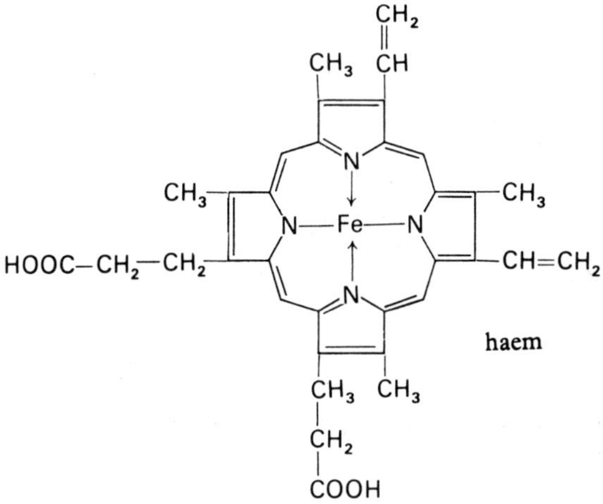

haem

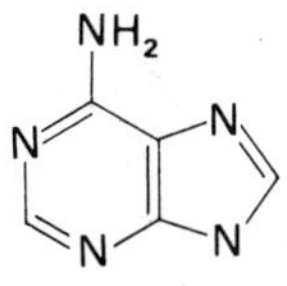

The parent unsaturated heterocyclic rings are planar and tend to be both sparingly soluble in water and unreactive. However, many molecules containing such ring systems are quite soluble and reactive by virtue of the polar groups attached to the ring. Thus whilst adenine is soluble only to the extent of 90 mg/100 ml at 25° and is rather inert, ATP is readily soluble (about 70 gm/100 ml at 25°) and highly reactive.

The non-polar nature of many ring systems encourages them to enter oily environments or to cluster together. For example, haem fits snugly into a hydrophobic cleft in the protein globin to form haemoglobin. Similarly, successive

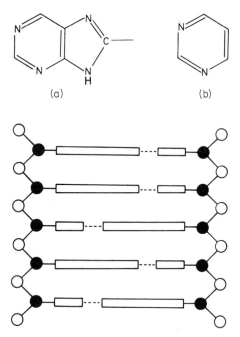

Figure 1.18 Two-dimensional diagram of the DNA double helix showing the vertical stacking of purine and pyrimidine bases between the ribose phosphate backbones. (a) Purine ring system as in adenine and guanine. (b) Pyrimidine ring system as in cytosine and thymine. ○ Phosphate; ● Ribose; ⬜⬜ Purine base: adenine or guanine; ⬜ Pyrimidine base: thymine or cytosine

pairs of sparingly soluble purine and pyrimidine bases present in the two polynucleotide chains forming DNA stack together and so help to stabilize the double helix (see also p. 169). The conjugated double bond system (i.e. alternating single and double bonds) found in many heterocyclic rings is an expression of the fact that the π electrons are delocalized and consequently more mobile than electrons in saturated systems. The ring is therefore able to provide a pathway for the displacement of electrons from one region of the molecule to another during the course of a reaction. An example of this is provided by the removal of an amino group from an amino acid in a reaction involving pyridoxal phosphate.

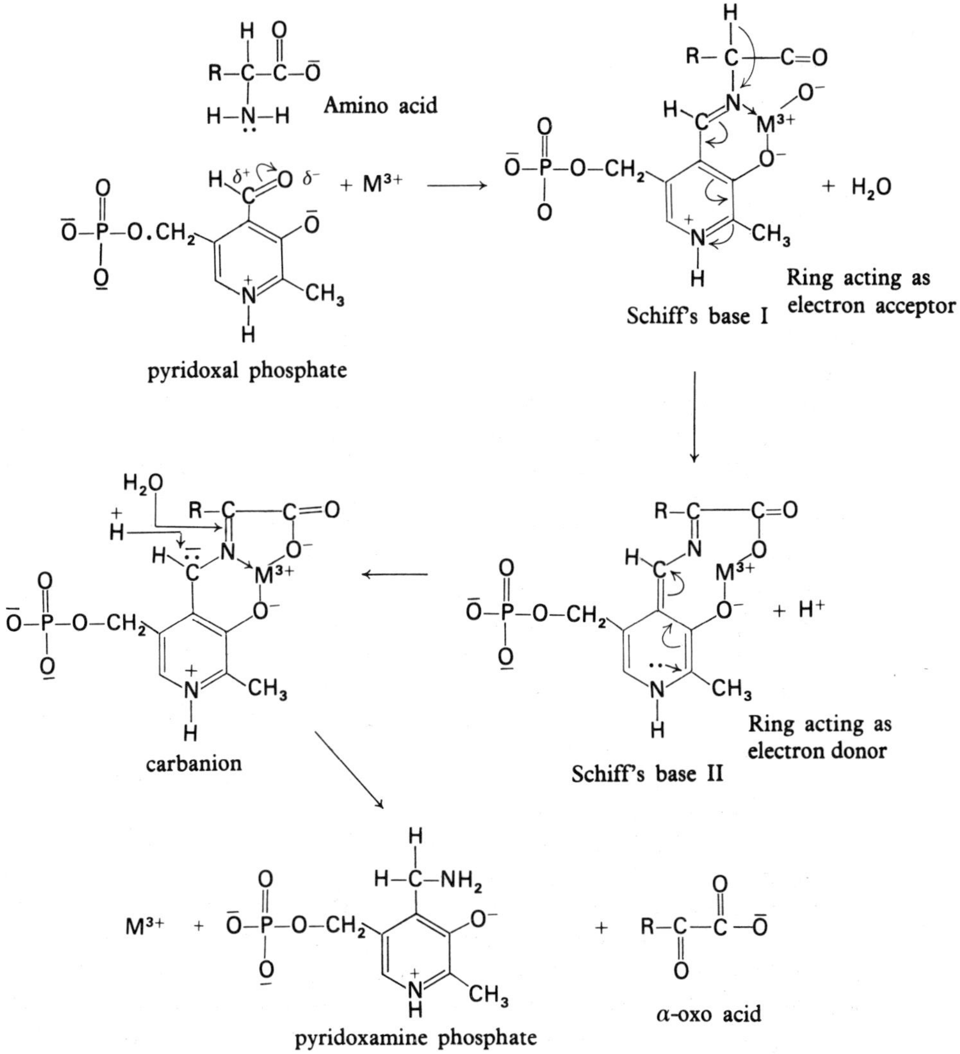

Figure 1.19 Transamination between an amino acid and pyridoxal phosphate. NB. The small curly arrows ⌢ denote shifts of electrons within the system of conjugated double bonds

CHAPTER 3

Bioenergetics

It is a fact of experience that some things happen and some do not. For example, owing to the ravages of time and fortune, buildings eventually fall down; the pile of rubble so formed does not spontaneously reassemble itself into a house. This type of experience has led to the formulation of one of the basic statements about Nature, namely the second Law of Thermodynamics. This states that a process occurs spontaneously only if it is accompanied by an overall increase in the disorder or *entropy* of the Universe at large. In other words, things only happen if the Universe becomes more chaotic as a result.

Life seems to have arisen spontaneously on Earth and is characterized by being a most subtle and intricately ordered state of matter. We are, therefore, faced with a paradox. How can the order of some parts of the Earth spontaneously increase to such a high level that the chemical elements there can exist in the living state and constitute an organism when spontaneous processes must decrease the overall order of the Universe? Like most paradoxes, this one is more apparent than real. We all know that piles of bricks and timber can be converted into a house if sandwiches and petrol are converted to carbon dioxide, water and heat by the men and machines that toil on the building site. The disorder or entropy accompanying the consumption of food and fuel has to be greater than the increase in order achieved by erecting the house. Thus a local increase in order and complexity can be obtained at the price of producing greater disorder somewhere else. So far as life is concerned, order is gained by capturing light energy from the sun and using it to drive the reduction of CO_2 into plant carbohydrates. The disordering which must necessarily accompany this photosynthetic process is the decomposition of liquid water to gaseous oxygen and the conversion of some of the absorbed light to heat. Heterotrophs, such as animals, disorder plants and other animals to increase and maintain their own organization; they must, of course, destroy more order externally than they gain internally. To maintain himself an adult man has to disorder the equivalent of about 24 sheep a year. Living organisms are, then, open systems continually exchanging materials and energy with their surroundings. Animals take in highly ordered chemicals and return less-ordered refuse such as CO_2, H_2O, urea and heat to the surroundings. The flow and progressive disordering of material through the system we call an organism leads to the continual renewal of that system. Hence order is built and maintained on a stream of disordering. Once the stream of disordering ceases, the order of the system begins to disintegrate.

The question now arises as to how the spontaneous, entropy producing, processes involved in food breakdown are coupled to the production of the ordered, self-reproducing, constantly changing systems characteristic of the living state. To answer this question we must examine the concept of chemical equilibrium and introduce some elementary thermodynamics.

Equilibria

In a reaction such as

$$A + B \rightleftharpoons C + D$$

the formation of C and D proceeds at a rate proportional to the product of the concentrations of A and B. Conversely the rate of reconversion of C and D to A and B is proportional to $[C] \times [D]$. However, this proportionality does not tell us exactly how many molecules of A and B or of C and D react in unit time, that is, it does not tell us how fast the two reactions proceed. For this purpose rate constants k_1 and k_2 are required. k_1 is defined so that $k_1 [A][B]$ gives the absolute number of moles of A and B/litre reacting in unit time and $k_2 [C][D]$ gives the absolute number of moles of C and D/litre reacting in unit time. At equilibrium the number of moles/litre of A and B reacting in unit time equals the number of moles/litre of C and D reacting to give A and B in unit time, i.e.

$$A + B \underset{k_2}{\overset{k_1}{\rightleftharpoons}} C + D$$

at equilibrium therefore

$$k_1 [A][B] = k_2 [C][D]$$

then

$$\frac{[C][D]}{[A][B]} = \frac{k_1}{k_2} = K$$

where K is the equilibrium constant.

Entropy and Free Energy Changes During Reactions

In addition to describing the equilibrium position of a reaction (by means of the equilibrium constant K) it is also desirable to be able to make predictions about the probability of a reaction occurring under particular conditions. As we have seen, a reaction is only able to proceed if it leads to an *overall* increase in the entropy or disorder of the reacting system and its surroundings. It is clearly difficult in many instances to determine precisely the overall entropy increase in the Universe produced by a particular reaction. Instead another parameter indicating the possibility of a process occurring spontaneously has been constructed, this is the free energy change of the process, ΔG. The free energy change is so called because it is the amount of energy available from a reaction for the performance of work.

At constant pressure and temperature (the normal physiological situation) ΔG is defined thus

$$\Delta G = \Delta H - T\Delta S$$

Where ΔH is the heat given out or taken in when bonds in the reactants are broken and new bonds are formed to give products, T is the absolute temperature and ΔS is the entropy change of the *reacting system** or the change in the randomness of the reacting materials.

If the free energy change (ΔG) of a reaction such as $A + B \rightleftharpoons C + D$ is negative, (e.g. -5 kcal/mole) the forward reaction can occur with the net formation of C and D. If it is positive (e.g. $+5$ kcal/mole) the reverse reaction will be the spontaneous one. If ΔG is zero, the system is at equilibrium.

ΔG would, of course, be negative if ΔH were negative and ΔS were positive. ΔG could also be negative, despite a negative value for ΔS, if ΔH were large and negative. Similarly a positive value for ΔH would still allow a negative value of ΔG if ΔS were large and positive. Thus an endothermic reaction (one that takes up heat from the surroundings) can proceed if it is accompanied by a large increase in the entropy or disorder of the system. An example is the dissolution of many salts. The freezing of water is an example of a large negative value of ΔH leading to a negative value of ΔG despite the large decrease in the entropy or disorder of the system occurring as water molecules have their freedom of movement restricted in the ice lattice.

The free energy change in a reaction is related to the concentrations of products and reactants in a manner described by the equation

$$\Delta G = \Delta G^\circ + RT \ln \frac{[C][D]}{[A][B]}$$

Since $\Delta G = 0$ at equilibrium, it follows that at equilibrium

$$\Delta G^\circ = -RT \ln \frac{[C][D]}{[A][B]}$$

and therefore that

$$\Delta G^\circ = -RT \ln K$$

because the concentrations of reactants and products are the equilibrium concentrations.

ΔG° at a given temperature is called the standard free energy change. For biological systems, values of ΔG° are often quoted at 25 °C and, since H^+ ions often take part in biochemical reactions, at pH 7·0. Such values of ΔG° adjusted to pH 7·0 are often written $\Delta G^{\circ\prime}$. Unless stated otherwise, it should be assumed that values of $\Delta G^{\circ\prime}$ apply to 25°. If the value of $\Delta G^{\circ\prime}$ is known then the value of ΔG for any concentration of products and reactants can be readily calculated. Note that if $\Delta G^{\circ\prime}$ is positive, K is less than 1; if it is zero, K is equal to 1; and if it is negative, K is greater than 1.

* Not of the Universe as a whole.

Table 1.1 Numerical relationship between K and $\Delta G^{\circ\prime}$ at $25°$

K	$\Delta G^{\circ\prime}$ (kcal/mole)
0·001	+4·089
0·01	+2·726
0·1	+1·363
1·0	0
10·0	−1·363
100·0	−2·726
1000·0	−4·089

Illustration of the Dependence of Free Energy Changes on the Concentration of Reactants

The importance of the concentration of reactants in determining the free energy change in a reaction can be illustrated with the conversion of glucose-1-phosphate to glucose-6-phosphate for which the equilibrium constant at $30°$ is 19 (i.e. at equilibrium there are 95 parts of glucose-6-phosphate and 5 parts of glucose-1-phosphate) so that

$$\Delta G^{\circ\prime} = -1400^* \times \log 19 = -1800 \text{ cal/mole}$$

Now if we mix glucose-1-phosphate (0·01 M) and glucose-6-phosphate (0·001 M) we can calculate ΔG as follows

$$\Delta G = \Delta G^{\circ\prime} + RT \ln \frac{[\text{glucose-6-P}]}{[\text{glucose-1-P}]}$$

$$= -1800 + 1400 \log \frac{0\cdot001}{0\cdot01}$$

$$= -1800 + 1400 \times -1$$

$$= -3200 \text{ cal/mole}$$

Thus the amount of glucose-6-phosphate will increase and that of glucose-1-phosphate will decrease with a free energy yield of 3200 cal/mole transformed. Conversely, if the initial concentration of glucose-1-phosphate is 0·0001 M and of glucose-6-phosphate is 0·01 M, ΔG will equal +1000 cal/mole (i.e. the reverse reaction proceeds and the amount of glucose-1-phosphate increases with a yield of 1000 cal/mole).

The effect of concentration is often important in biological systems as materials are constantly being formed and used. Nevertheless the range of concentrations found in living matter is limited, with most metabolites having a concentration of about 1 mM (0·001 M). If $\Delta G^{\circ\prime}$ for a reaction is large and negative (about −4 kcal/mole or more), then the reaction is, for all practical purposes, irreversible. This follows because, to reverse the reaction, ΔG must be negative

* $RT \ln K = 2\cdot3 \, RT \log K = 1400 \log K$ at $30°$.

for the back reaction (i.e. positive for the forward reaction)

$$\Delta G = \Delta G^{\circ\prime} + RT \ln \frac{[C][D]}{[A][B]}$$

Thus, in order to reverse the reaction (i.e. make ΔG forward positive) when $\Delta G^{\circ\prime} = -4000$ cal/mole

$$1400 \log \frac{[C][D]}{[A][B]} \text{ must exceed } 4000$$

Therefore

$$\log \frac{[C][D]}{[A][B]} \text{ must be greater than } \frac{4000}{1400} = 2 \cdot 86$$

Thus

$$\frac{[C][D]}{[A][B]} \text{ must be greater than antilog } 2 \cdot 86 \text{ or } 724$$

Therefore to reverse the reaction, $[C] \times [D]$ must be at least $724 \times [A] \times [B]$. Thus if A and B are about millimolar [C] and [D] would have to be about 27 mM which is most unlikely.

After this brief excursion into thermodynamics, we are now in a position to answer the problem posed earlier of how living cells couple the spontaneous processes involved in food breakdown with the production of order and the performance of work.

Energy Supply by ATP

All living processes depend in some way upon a continuing supply of energy. In purely chemical terms, living organisms have evolved in such a way that this requirement for energy is manifested as a continuing necessity to carry out phosphorylation reactions (i.e. the transfer of phosphoryl groups

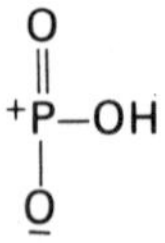

$$\overset{O}{\underset{O^-}{\overset{\parallel}{+P}-OH}}$$

from one compound to another). These phosphorylation reactions are vital to the chemical economy of the cell as they ensure that certain key reactions proceed with high yields of their product, that is, with a favourable equilibrium position. It follows that cells must be able to generate an efficient phosphorylating reagent.

The organic chemist, faced with the problem of performing a synthesis involving the transfer of an acyl group, commonly uses the appropriate acid anhydride; thus for acetylations (the transfer of acetyl, $CH_3\overset{O}{\overset{\parallel}{C}}-$, groups) reagents like acetic anhydride

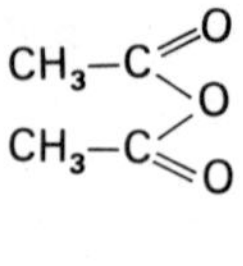

$$\begin{array}{c} CH_3-C\overset{O}{\underset{O}{\diagdown}} \\ CH_3-C\overset{O}{\diagup} \end{array}$$

or acetylchloride

$$CH_3-C\underset{Cl}{\overset{O}{\diagup}}$$

may be used. Living cells have adopted the same chemical principle, as they use an anhydride of phosphoric acid for the essential phosphorylation reactions. The particular phosphorylating agent is ATP. We may consider it as having a chemical grouping serving as a chemical handle (AMP) to which is attached the

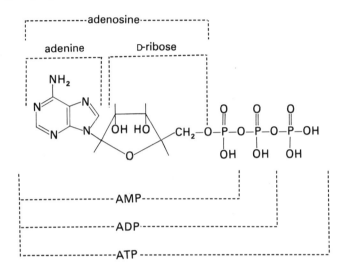

Figure 1.20 Adenosine mono-, di- and triphosphates

phosphoric anhydride. The essential function of this compound can be illustrated by considering the formation of glucose-6-phosphate from glucose.

The equilibrium constant at 25° for the reaction glucose + phosphoric acid ⟶ glucose-6-phosphate + water is 0·0045 whereas that for the reaction glucose + ATP ⟶ glucose-6-phosphate + ADP is 5500. Thus it is clear that the yield of the sugar phosphate is greater when the donor of the phosphoryl group is ATP than when it is phosphoric acid. This property of ATP is sometimes described by saying that it has a relatively high 'phosphoryl group transfer potential'; higher than that of phosphoric acid. Cells contain a great number of other phosphate compounds, some of which are more efficient phosphorylating agents than ATP, some less. Because of the importance of phosphoryl groups in living processes, biochemists have found it convenient to draw up a relative order of phosphorylating ability which enables them to make predictions about the direction of transfer of phosphoryl groups between various compounds. This can be done quantitatively by measuring the ease of transfer to a standard acceptor, namely water: and, for arithmetical convenience, instead of using the equilibrium constant, the standard free energy changes for the transfer reactions to water (standard free energy of hydrolysis $\Delta G°$) are preferred. Table 1.2 illustrates the

Table 1.2 Free energies of hydrolysis of some biologically important phosphates. Pi inorganic phosphate; PP pyrophosphate. Values are based on those given by M. R. Atkinson and R. K. Morton in *Comparative Biochemistry*, Vol. II (Eds. H. S. Mason and M. Florkin) Academic Press, New York, 1960

	$\Delta G^{\circ\prime}$ (kcal)
Phosphoenol pyruvic acid + H_2O → pyruvic acid + Pi	−13
1,3-Diphosphoglyceric acid + H_2O → 3-phosphoglyceric acid + Pi	−12
Creatine phosphate + H_2O → creatine + Pi	−11
ATP + H_2O → AMP + PP	−8
ATP + H_2O → ADP + Pi	−7
ADP + H_2O → AMP + Pi	−6
Glucose-1-phosphate + H_2O → glucose + Pi	−5
Fructose-1-phosphate + H_2O → fructose + Pi	−4
Glucose-6-phosphate + H_2O → glucose + Pi	−3
Glycerol-1-phosphate + H_2O → glycerol + Pi	−2

range of ΔG° values found. These values can be used to make predictions about phosphorylations, for example that (a) ATP will phosphorylate glucose rather than glucose-6-phosphate phosphorylating ADP. Similarly (b) 1,3-diphosphoglyceric acid will phosphorylate ADP to give ATP.

(a) glucose-6-phosphate + H_2O ⇌ glucose + phosphate; ΔG° −3 kcal/mole

ATP + H_2O ⇌ ADP + phosphate; ΔG° −7 kcal/mole

i.e.

glucose + ATP ⇌ glucose-6-phosphate + ADP; $\Delta G^\circ = -7 - (-3)$
$= -4$ kcal/mole

(b) 1,3-diphosphoglyceric acid + H_2O ⇌ 3-phosphoglyceric acid + phosphate;

$\Delta G^\circ = -12$ kcal/mole

ATP + H_2O ⇌ ADP + phosphate; $\Delta G^\circ = -7$ kcal/mole

i.e.

1,3-diphosphoglyceric acid + ADP ⇌ ATP + 3 phosphoglyceric acid;
$\Delta G^\circ = -12 - (-7) = -5$ kcal/mole

Living organisms are arranged so that spontaneous reactions occurring during degradation of foodstuffs involve the production of compounds whose use in synthetic and mechanical processes ensure that those vital functions also proceed spontaneously. Compounds with high phosphoryl group donor potential such as ATP, which are both products of degradative processes and reactants in syntheses, act as chemical coupling agents in biological machines. For example, as a preliminary to protein synthesis, amino acids have to be attached to molecules of transfer RNA.

$$\text{aa} + t\text{-RNA} \longrightarrow \text{aa-}t\text{-RNA} + H_2O; \quad \Delta G^{\circ\prime} = +6 \text{ kcal/mole}$$

This energy-requiring process is carried out by a set of spontaneous reactions linked by common intermediates, at least one of which is generated during the breakdown of foodstuffs.

$$\text{aa} + \text{ATP} \rightleftharpoons \text{aa-AMP} + 2\text{Pi}; \quad \Delta G^{\circ\prime} = -5 \text{ kcal/mole}$$

$$\text{aa-AMP} + t\text{-RNA} \rightleftharpoons \text{aa-}t\text{-RNA} + \text{AMP}; \quad \Delta G^{\circ\prime} = -1 \text{ kcal/mole}$$

$$\text{AMP} + \text{ATP} \rightleftharpoons 2 \text{ ADP}; \quad \Delta G^{\circ\prime} = -1 \text{ kcal/mole}$$

$$2 \text{ ADP} + 2 \text{ PEP*} \rightleftharpoons 2 \text{ ATP} + 2 \text{ pyruvate}; \quad \Delta G^{\circ\prime} = -12 \text{ kcal/2 mole}$$

$$\text{glucose} + 2 \text{ Pi} \rightleftharpoons 2 \text{ PEP}; \quad \Delta G^{\circ\prime} = -9 \text{ kcal/mole}$$

The overall reaction is

$$\text{aa} + t\text{-RNA} + \text{Glucose} \longrightarrow \text{aa-}t\text{-RNA} + \text{Pyruvate}; \quad \Delta G^{\circ\prime} - 28 \text{ kcal/mole}$$

Thus the biological method of synthesizing aa-t-RNA, by a set of reactions involving glucose degradation to pyruvate, ensures that the process is driven by an overall decrease in standard free energy. A large negative free energy change ensures that the equilibrium position of the reaction is very much in favour of the product since

$$\Delta G^{\circ\prime} = -RT \ln K$$

Thus in the example given above K is approximately 10^{20}; therefore at equilibrium [aa-t-RNA][pyruvate]^2 would be about 10^{20} times greater than [aa][t-RNA][glucose].

Catalysis and Enzymes

A negative value for ΔG indicates that a reaction can proceed but it does not give any indication of the rate at which the change will occur, or the mechanism by which it will occur.

One of the characteristic properties of living cells is their ability to permit complex reactions to proceed quickly at relatively low temperatures, pressures and concentrations, despite the fact that most organic compounds are remarkably stable under those conditions. For example, urea, the chief form of waste

* PEP = phosphoenolpyruvic acid, generated in glycolysis during the breakdown of glucose (See p. 112).

nitrogen in the urine, does not react with water at an appreciable rate at physiological temperatures despite the favourable standard free energy change involved.

$$\begin{array}{c} NH_2 \\ | \\ C=O \\ | \\ NH_2 \end{array} + H_2O \longrightarrow CO_2 + NH_3; \qquad \Delta G^{\circ\prime} = -13\cdot 8 \text{ kcal/mole}$$

However, some bacteria and plants are able to break down urea very rapidly because they possess an enzyme (a protein catalyst) called urease which increases the rate of the reaction under physiological conditions without being consumed in the process.

Almost every reaction occurring in cells is catalysed by its own specific enzyme, thus cells possess several thousands of different enzymes. These do not alter the equilibrium position of their respective reactions, they merely facilitate the attainment of, or approach towards, equilibrium. Much of biochemistry centres around the role of enzymes so we must discover why reactions which occur slowly in the absence of a catalyst, can be speeded up by enzymes.

If two molecules are to react together they must collide; however, that is not to say that each collision involving appropriate molecules will lead to reaction. This is, in fact, far from being the case. In order to react, the colliding molecules must possess between them sufficient kinetic energy to overcome the mutual repulsion of their electron clouds and to redistribute electrons so that an intermediate activated complex is formed. This complex then decomposes, to its appropriate products.

Thus the reaction

$$AB + C \longrightarrow A + BC$$

should be written more correctly as

AB + C	$\longrightarrow$	A - - - -B - - - - C	$\longrightarrow$	A + BC
reactants		activated complex with loosened and rearranged bonds		products

Under normal conditions, few colliding molecules have sufficient kinetic energy to form an activated complex. Hence, the concentration of the complex is low and the rate of product formation (which is proportional to [activated complex]) is correspondingly low. One way in which the rate of reaction can be increased is by heating the reactants so that more colliding molecules have sufficient kinetic energy to form an activated complex. Another way is to introduce a catalyst, such as an enzyme, into the reacting system. This allows the reaction to proceed via an activated complex (or series of activated complexes) involving not only the reactants but also the enzyme. The kinetic energy required to form such a complex is less than that needed to form the complex in the uncatalysed reaction, partly because binding to the enzyme surface facilitates the approach of reactants and redistributes electrons within them in a manner favouring reaction. Hence, at a given temperature, more colliding molecules will be able to form activated complexes than in the absence of the catalyst. The concentration of activated complex will thus be higher than in the uncatalysed reaction and the rate of product formation faster. The role of an enzyme is thus to provide an alternative reaction pathway involving a different activated complex.

A simple representation of the course of an enzyme-catalysed reaction is as shown below:

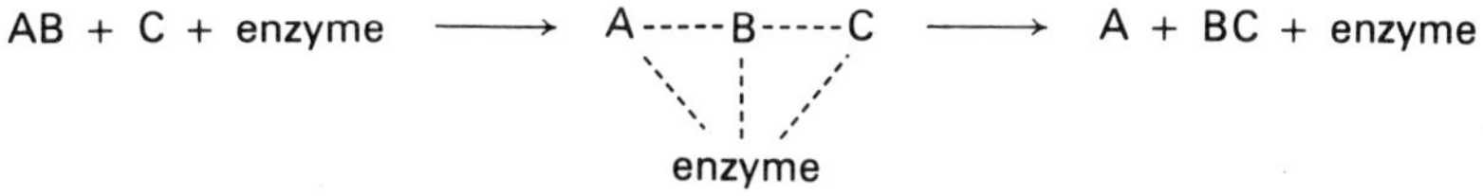

$$AB + C + enzyme \longrightarrow \underset{\underset{\text{enzyme}}{\text{(activated complex)}}}{A\text{-----}B\text{-----}C} \longrightarrow A + BC + enzyme$$

Redox Reactions

The biosphere converts sunlight into heat by a cycle of oxidation–reduction reactions. Carbon dioxide is reduced during photosynthesis to the level of carbohydrate, using light as the energy source and water as the reductant. The resulting carbohydrate is oxidized with the evolution of heat to yield carbon diox-

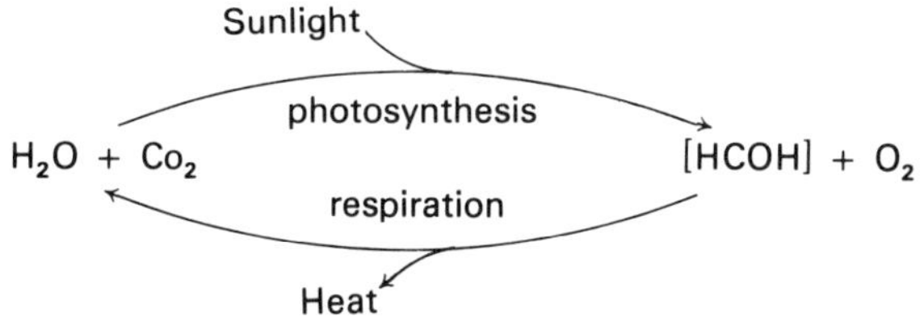

$$H_2O + CO_2 \underset{\underset{\text{Heat}}{\text{respiration}}}{\overset{\overset{\text{Sunlight}}{\text{photosynthesis}}}{\rightleftarrows}} [HCOH] + O_2$$

ide and water during respiration. Oxidations and reductions are thus processes of fundamental importance in living systems.

Oxidations are reactions in which the material oxidized gains oxygen or loses electrons or hydrogen. The following transformations are all examples of oxidations.

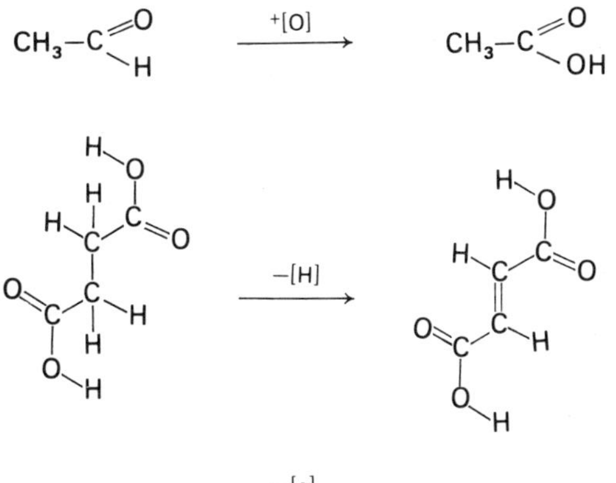

$$CH_3-CHO \xrightarrow{+[O]} CH_3-COOH$$

[structure of tartaric acid-like molecule] $\xrightarrow{-[H]}$ [structure of fumaric/maleic acid-like molecule]

$$Fe^{2+} \xrightarrow{-[e]} Fe^{3+}$$

Conversely, reductions involve a loss of oxygen or a gain of electrons or hydrogen. From this it follows that all oxidations must be accompanied by reduction of the oxidant; thus it is more correct to talk of oxidation–reduction or redox reactions than simply of oxidations or reductions. Some redox reactions representative of those encountered in biochemical systems are indicated below.

$$\text{alanine} + O_2 + H_2O \rightleftharpoons \text{pyruvic acid} + H_2O_2 + NH_3$$

$$\text{fatty acid} + FAD \rightleftharpoons \alpha{:}\beta \text{ unsaturated fatty acid} + FAD \cdot 2H$$

$$\text{cytochrome } b \text{ Fe}^{2+} + \text{cytochrome } c_1 \text{ Fe}^{3+} \rightleftharpoons \text{cytochrome } b \text{ Fe}^{3+} + \text{cytochrome } c_1 \text{ Fe}^{2+}$$

In such reactions the pairs of oxidized and reduced materials such as FAD/FAD.2H and cytochrome b Fe^{2+}/cytochrome b Fe^{3+} are called redox couples. Frequently it is useful to be able to decide whether a reaction such as

$$NADH + H^+ + FAD \rightleftharpoons NAD^+ + FAD \cdot 2H$$

will proceed with the oxidation of NADH + H$^+$ or with the oxidation of FAD.2H. This can be ascertained from a knowledge of the standard redox potential ($E^{\circ\prime}$) of the couples concerned at 25° and pH 7·0.

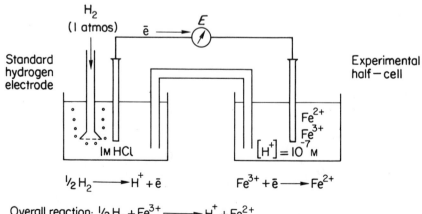

Overall reaction: $\frac{1}{2}H_2 + Fe^{3+} \longrightarrow H^+ + Fe^{2+}$

With a hydrogen electrode as the reference standard, it is possible to determine the electrode potentials (E) of half-cells containing an inert electrode in a solution of a redox couple such as Fe^{2+}/Fe^{3+} at pH 7·0. The standard electrode potential of the hydrogen electrode is arbitrarily defined as zero when $[H^+]$ is 1 M. The potential (E) recorded by the potentiometer is related to the standard electrode potential $E^{o\prime}$ of the Fe^{2+}/Fe^{3+} couple in the experimental half-cell by the expression

$$E = E^{o\prime} + \frac{RT}{nF} \ln \frac{[Fe^{3+}]}{[Fe^{2+}]}$$

where F is the Faraday and n is the number of electrons transferred per mole. Now if the experimental half-cell were to contain equal concentrations of Fe^{2+} and Fe^{3+} ions

$$\frac{RT}{nF} \ln \frac{[Fe^{3+}]}{[Fe^{2+}]} \text{ would equal } 0$$

thus under these conditions $E = E^{o\prime}_{Fe^{2+}/Fe^{3+}}$

The standard redox potentials of some biologically interesting couples at 25° and pH 7·0 are given in Table 1.3. It should be noted that these values are all

Table 1.3 Standard electrode potentials of some redox couples of biological interest. (25°; pH 7·0)

Couple	$E^{o\prime}$
$\frac{1}{2} H_2/H^+$	−0·42 V
$NADH + H^+/NAD^+$	−0·32 V
$Lactate^{2-}/Pyruvate^{2-}$	−0·19 V
$Malate^{2-}/Oxaloacetate^{2-}$	−0·17 V
$Succinate^{2-}/Fumarate^{2-}$	+0·03 V
Coenzyme Q2H/Coenzyme Q	+0·09 V
Cytochromes	
$b\ Fe^{2+}/Fe^{3+}$	+0·10 V
$c\ Fe^{2+}/Fe^{3+}$	+0·21 V
$c\ Fe^{2+}/Fe^{3+}$	+0·23 V
$a/a_3\ Fe^{2+}/Fe^{3+}$	+0·29 V
H_2O/O_2	+0·82 V

relative to a standard redox potential of zero for hydrogen at pH 1; hence the value of $E^{o\prime}$ for the H_2/H^+ couple of −0·42 refers to a hydrogen cell at pH 7·0 relative to one in 1 M HCl. The convention is that the reducing power of a couple is greater the more negative its standard electrode potential.

An important system of redox couples present in mitochondria (see p. 135) permits compounds such as malic acid ($E^{o\prime}$ −0·17 V) to transfer some of their

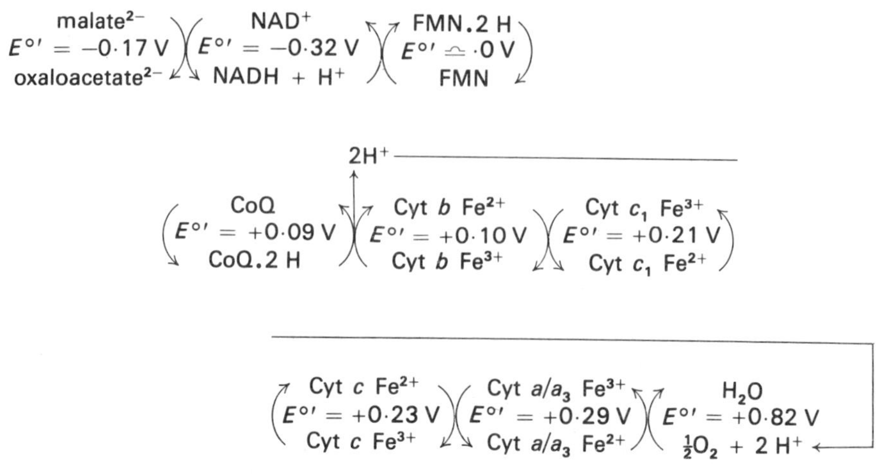

$$\text{malate}^{2-} \underset{\text{oxaloacetate}^{2-}}{\overset{}{\searrow\nearrow}} \underset{\text{NADH} + \text{H}^+}{\overset{\text{NAD}^+}{\searrow\nearrow}} \underset{\text{FMN}}{\overset{\text{FMN.2H}}{\searrow\nearrow}}$$
$$E^{o\prime} = -0.17\text{ V} \quad E^{o\prime} = -0.32\text{ V} \quad E^{o\prime} \simeq .0\text{ V}$$

$$\overset{2\text{H}^+}{\underset{}{\longrightarrow}}$$

$$\underset{\text{CoQ.2H}}{\overset{\text{CoQ}}{\searrow\nearrow}} \underset{\text{Cyt }b\text{ Fe}^{3+}}{\overset{\text{Cyt }b\text{ Fe}^{2+}}{\searrow\nearrow}} \underset{\text{Cyt }c_1\text{ Fe}^{2+}}{\overset{\text{Cyt }c_1\text{ Fe}^{3+}}{\searrow\nearrow}}$$
$$E^{o\prime} = +0.09\text{ V} \quad E^{o\prime} = +0.10\text{ V} \quad E^{o\prime} = +0.21\text{ V}$$

$$\underset{\text{Cyt }c\text{ Fe}^{3+}}{\overset{\text{Cyt }c\text{ Fe}^{2+}}{\searrow\nearrow}} \underset{\text{Cyt }a/a_3\text{ Fe}^{2+}}{\overset{\text{Cyt }a/a_3\text{ Fe}^{3+}}{\searrow\nearrow}} \underset{\tfrac{1}{2}\text{O}_2 + 2\text{H}^+}{\overset{\text{H}_2\text{O}}{\searrow\nearrow}}$$
$$E^{o\prime} = +0.23\text{ V} \quad E^{o\prime} = +0.29\text{ V} \quad E^{o\prime} = +0.82\text{ V}$$

Figure 1.21 The system of redox couples linking malate oxidation to oxygen reduction. N.B. Although the standard electrode potential ($E^{o\prime}$) of the malate/oxaloacetate couple is more positive than that of the NADH + H$^+$/NAD$^+$ couple, malate is able to pass its hydrogens to NAD$^+$ because $E_{\text{malate-oxaloacetate}}$ is more negative than $E_{\text{NADH+H}^+/\text{NAD}^+}$ by virtue of the concentration terms in the equation

$$E = E^{o\prime} + \frac{RT}{nF} \ln \frac{[\text{oxidant}]}{[\text{reductant}]}$$

The concentration of NADH + H$^+$ is kept low because its hydrogens are passed on to FMN and eventually to oxygen

hydrogen atoms to molecular oxygen ($E^{o\prime} = +0.82$ V). The standard free energy change associated with the transfer of hydrogen from malate to oxygen is related to the difference in standard electrode potential between the couples malate^{2-}/ox-

$$\begin{array}{c}\text{O}\diagdown_{\text{C}}\diagup\text{O}^- \\ | \\ \text{H}-\text{C}-\text{OH} \\ | \\ \text{H}-\text{C}-\text{H} \\ | \\ \text{O}\diagdown_{\text{C}}\diagup\text{O}^- \\ \text{malate}^{2-}\end{array} + \tfrac{1}{2}\text{O}_2 \longrightarrow \begin{array}{c}\text{O}\diagdown_{\text{C}}\diagup\text{O}^- \\ | \\ \text{C}=\text{O} \\ | \\ \text{H}-\text{C}-\text{H} \\ | \\ \text{O}\diagdown_{\text{C}}\diagup\text{O}^- \\ \text{oxaloacetate}^{2-}\end{array} + \text{H}_2\text{O}$$

aloacetate^{2-} and oxygen/H$_2$O by the expression $\Delta G^{o\prime} = -\Delta E^{o\prime} nF$ as shown below

$$E_{\text{malate/oxaloacetate}} = E^{o\prime}_{\text{malate/oxaloacetate}} + \frac{RT}{nF} \ln \frac{[\text{oxaloacetate}]}{[\text{malate}]}$$

$$E_{\text{O}_2/\text{H}_2\text{O}} = E^{o\prime}_{\text{O}_2/\text{H}_2\text{O}} + \frac{RT}{nF} \ln \frac{[\text{O}_2]^{\frac{1}{2}}}{[\text{H}_2\text{O}]}$$

At equilibrium
$$E_{\text{malate/oxaloacetate}} = E_{\text{O}_2/\text{H}_2\text{O}}$$

or

$$E^{\circ\prime}_{\text{malate/oxaloacetate}} + \frac{RT}{nF} \ln \frac{[\text{oxaloacetate}]}{[\text{malate}]} = E^{\circ\prime}_{\text{O}_2/\text{H}_2\text{O}} + \frac{RT}{nF} \ln \frac{[\text{O}_2]^{\frac{1}{2}}}{[\text{H}_2\text{O}]}$$

Therefore

$$E^{\circ\prime}_{\text{O}_2/\text{H}_2\text{O}} - E^{\circ\prime}_{\text{malate/oxaloacetate}} = \Delta E^{\circ\prime} = \frac{RT}{nF} \ln \frac{[\text{oxaloacetate}]}{[\text{malate}]} - \frac{RT}{nF} \ln \frac{[\text{O}_2]^{\frac{1}{2}}}{[\text{H}_2\text{O}]}$$

$$= \frac{RT}{nF} \ln \frac{[\text{oxaloacetate}][\text{H}_2\text{O}]}{[\text{malate}][\text{O}_2]^{\frac{1}{2}}}$$

i.e.
$$\Delta E^{\circ\prime} = \frac{RT}{nF} \ln K_{\text{equib}}$$

Now we have seen previously that
$$-RT \ln K_{\text{equib}} = \Delta G^{\circ\prime}$$

Therefore
$$\Delta G^{\circ\prime} = -\Delta E^{\circ\prime} nF$$

N.B. To obtain $\Delta G^{\circ\prime}$ in calories $-\Delta E^{\circ\prime} nF$ must be divided by 4·18 to convert joules to calories.

Thus the standard free energy change for the oxidation of malate is

$$\frac{-0\cdot 99 \times 2 \times 96,500}{4\cdot 18} \text{ cal/mole} = -46 \text{ kcal/mole}$$

Now, mitochondria are constructed so that the flow of electrons from malate to oxygen can be coupled to the synthesis of 3 molecules of ATP from 3 of ADP. The standard free energy for the phosphorylation of ADP is 3×7 kcal = 21 kcal. Hence coupled electron transport and ADP phosphorylation has a net standard free energy change of 25 kcal/mole malate oxidized and is therefore a spontaneous process.

CHAPTER 4

Water

The organization of living organisms is determined to a large extent by the physical properties of water; this is perhaps hardly surprising since it is the chief component of living systems. Water is an extremely polar molecule derived from sp^3 hybridized oxygen; two of the four sp^3 hybrid orbitals of oxygen (which is strongly electronegative) form σ bonds with electropositive hydrogen atoms whilst the two remaining sp^3 orbitals are occupied by lone electron pairs. The molecule is thus tetrahedral in shape with a partial positive charge on both hydrogens and a partial negative charge on the oxygen

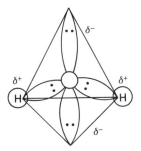

The anomalous and, from a biological standpoint, the significant properties of water are consequences of this extreme polarity.

Water is a liquid at temperatures at which related compounds, such as H_2S, are gases because its molecules are able to form short-lived hydrogen bonds with one another.

$$\overset{\delta+}{H}\diagdown\underset{\underset{H}{\overset{\delta+}{\diagup}}}{\overset{\delta-}{O}}\cdots\overset{\delta+}{H}\diagdown\underset{\underset{H}{\overset{\delta+}{\diagup}}}{\overset{\delta-}{O}}\cdots\overset{\delta+}{H}\diagdown\underset{\underset{H}{\overset{\delta+}{\diagup}}}{\overset{\delta-}{O}}$$

These are weak electrostatic interactions ($\Delta G^{\circ\prime}$ for formation about 4·5 kcal/mole, compared with about 50 kcal/mole for most covalent bonds) between the electropositive hydrogens and electronegative oxygens of adjacent molecules. The energy required to break intermolecular hydrogen bonds accounts for the high latent heats of fusion and evaporation possessed by water. The high latent heat of evaporation (540 cal/g) provides organisms with an effective temperature buffer since it allows heat to be absorbed by the body fluids from the surroundings or to be generated internally without a significant rise in body temperature. Thus organisms are protected from having their metabolic rates grossly disturbed by changes in ambient temperature. On a wider scale, the high latent heat of evaporation and large heat capacity of water permit the sea to act as a buffer so that the diurnal and seasonal variations of temperature on Earth are not too hostile to living systems.

When water freezes to become ice, each water molecule becomes linked to four others by hydrogen bonds so that a regular, open lattice is formed. On average, even at 37° about 15 per cent of water molecules are bonded to four others and liquid water can be regarded as having a flickering structure in which molecules are continually changing their degree of hydration and aggregation.

On account of its polar nature, water is able to act as a solvent for many salts and polar organic molecules. In crystals of a salt, such as NaCl, the distances between Na^+ and Cl^-, Na^+ and Na^+ and Cl^- and Cl^- are defined by the balance between attractive and repulsive electrostatic forces. Polar water molecules, by clustering around the ions, diminish the attractive electrostatic interactions between oppositely charged ionic species and so allow them to enter solution.

Polar organic molecules are brought into solution through the formation of hydrogen bonds with water which effectively build the solute into the structure of the liquid.

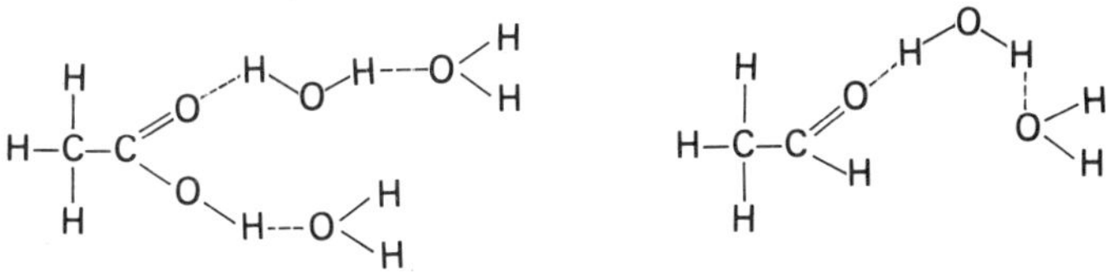

Non-polar hydrocarbons are not readily soluble because they are unable to engage in hydrogen bonding with water. Moreover, if non-polar groups are introduced into a watery medium, water molecules tend to cluster around them to form a hydrogen-bonded cage which allows van der Waals bonds to be formed between the atoms of the non-polar group and the oriented water molecules. This arrangement is energetically less favourable than one in which the non-polar groups cluster together, united by weak van der Waals bonds and the water molecules are allowed to maintain a less ordered conformation. Hence molecules containing both non-polar, hydrophobic groups and polar, hydrophilic groups tend to form micelles in water with the non-polar parts packed together to form an oily phase and the hydrophilic parts projecting into the water. This tendency of hydrocarbon groups to keep out of a watery environment is termed hydrophobic bonding; energetically the most important contribution to this type of interaction is made by maximizing the entropy of water.

Acids and Buffers

A solution is acid when its concentration of hydrogen ions is greater than that of hydroxyl ions and is alkaline when its OH^- ion concentration exceeds that of H^+ ions. Water dissociates to a limited extent into H^+ and OH^- ions as shown below:

$$H_2O \rightleftharpoons H^+ + OH^-$$

since this yields equal numbers of H⁺ and OH⁻ ions, water is neutral (that is neither acid nor alkaline). The dissociation constant of water (K_a) is given by the expression

$$K_a = \frac{[H^+][OH^-]}{[H_2O]}$$

but, as the concentration of water in water (1000/18 = 55·5 M) is very much larger than that of H⁺ and OH⁻ ions, it may be regarded as a constant. Therefore we can write

$$K_a[H_2O] = K_w = [H^+][OH^-]$$

K_w is called the ion product of water and has a value of 10^{-14} at 25°. Thus

$$[H^+][OH^-] = 10^{-14}$$

since

$$[H^+] = [OH^-], [H^+] = 10^{-7} \text{ M}$$

That is, neutral solutions have a hydrogen ion concentration of 10^{-7} M.

It has become a widespread convention to express acidity not in terms of [H⁺] but by means of a scale of pH, where pH is defined as $-\log[H^+]$. Thus an acidic solution containing H⁺ ions at a concentration of 10^{-2} M is at a pH of 2. Neutral solutions whose $[H^+] = 10^{-7}$ M have a pH of 7 whilst an alkaline solution with a [H⁺] of 10^{-10} M (and therefore an [OH⁻] of 10^{-4} M) has a pH of 10.

	←——————Acidic——————						→Neutral←			——————Alkaline——————				
pH	0	1	2	3	4	5	6	7	8	9	10	11	12	13	14
[H⁺]	10^0	10^{-1}	10^{-2}	10^{-3}	10^{-4}	10^{-5}	10^{-6}	10^{-7}	10^{-8}	10^{-9}	10^{-10}	10^{-11}	10^{-12}	10^{-13}	10^{-14}
[H⁺]	1 M	0·1 M	0·01 M etc.												
[OH⁻]	10^{-14}	10^{-13}	10^{-12}	10^{-11}	10^{-10}	10^{-9}	10^{-8}	10^{-7}	10^{-6}	10^{-5}	10^{-4}	10^{-3}	10^{-2}	10^{-1}	10^0

A useful definition of an acid is that given by Brønsted, namely that acids can donate protons whilst bases can accept them. Thus acetic acid

$$H-\underset{\underset{H}{|}}{\overset{\overset{H}{|}}{C}}-C\overset{O}{\underset{OH}{\diagdown}}$$

is an acid because it can release a proton or hydrogen ion, thus

$$H-\underset{\underset{H}{|}}{\overset{\overset{H}{|}}{C}}-C\overset{O}{\underset{OH}{\diagdown}} \rightleftharpoons H-\underset{\underset{H}{|}}{\overset{\overset{H}{|}}{C}}-C\overset{O}{\underset{O^-}{\diagdown}} + H^+$$

whilst the acetate ion

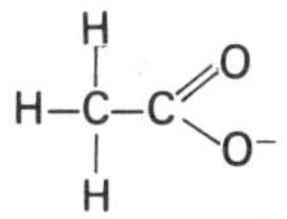

$$\text{H-C(H)(H)-C}(\!=\!\text{O})\text{O}^-$$

is the conjugate base of acetic acid since it can accept a proton to form the acid.

Similarly the ammonium ion, NH_4^+, is an acid because it can donate a proton thus

$$NH_4^+ \rightleftharpoons NH_3 + H^+$$

clearly ammonia (NH_3) is the conjugate base in this instance. Many of the compounds encountered in biochemical systems are weak acids whose degree of dissociation varies with changes in ambient hydrogen ion concentration and this exerts a profound influence upon their chemical behaviour.

Let us examine the process of dissociation in more detail by considering a weak acid, HA, which dissociates to give a proton and conjugate base;

$$HA \rightleftharpoons H^+ + A^-$$

the dissociation, or equilibrium, constant is given by the expression,

$$K_a = \frac{[H^+][A^-]}{[HA]}$$

Rearranging gives,

$$[H^+] = K_a \frac{[HA]}{[A^-]}$$

Hence when the acid is half dissociated, i.e. [HA] = [A$^-$]

$$[H^+] = K_a$$

Thus the dissociation constant of a weak acid is equal to the hydrogen ion concentration at which the acid is half dissociated, i.e. 50 per cent exists as acid [HA] and 50 per cent as the conjugate base [A$^-$]. Now if

$$[H^+] = K_a \frac{[HA]}{[A^-]}$$

$$\log [H^+] = \log K_a + \log \frac{[HA]}{[A^-]}$$

then

$$-\log [H^+] = -\log K_a - \log \frac{[HA]}{[A^-]}$$

or

$$pH = pK_a + \log \frac{[A^-]}{[HA]}$$

which is known as the Henderson–Hasselbalch equation. When the acid is half dissociated,

$$[A^-] = [HA]$$

and

$$pH = pK_a$$

Thus the pK value for a weak acid gives the pH at which the acid is half dissociated.

The Henderson–Hasselbalch equation is useful because it can be used to calculate the composition of buffer solutions of defined pH with a resistance to changing pH when acid or alkali is added. For example, a buffer solution at pH 5·1 can be prepared for acetic acid and sodium acetate.

$$pK_a \text{ for acetic acid} = 4\cdot76$$
$$\text{Sodium acetate concentration} = 0\cdot14 \text{ M}$$
$$\text{Acetic acid concentration} = 0\cdot06 \text{ M}$$
$$pH = 4\cdot76 + \log \frac{0\cdot14}{0\cdot06}$$
$$= 4\cdot76 + \log 2\cdot33$$
$$= 4\cdot76 + 0\cdot37$$
$$= 5\cdot13$$

Such buffer solutions provide controlled environments for biochemical reactions in the laboratory, whilst in living organisms buffer systems form part of the mechanism for regulating the internal environment. Buffer action is illustrated by Figure 1.22 which shows the effect of titrating 100 millimoles of a weak

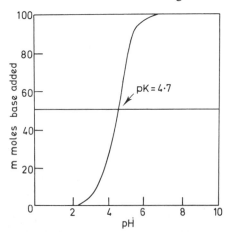

Figure 1.22 Titration curve of a weak acid with a strong base

monobasic acid with a strong base. It can be seen that the pH change per millimole of base added is least when the acid is half neutralized, (i.e. when pH = pK_a) and is low in the pH range $pK_a \pm 1$. Thus buffer systems can be constructed using the Henderson–Hasselbalch equation, such that the pH can have any value from $pK_a -1$ to $pK_a +1$ by adjusting the relative proportion of acid HA and its conjugate base A^-. The more concentrated the buffer, i.e. the higher the value of $[HA] + [A^-]$, the greater will be its capacity to maintain its pH when acid or alkali is added. Thus a 0·2 M acetate buffer at pH 5·1 will be able to mop up more acid or alkali than a 0·02 M acetate buffer at the same pH.

Buffering is important, both experimentally and in living organisms, because once the pH is defined it follows that the state of ionization of acidic and basic groups will also be defined. For example, an acidic group with $pK_a = 3$ will be 50 per cent dissociated at pH 3 and almost completely dissociated at pH 7. Similarly an acid with a pK_a of 9 will be 50 per cent protonated at pH 9 and virtually completely protonated at pH 7. Amino acids such as glycine

$$H-\underset{\underset{NH_2}{|}}{\overset{\overset{H}{|}}{C}}-C\underset{OH}{\overset{O}{\diagup\!\!\!\diagdown}}$$

carry no net charge at about pH 6·5 because their carboxyl groups are dissociated and amino groups are protonated

$$H-\underset{\underset{NH_3^+}{|}}{\overset{\overset{H}{|}}{C}}-C\underset{O^-}{\overset{O}{\diagup\!\!\!\diagdown}}$$

This is because pK_as for their carboxyl and amino groups are around 3 and 9 respectively. Figure 1.23 shows the titration curve for glycine.

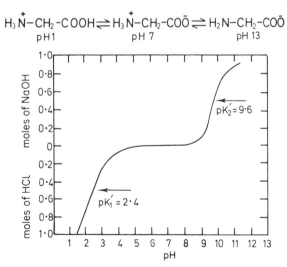

Figure 1.23 The titration curve of glycine

The imposition of certain states of ionization on acidic groups in enzymes and substrates has profound implications for the reactivity, membrane solubility and form of biologically important molecules. For example, dissociation of side chain carboxyl groups in some polypeptides as the pH increases (i.e. [H$^+$] falls) can lead to intramolecular repulsions which may cause a formerly globular molecule to assume the form of a random coil. Many drugs can pass through cell membranes

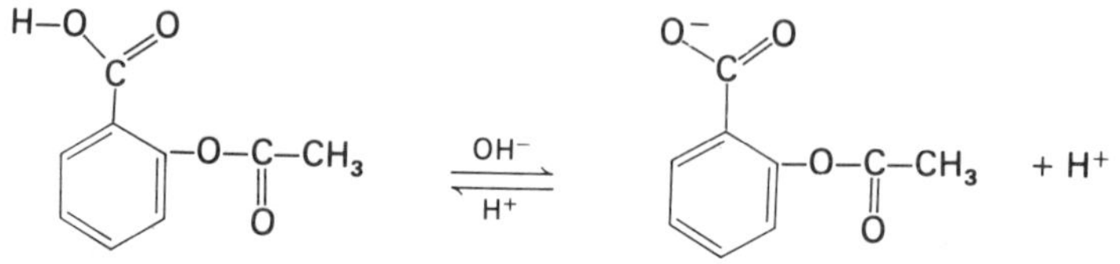

$$\text{aspirin-COOH} \underset{H^+}{\overset{OH^-}{\rightleftharpoons}} \text{aspirin-COO}^- + H^+$$

only when they are uncharged, for example aspirin is absorbed more readily from the stomach than from the small intestine because the low pH of the stomach contents (high [H$^+$]) ensures that a high proportion of the molecules are in the protonated state.

CHAPTER 5

Proteins

Proteins are so varied in shape, size, composition, physical properties and function that it is difficult to make many statements about them as a class that do not have to be hedged about with qualifications and exceptions. Indeed, it is their diversity that gives them such significance in living systems where they act, amongst other things, as enzymes, transducers, structural elements, receptors, antibodies, hormones and toxins.

The Isolation of Proteins

As the detailed study of an enzyme (or other protein) often involves its isolation in pure form, a variety of special procedures has been designed for this purpose. One of the commonest is the stepwise addition of ammonium sulphate to a solution of protein, which, by removing water from the surface of the protein molecules, allows them to aggregate and precipitate. This process, known as 'salting out', can be used to separate those proteins which precipitate in different concentrations of ammonium sulphate. Proteins can also be precipitated selectively by altering the pH of the solution as they often have different 'isoelectric points', i.e. pH values at which the protein will not move in an electric field and at which the solubility is minimal. Finally, proteins may be selectively adsorbed onto materials such as calcium phosphate gels or columns of cellulose derivatives (carboxymethylcellulose and diethylaminoethylcellulose) from which they can be eluted by solutions of different salt concentration or pH.

Another technique of considerable value, gel filtration, depends on the separation of proteins from one another or from other compounds on the basis of their molecular dimensions. An artificially cross-linked dextran provides a suitable filter. The holes within the dextran network allow the entry of small molecules and exclude the large; therefore smaller molecules are entrapped in the dextran while larger molecules pass through. A number of grades of suitable dextrans (marketed under the trade name of Sephadex) are available for separating compounds of molecular weight from about 4000 to 200,000.

Testing the Physical Homogeneity of Isolated Proteins

A number of techniques may be employed to test the purity of an isolated protein. One method used extensively is to see whether the protein sediments in an homogeneous way in a high centrifugal field developed in an ultracentrifuge;

this instrument can also give information about the molecular weight, shape, size and density of the protein. The centrifuge cell spins fast enough to sediment the larger dissolved molecules under forces of 5×10^5 g and by measuring the change in refractive index across the cell it is possible to detect the point at which a solute 'boundary' is separating from the solvent. If the protein solution has a number of components they will sediment at different rates and a number of peaks (corresponding to the changes in the refractive index) will be observed. Although a pure protein solution should sediment as one peak, it does not follow that if a single peak is found the protein must be pure, as proteins sometimes associate and sediment as a complex.

Another technique for testing the purity of a protein depends on the fact that all protein molecules are charged, the sign of the net charge depending on the pH of the solution. Thus, if a protein molecule in solution is subjected to an electric field it will tend to move towards one or the other pole, depending on the sign of its charge. Hospital laboratories commonly perform electrophoresis using a gel (of starch or polyacrilamide) as the medium through which the protein migrates. Figure 1.24 illustrates the separation of various forms (termed 'isoenzymes' or 'isozymes') of the enzyme lactic dehydrogenase in human plasma; these isozymes vary in proportion in different diseases. The illustration shows two characteristic patterns of isozymes which occur in infectious hepatitis and myocardial infarction.

Immunological Tests

Proteins purified until they give only one peak in the ultracentrifuge under a variety of conditions and which migrate as a single peak during electrophoresis in a range of buffers are usually considered as pure proteins. However, immunological examination of preparations from different organs can reveal differences in proteins which behave in the same way in all these physical tests. These procedures measure the overall statistical properties of the molecule—for example, a protein of moderate molecular weight may contain about 300 amino acid residues selected from the 20 common amino acids. Even if exactly the same numbers and sorts of amino acids were selected, the arrangement within the molecule could be tremendously varied and still produce the same physical properties (sedimentation velocity, net charge). Nevertheless, two apparently identical proteins of this type injected separately into animals may produce two separate sets of antibodies, which can be detected by precipitation when antigen and antisera are mixed. Such a test is probably the most sensitive method of distinguishing between proteins which differ only slightly in their structure.

The Structure of Proteins

All proteins contain one or more polypeptide chains each of whose amino acid sequence is defined by the store of genetic information in the cell where they were synthesized. In addition to their polypeptides, many proteins contain other components of low molecular weight and varied nature.

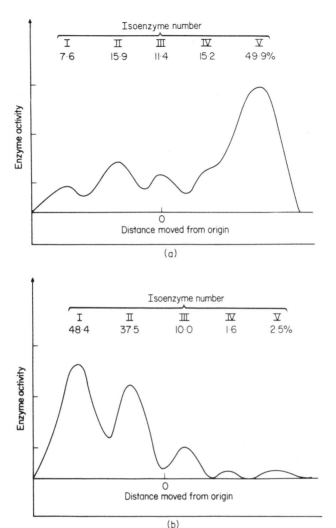

Figure 1.24 The isoenzymes of serum lactic dehydrogenase separated by electrophoresis in agar gel. After electrophoresis each pattern contains 5 isoenzymes, their activities being detected by allowing the enzyme in the gel to react with pyruvate NADH + H$^+$, and measuring the decrease in optical density (at 366 nm) resulting from the oxidation of the NADH + H$^+$. Pattern (a) infectious hepatitis; pattern (b) myocardial infarction. From J. Kamaryt and Z. Zazvorka, *Science Tools*, 1963

The three-dimensional form of most proteins in aqueous solutions probably represents, within limits set by their covalent structure and the prevailing physical conditions, the spatial arrangement that confers the lowest possible free energy on the protein–water system as a whole. It is, however, conceivable that the conformation of some proteins represents the structure that was most stable at the time of their synthesis and original coiling but is metastable under later conditions.

In describing protein structure it is most convenient to consider the covalent or primary structure of the polypeptide chains before dealing with the more numerous, but weaker, interactions which are of particular importance in determining their overall arrangement in space.

The Covalent Structure of Proteins

The polypeptide chains of proteins consist of amino acid residues joined together in a linear sequence; the linkage between successive residues (the peptide bond

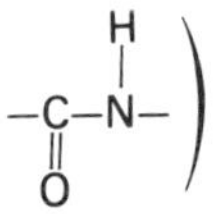

$$\left(-\underset{\underset{O}{\|}}{C}-\underset{\underset{}{|}}{\overset{H}{N}}- \right)$$

is derived from the α-carboxyl group of one amino acid and the α-amino group of the next by the elimination of water. Thus all the polypeptides have the general structure

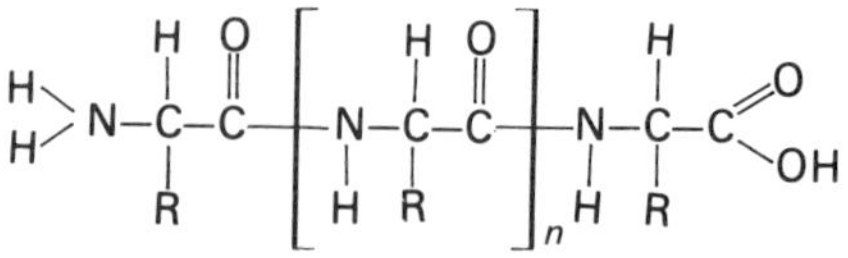

$$H_2N-\underset{R}{\overset{H}{\underset{|}{C}}}-\underset{}{\overset{O}{\underset{\|}{C}}}-\left[-N-\underset{R}{\overset{H}{\underset{|}{C}}}-\underset{}{\overset{O}{\underset{\|}{C}}}- \right]_n -N-\underset{R}{\overset{H}{\underset{|}{C}}}-C\underset{OH}{\overset{O}{\diagup}}$$

where n may have any value between about 50 and 2000; R represents the side group of any of the amino acids described in Figure 1.14. All polypeptides of this type have one terminal α-amino group and one terminal α-carboxyl group but differ from each other in the number and sequence of their constituent amino acids.

Some proteins, myoglobin for instance, possess a single polypeptide chain whilst others contain several chains which are not linked to one another by covalent cross-bridges.

The presence of two or more cysteine residues

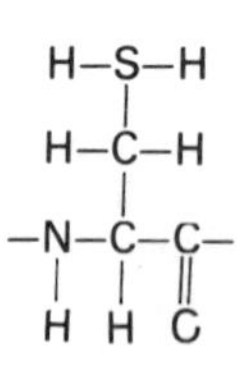

$$\begin{array}{c} H-S-H \\ | \\ H-C-H \\ | \\ -N-C-C- \\ | \; | \; \| \\ H \; H \; O \end{array}$$

in a polypeptide chain introduces the possibility of forming a loop in the chain by means of a disulphide (S—S) bridge (see Figure 1.11); four such loops occur in the enzyme lysozyme (Figure 1.25). Interchain disulphide bridges are frequently found in proteins such as insulin with two or more polypeptide chains.

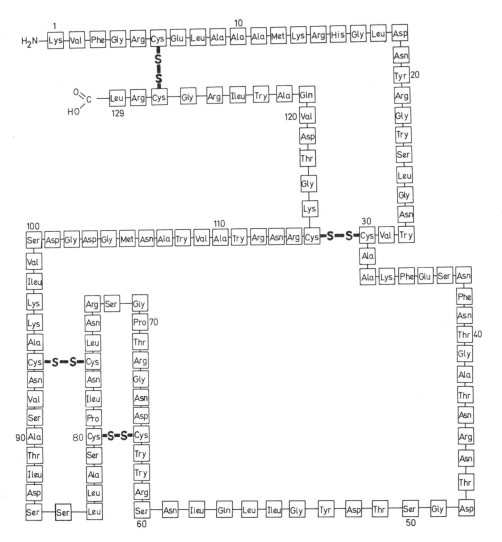

Figure 1.25 The covalent structure of lysozyme. Gln = glutamine; Asn = asparagine

Determination of the Sequence of Amino Acids in Proteins

Once a protein has been isolated in a highly purified form, it is possible to determine the sequence of component amino acids. This was done first for insulin. Obviously, simple hydrolysis of the protein into its constituent amino acids tells nothing about their arrangement, though if the protein is first reacted with some compound which labels an amino acid by virtue of its position in the molecule, subsequent hydrolysis and identification of the labelled amino acid will be useful. The compound 1-fluoro-2,4-dinitrobenzene reacts with free amino

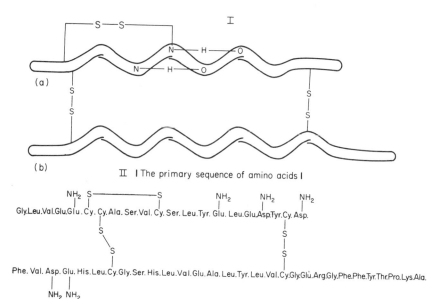

Figure 1.26 A diagrammatic representation of the primary structure of the protein molecule insulin. The protein is composed of two polypeptide chains, designated (a) and (b). The two polypeptide chains are held together by two prominent sulphur-to-sulphur (disulphide) bridges. A similar disulphide bridge exists between two amino acids on the (a) chain. (The abbreviations are the initial letters of the names of the constituent amino acids: Glu—NH$_2$, Asp—NH$_2$, Cy—S—S—Cy represent glutamine, asparagine and cystine, respectively.) Modified from C. B. Anfinsen, *The Molecular Basis of Evolution*, Wiley, New York, 1963

groups in proteins, giving the coloured dinitrophenyl (DNP) derivatives which after hydrolysis can be isolated and identified by paper chromatography.

Such a procedure indicated that the insulin molecule had an N-terminal glycine and an N-terminal phenylalanine residue and that there was one lysine residue. Thus each molecule consisted of two linked polypeptide chains (the A and the B chain). As the protein contained cystine and, when treated with a mixture of formic acid and hydrogen peroxide (which splits disulphide bridges), broke down into A and B chains (which could be isolated and purified) it was apparent that these two chains were linked by disulpide bridges.

The sequence of amino acids within the two separated chains was determined by partial hydrolysis, identification of the N-terminal residues of each resulting polypeptide chain as the dinitrophenyl derivative followed by complete hydrolysis to give the amino acid composition of the polypeptides. The results obtained, together with the proportions of amino acids in the entire chain and its molecular weight, could be used to deduce the particular sequence.

The use of specific hydrolytic enzymes for the initial cleavage of the protein has made such analyses somewhat easier. The hydrolytic enzymes used are of two kinds, firstly the exopeptidases, which remove amino acids from either the N-terminal end (amino peptidases) or the carboxyl end (carboxypeptidase) of the

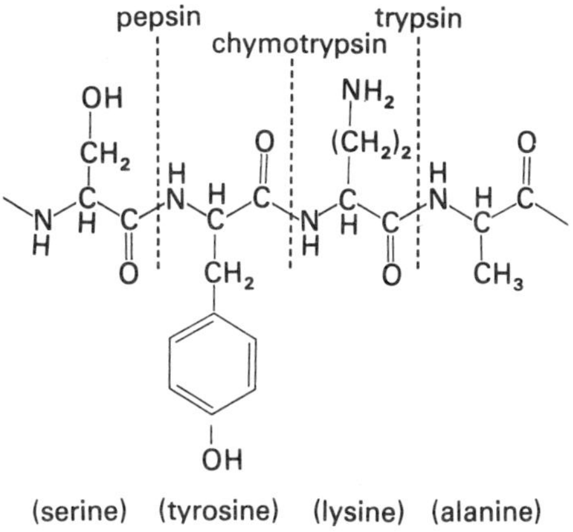

Figure 1.27 The bond specificities of pepsin, trypsin and chymotrypsin

protein; and secondly endopeptidases which attack the protein specifically wherever certain groupings are present in the molecule. Thus pepsin cannot act at positively charged residues in the chain (lysine or arginine) whereas trypsin acts only in this situation. Pepsin and chymotrypsin both require an aromatic amino acid at the linkage; for pepsin the aromatic residue must provide the —NH group of the peptide bond, for chymotrypsin it must provide the C=O group.

The Three-dimensional Conformation of Proteins

Peptide linkages have considerable double-bond character (usually indicated thus:

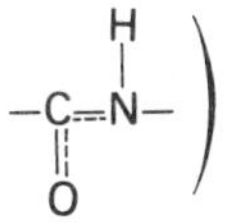

$$-\underset{\underset{O}{\|}}{C}=\underset{}{N}-\underset{}{H}$$

Consequently rotation about the axis does not occur to any significant extent and the C, O, N and H atoms are essentially coplanar. Furthermore, in almost all known instances the C=O and N—H groups are in the *trans* conformation relative to one another. Rotation is possible about the N—C bond ($\phi°$) and about the C—C bond ($\psi°$) and the three-dimensional conformation of the chain would be defined if all the rotations were known.

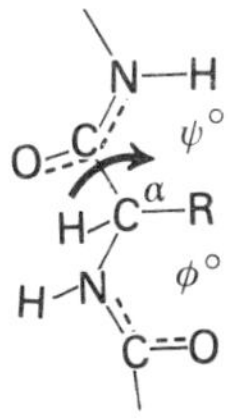

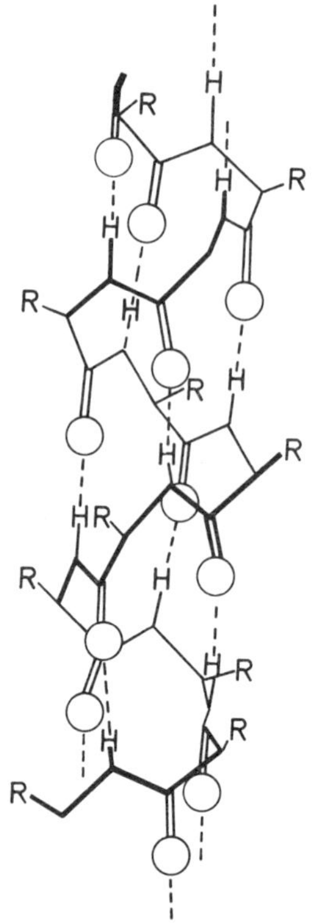

Figure 1.28 A right-handed α-helix. There are 3·6 amino acids per turn, a pitch of 5·4 Å, and a diameter of 10·5 Å. Notice that all —C=O and —NH groups form hydrogen bonds. In the right-handed α-helix, all R-groups point away from the helix. $\phi = 48°$; $\psi = 57°$

If the rotations $\phi°$ and $\psi°$ have constant values throughout a polypeptide, the chain assumes a regular helical pattern whose pitch and number of repeating units depends on the precise values of ϕ and ψ. The best known of these helices is the α-helix. On the other hand, if there is a variety of values for ϕ and ψ, the chain assumes a much more irregular conformation. Some values of ϕ and ψ do not occur as they lead to very unfavourable steric interactions.

The values of ϕ and ψ which exist in the polypeptides chains of proteins are determined by the natural tendency of the protein–water system to achieve the lowest possible state of free energy. Generally this involves maximizing favourable weak interactions between various groups in the chains and minimizing unfavourable interactions between the polypeptides and water. The reason for this is that the greatest loss of free energy and consequent gain in stability is often achieved by maximizing entropy or disorder of water. Clearly maximizing inter- and intrachain weak links diminishes the number of polypeptide groups available for binding with, and hence ordering, water.

The examples that follow indicate the various factors contributing to the overall three-dimensional structure of proteins. The structures concerned have been determined by X-ray crystallography. X-rays passing through crystals of a protein are diffracted and can be made to yield diffraction patterns of spots of differing intensity on photographic films. From the position and intensities of the spots it is possible to calculate the positions of the diffracting atoms in the crystal and hence the structure of the individual protein molecules.

Silk

About 86 per cent of the amino acids of silk are provided by glycine, alanine and serine in the ratio 3 : 2 : 1. Fibres of silk are constructed from extended polypeptide chains lying along the axis of the fibre. Neighbouring chains run in opposite directions and are hydrogen bonded together through the C=O and N—H groups of their peptide bonds to form a pleated, sheet-like, structure (Figure 1.29). The fibre consists of many layers of such sheets. The silk polypeptide is unusual in that it possesses repeating units of six amino acids along much of its length.

$$-(Gly-Ser-Gly-Ala-Gly-Ala)_n-$$

As a consequence of this, all the glycine residues extend from one side of the sheet and all the alanine and serine residues from the other. The sheets are stacked together with glycine/glycine faces and serine/serine faces in contact. The extended covalent bonds along the axis of the polypeptides render the silk fibre relatively inextensible whilst the weaker van der Waals forces between the sheets permit it to be flexible.

Tropocollagen

Tropocollagen, the protein subunit of collagen, is about 15 Å in diameter, 2800 Å long and has a molecular weight of about 300,000. Its composition is peculiar because glycine accounts for 33 per cent of its amino acid content and the α-imino acids proline and hydroxyproline together for another 21·6 per cent. The molecule consists of three polypeptide chains each of which is an extended, left-hand, helix with three amino acid residues per turn. These three helices are coiled around each other to form a right-handed, three-fold, super helix. Two-thirds of the amide and carbonyl groups in any one chain are involved in hydrogen bonding with complementary groups in the other two chains. Every third amino acid side group in a given chain projects into the centre of the three-fold super helix and can only be accommodated if it is an α-hydrogen atom of

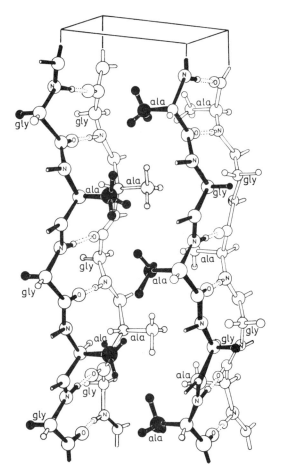

Figure 1.29 The three-dimensional structure of silk. The side chains of one sheet nestle between those of neighbouring sheets. The cut bonds extend to neighbouring chains in the same sheet. $\phi = 60°$, $\psi = 135°$ and there are two amino acid residues per turn of the chains

glycine: this steric restraint accounts for glycine providing 33 per cent of the amino acid residues in the molecule.

The favoured form of chains with a high proportion of the α-imino acids, proline and hydroxyproline, which do not fit into α-helices, is an extended one maintained by interchain hydrogen bonding. These structural features make the triple chain, coiled-coil molecule of tropocollagen very inextensible and rigid. Individual tropocollagen molecules are only sparingly soluble and have a strong tendency to aggregate to form insoluble collagen fibres (see p. 264).

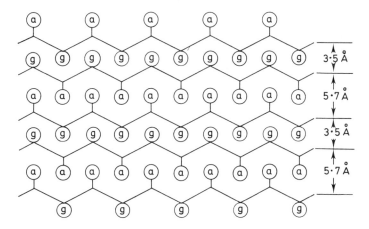

Figure 1.30 The sheets in silk are packed with alanine (a) against alanine and glycine (g) against glycine. The spacing between sheets therefore alternates between 5·7 Å and 3·5 Å. For clarity serine residues are not shown, in reality they would occupy some of the places designated by a

Figure 1.31 Tropocollagen showing the three left-handed polypeptide helices coiling together to form a right-handed super helix

Myoglobin

Unlike the two structural proteins discussed above which were very elongated and insoluble, myoglobin is a soluble, globular, protein whose single polypeptide chain is tightly coiled upon itself in a complex manner. The molecule consists of eight lengths of α-helix linked by non-helical regions; the whole being arranged to form a box into which fits the oxygen-carrying haem group. Some of the non-helical regions contain a proline residue which prevents the assumption of an α-helical form. The stability of an α-helix, in the absence of any other stabilizing

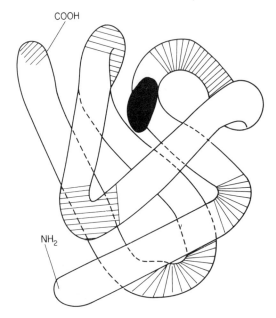

Figure 1.32 Diagrammatic representation of helical (unshaded) and non-helical (shaded) parts of the molecule of myoglobin. Also shown are N-terminal (NH_2) and C-terminal (COOH) ends of the structure and the position of the haem group (black)

factor, depends upon whether the formation of —C=O---H—N— hydrogen bonds between successive turns of the helix (see Figure 1.28) or hydrogen bond formation between water and the —C=O and —N—H groups leads to the greater loss of free energy. In general there is little to choose between these alternatives so that the α-helix is not intrinsically very stable. It may, however, become a stable conformation if further stability can be gained by interactions of the side chains with other groups in the molecule. The α-helices of myoglobin are stabilized because their hydrophobic amino acid residues interact through van der Waal's forces with hydrophobic residues in other α-helical segments. Myoglobin also conforms to a pattern common in globular proteins—namely that of having most of its hydrophobic residues packed together with the centre of the molecule and the hydrophilic ones at the surface. The removal of the hydrophobic groups from contact with water is largely driven by the favourable free energy loss that accompanies the freeing of water molecules that otherwise tend to be in an ordered state around hydrophobic residues. Free energy is also lost by virtue of the enforced van der Waals interactions between the packed hydrophobic residues forming the core of the protein.

The haem residue fits into the cleft in the molecule so that its non-polar edge interacts with non-polar amino acid residues and the iron atom is situated between two histidine residues.

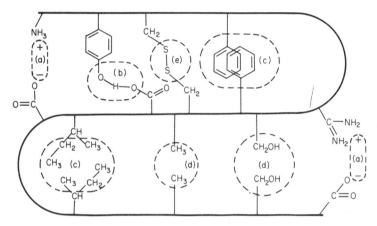

Figure 1.33 Some types of non-covalent bonds which stabilize protein structure: (a) electrostatic interaction; (b) hydrogen bonding between tyrosine residues and carboxylate groups on side chains; (c) interaction of non-polar side chains caused by their mutual exclusion from water; (d) van der Waals interactions; (e) a disulphide linkage, a covalent bond. From C. B. Anfinsen, *The Molecular Basis of Evolution*, Wiley, New York, 1959

Although myoglobin contains a high proportion of α-helix this feature is not found in all globular proteins. α-Chymotrypsin contains very little α-helix but has extensive regions where the polypeptide chain is extended and folded back upon itself to form a structure not unlike the β-pleated sheet found in silk.

This review of protein structure indicates that the conformation which is adopted in any instance is the result of the interplay of so many subtle interactions that it cannot be predicted *a priori* at the present time. A summary of the weak bonds maintaining the conformation of proteins is shown in Figure 1.33.

Macro-molecular Aggregates and Self-Assembly

Many functionally active proteins are oligomers consisting of two or more protein subunits or protomers; the forces uniting the subunits are usually weak, non-covalent, bonds that can be broken by altering the pH or ionic strength or by adding dissociating agents such as urea.

Adult human haemoglobin is an oligomer constructed from two α- and two β-subunits closely related to myoglobin in composition and conformation. Solutions of α- and β-subunits mixed under suitable conditions aggregate spontaneously into tetrameric haemoglobin oligomers. The driving force in this process is the natural tendency of all systems (in this case the water/protein system) to lose the largest possible amount of free energy. The largest factor in the free energy loss probably being the gain in entropy achieved as water molecules are freed from

Figure 1.34 Model of haemoglobin molecule at 5·5 Å resolution, showing the two α-chains (white) and two β-chains (black) tetrahedrally arranged relative to each other (Perutz and coworkers, 1960)

the ordering effects of groups in the subunits able to engage in interprotomer bonds. There are some 19 amino acid residues involved in linking unlike subunits having neighbouring haem groups (i.e. $\alpha_1\beta_2$ and $\alpha_2\beta_1$ (see Figure 1.34) and about 34 joining unlike subunits having more widely separated haems ($\alpha_1\beta_1$ and $\alpha_2\beta_2$). Most of the intersubunits bonds are hydrophobic but some hydrogen bonds and charged group interactions are also involved. There is very little contact between α_1 chains and between pairs of β_2 chains.

The formation of the oligomer introduces a subtlety into oxygen binding and release not shared by myoglobin or by the free α- and β-protomers. The binding of oxygen by the haem group of one protomer in the oligomer facilitates the binding of oxygen by the others. This cooperative effect is a consequence of the conformational changes induced by oxygen binding to one protomer being transmitted to the others by changes in the orientation of groups involved in interprotomer bonding.

This brief survey of the structure of haemoglobin indicates that protomers having surface regions with sufficient complementarity of shape and grouping can interact with one another to form precisely ordered polymeric aggregates held together by the summation of many weak forces. The orientation of the protomers in an oligomer is determined by the amino acid sequence of the protomers since that determines their three-dimensional forms and hence the dis-

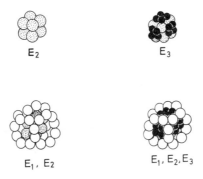

Figure 1.35 The structure of the pyruvic dehydrogenase complex. $E_1 =$ Pyruvic dehydrogenase, $E_2 =$ Lipoyl acetyl transferase, $E_3 =$ Lipoyl dehydrogenase

tribution of interacting (protomer/protomer binding) groups on their surfaces. The oligomers produced in this way can display more complex behaviour than their isolated protomers. Hence a process of spontaneous self-assembly based on weak interactions between complementary surfaces can lead to the formation of systems showing complex physiological behaviour. For example, aspartate transcarbamylase (see p. 187) is an oligomeric enzyme consisting of six catalytic subunits and six regulatory subunits. The activity of the enzyme is modified when small molecules, termed heterotropic effectors (see p. 87), bind to the regulatory units and induce a conformational change that is transmitted to the catalytic units. Pyruvate dehydrogenase is a more complex oligomer composed of three different types of enzymically active protomer each of which catalyses a different reaction. The multienzyme system contains a total of 72 subunits (24 of enzyme 1, 24 of enzyme 2 and 24 of enzyme 3) and converts pyruvate to acetyl coenzyme A via three sequential reactions in which the substrate is passed directly from one enzyme to the next (see p. 122).

CHAPTER 6

Enzymes

The Specificity of Enzyme Action

From the preceding discussion of the structure of proteins it is clear that the organization of an enzyme molecule in space will be extremely precise, with the polypeptide chain being folded in a certain way and, in consequence, having particular amino acids present at particular positions on its surface. Thus the enzyme in its cellular environment will have a certain characteristic shape or conformation which in a certain region provides those chemical groupings essential for specific interaction with its 'substrate' (the compound whose transformation is being catalysed). These points of interaction are called 'active sites' and a most important and striking property of enzymes is the selectivity of the binding of substrate in relation to these active sites, i.e. the *specificity* of the reaction catalysed by the enzyme. Commonly a small group of substances of similar chemical nature act as substrates for the enzyme though one compound reacts at a greater rate than the others and is often regarded as *the* substrate utilized by the enzyme *in vivo*. For example, the enzyme alcohol dehydrogenase will react with a wide variety of alcohols, transforming them to the corresponding aldehydes, but *ethanol* is the substrate utilized most rapidly.

$$\begin{array}{c} CH_2OH \\ | \\ CH_3 \end{array} + NAD^+ \rightleftharpoons \begin{array}{c} H\diagdown \diagup O \\ C \\ | \\ CH_3 \end{array} + NADH + H^+$$

With some enzymes, the range of substrates is very wide; for example, the enzyme esterase hydrolyzes any carboxylic ester to the corresponding alcohol and acid.

Many organic compounds have a centre of asymmetry and exist in D- and L-forms (see p. 12). Enzymes acting on such compounds usually exhibit complete optical specificity, reacting with only one of the two isomers. Thus kidney contains two enzymes each oxidizing a wide range of amino acids and producing α-keto acids, ammonia and hydrogen peroxide. One enzyme attacks only the D-amino acids, the other only the L-forms. Enzymes in mammals attacking the D-configuration of amino acids are rare, possibly because mammalian protein is built up from the L-amino acids series, the D-series being confined almost entirely to some peptides in bacteria. There is, however, a small group of enzymes that has both the optical isomers as substrates. These enzymes, the racemases, convert one isomer into the other.

Sometimes an enzyme may act on only one of a pair of *cis–trans* [isomers]. [For] example, fumarase catalyses the formation of malic acid from fuma[ric acid, the] *trans* isomer) by adding water across the double bond; this enzyme wi[ll not react] with the *cis* isomer, *maleic* acid.

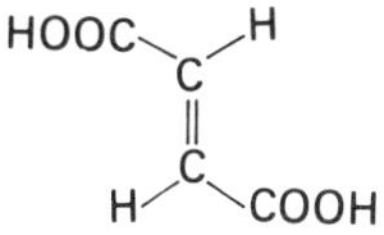

$$\text{fumaric acid} \qquad \text{maleic acid}$$

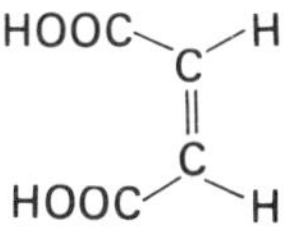

Some enzymes appear to have such a restricted specificity that only one substance will serve as a substrate. Thus *urease*, found in the plant kingdom and also in the mammalian gut because of the resident population of microorganisms, attacks only urea to form carbon dioxide and ammonia.

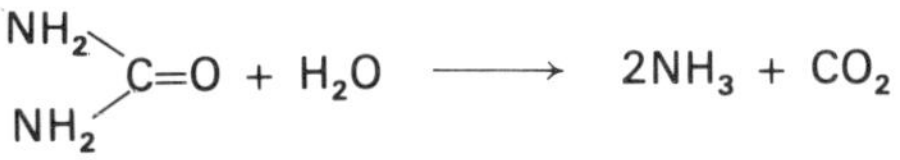

$$\begin{matrix} NH_2 \\ \diagdown \\ C=O + H_2O \longrightarrow 2NH_3 + CO_2 \\ \diagup \\ NH_2 \end{matrix}$$

Another enzyme arginase, essential for urea formation, reacts only with arginine.

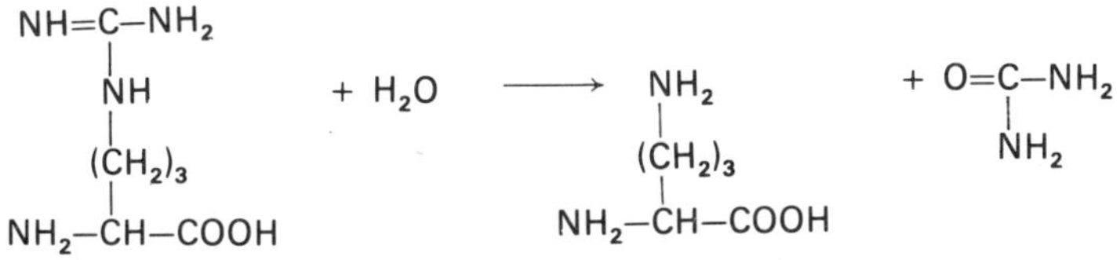

arginine $+ H_2O \longrightarrow$ ornithine $+ O=C-NH_2$ urea

Some enzymes previously regarded as specific for one naturally occurring substrate have been found to act also on synthesized analogues. Acetylcholinesterase, which is responsible for hydrolyzing acetylcholine to acetic acid and choline (thereby ensuring that the transmission of impulses from nerve to muscle lasts for a brief time only) also hydrolyses synthetic analogues such as carbamylcholine, although at a much slower rate. These analogues are very useful when a prolonged parasympathetic effect is required.

The Influence of Temperature on Enzyme Action

The rate of chemical reactions approximately doubles for each temperature rise of 10° ($Q_{10} = 2$). Within limits, enzyme-catalysed reactions behave in the same way but their behaviour is complicated by the protein nature of the catalyst. It will be recalled that the activity of the enzyme depends on the existence of a particular conformation of the protein. This conformation, which may not be the most probable one, is maintained by the elaborate cross-linkages of hydrophobic bonds, hydrogen bonds and covalent bonds between the amino acid side chains. Such a structure may be very delicately poised and heating the molecule may impart sufficient energy to it to break some of the bonds holding particular side chains in juxtaposition. The specific conformation of the enzyme molecule is not necessarily restored on cooling and the protein is then said to be 'irreversibly denatured'. In this condition it is catalytically inactive. The rate of denaturation of proteins is high and becomes progressively higher with increased temperature. At 70–80°, the Q_{10} for denaturation may be of the order of hundreds. Such behaviour produces the kind of change in the activity of an enzyme with

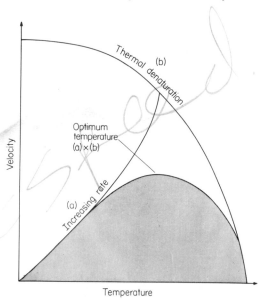

Figure 1.36 Effect of temperature on reaction rate of an enzyme-catalysed reaction: (a) represents the increasing rate of a reaction as a function of temperature; (b) represents the decreasing rate as a function of thermal denaturation of the enzyme protein. The shaded area represents the combination of (a) × (b). From E. E. Conn and P. K. Stumpf, *Outlines of Biochemistry*, Wiley, New York, 1963

temperature shown in Figure 1.36. Initially, as the temperature rises, the rate of reaction increases with the Q_{10} of about 2. When the temperature is high enough to permit conformational change in the protein the overall rate expresses a balance between the rate of the reaction at that temperature and the rate of removal of active enzyme by denaturation of the protein. Since the Q_{10} of denaturation increases so rapidly with temperature, the activity of the enzyme soon shows a precipitous fall. It is also clear that the optimum temperature (at which the rate of reaction is maximal) may well depend on the time for which the enzyme is kept at each temperature; if the exposure lasts only a few seconds the optimum temperature will be higher than if it is for hours or days.

Influence of pH on Enzyme Activity

The side chains of the amino acids of a protein may contain groups which act as acids or bases depending on the pH; these are commonly the carboxyl groups of glutamic and aspartic acids and the amino groups of lysine, the guanidino group of arginine and the imidazolyl group of histidine. Such groups may form part of the active site or may interact with each other to establish and maintain the characteristic conformation of the protein. Thus both the shape of the protein molecule and the efficiency of the active site may vary with the pH; it follows that the ease with which a substrate molecule is correctly oriented on the enzyme surface will also be influenced by the pH. Many of the substrates of enzymes are themselves substances which can ionize and the enzymes may interact only with the ionized or the unionized form. All these interactions will result in an influence of pH on the rate of the reaction. Finally, extremes of pH often result in denaturation of the enzyme, as does heating.

As a consequence of the interplay of these effects of pH most enzymes have an optimum pH, that is, a particular value of pH at which they work most rapidly. There is a wide range of these values (from about pH 2 to about pH 10) and it follows that many enzymes in the body cannot be working at their optimum pH since, except in the stomach, extremes of pH are not met; the pH range in the body is from about 6·4–7·5. There is surprisingly little information about the rates of enzyme reactions at physiological pH values. Although the pH of some tissues is known as an average it is possible that local variations in pH may occur at least for short times in the immediate vicinity of the enzyme molecule, producing large changes in rates in this circumscribed area. These localized pH changes may act as a type of control mechanism.

The Kinetics of Enzyme Catalysed Reactions

It is often necessary to measure the activities of enzymes as part of the study of metabolism and for diagnostic purposes in clinical medicine. Furthermore, detailed analyses of the kinetics of an enzyme-catalysed reaction can yield information concerning its mechanism.

The progress of an enzyme-catalysed reaction such as $S \rightleftharpoons P$ may be determined by following either the formation of the product (P) or the consumption of the substrate (S). The form of the progress curve obtained (see below) indicates that the rate of reaction diminishes as time proceeds. Several factors can contribute to the fall in rate, including (i) the increasing importance of the back reaction ($P \rightarrow S$) as the concentration of product (P) rises, (ii) the fall in the concentration of substrate (S) as the reaction proceeds and sometimes (iii) inhibition of the enzyme by products. Therefore, when an enzyme activity has to be measured it should be done by determining the *initial* rate of reaction; fortunately the initial portion of most progress curves is effectively linear so that its slope gives the rate.

The simplest way to study the rate of an enzymic reaction is to see what happens to the initial rate when the concentrations of enzyme and substrate are varied.

(A) If the substrate is present in excess and the quantity of enzyme is varied there is a strict proportionality between the amount of substrate transformed in unit time and the concentration of the enzyme.

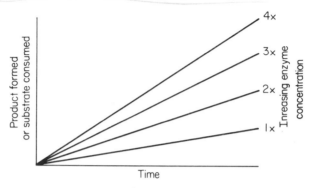

(B) The effect of increasing the substrate concentration [S] on the initial rate (v) at constant enzyme concentrations is shown in Figure 1-37. At first, as the substrate concentration is increased the rate of reaction rises but successive increments in substrate concentration lead to progressively smaller increases in rate

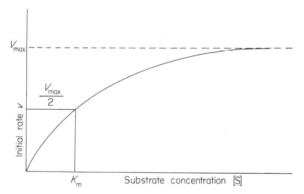

Figure 1.37 The effect of substrate concentration on the initial rate of an enzyme catalysed reaction

until eventually the reaction rate no longer changes with increasing substrate concentration. The enzyme is then said to be saturated and the rate is V_{max}. Since the curve obtained on plotting $v \times [S]$ is a rectangular hyperbola, v is given by the expression

$$v = \frac{V_{max}[S]}{[S] + K_m}$$

where K_m is termed the Michaelis Constant. As can be seen from Figure 1.37, K_m is equal to the substrate concentration at which the enzyme operates at half its maximum rate. A low value for K_m means that the enzyme will be working at its maximum rate with low substrate concentrations and, under natural conditions may always be working as fast as possible. An enzyme of this kind is not likely to be important in exerting control over metabolic processes by changing its activity in response to changes in substrate concentrations. Enzymes with higher K_m values have a wider scope as controlling elements.

In order to explain the observed kinetic behaviour of simple enzyme-catalysed reactions, Michaelis and Menten suggested that enzyme and substrate combine to form an enzyme–substrate (ES) complex which subsequently breaks down to yield product and free enzyme.

$$E + S \underset{k_{-1}}{\overset{k_1}{\rightleftharpoons}} ES \overset{k_2}{\longrightarrow} E + P$$

The enzyme molecules possess an active site where the substrate is bound and transformed; when the substrate concentration is high enough to ensure that all the active sites are occupied (i.e. when all the enzyme is present as ES complex) the enzyme is said to be saturated and the rate of reaction is $k_2[ES] = V_{max}$.

The significance of the Michaelis constant K_m in terms of the rate constants k_1, k_{-1} and k_2 depends upon what assumptions are made about the mechanism of the reaction. Michaelis and Menten assumed that k_2 was smaller than k_{-1} and consequently concluded that $K_m = k_{-1}/k_1 = K_s$, the dissociation constant of the enzyme substrate complex which at equilibrium equals [E][S]/[ES]. Briggs and Haldane made no assumptions about the relative sizes of k_{-1} and k_2 and expressed K_m as $(k_{-1} + k_2)/k_1$. Both of these treatments are less than perfect since the model chosen ignores the possibility of products reacting with enzyme to form ES complex; however, this is not an unreasonable assumption when initial rates are being considered. When this back reaction is included, the simplest form of the reaction has to be written

$$E + S \underset{k_{-1}}{\overset{k_1}{\rightleftharpoons}} ES \underset{k_{-2}}{\overset{k_2}{\rightleftharpoons}} EP \underset{k_{-3}}{\overset{k_3}{\rightleftharpoons}} E + P$$

Thus, although K_m is an experimentally measurable quantity and can be defined operationally as [S] when $v = \frac{1}{2}V_{max}$ for hyperbolic curves of $v \times [S]$, its significance can be interpreted in terms of rate constants (k_1, k_{-1}, k_2, etc.) only if the detailed mechanism of the reaction is known.

So far we have considered reactions in which a single substrate is transformed into products; this formulation is also applicable to two substrate reactions where one of the substrates is present at a saturating concentration. An example of this type of reaction would be one where water is a reactant, such as in an hydrolysis. However, the majority of reactions encountered in biological systems involve two or more substrates, none of which is present in saturating concentrations. In the case of a reaction such as

$$A + B + enz \rightleftharpoons enz\text{-}A\text{-}B \longrightarrow enz + C + D$$

K_m values for A and B (K_{mA} and K_{mB}) can be found by determining v as a function of [A] at a saturating concentration of B and as a function of [B] at a saturating concentration of A respectively.

If the K_m for A is determined in this way, the concentration of A giving half-maximal rate of reaction at a saturating concentration of B is obtained since $K_{mA} = $ [A] when $v = \frac{1}{2} V_{max}$ (see Figure 1.36). It is not permissible to assume that the concentration of A giving half-maximal rate of reaction *in vivo* is equal to the K_m determined *in vitro* unless B is known to be present *in vivo* at a saturating concentration and the products to be present at zero concentration. In many instances the concentration of B *in vivo* may not saturate the enzyme and products are usually present. Thus the concentration of A *in vivo* giving half-maximal rate of reaction at the standing concentrations of B and the products will be equal to a quantity properly called K_{mA} 'apparent' which may differ considerably from the true K_{mA}.

Not all plots of $v \times $ [S] give rise to hyperbolic curves; those obtained from some enzymes are sigmoidal (S-shaped) in form. Such curves can arise for a variety of

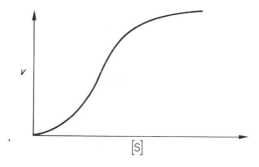

reasons; one popular model suggests the existence of several substrate binding sites on each enzyme molecule and of a cooperative interaction between them whereby the binding of substrate at one site facilitates the binding, and hence the transformation, of substrate at the others. This type of cooperative interaction between sites binding identical substrate molecules is termed a homotropic effect. Enzymes displaying such homotropic interactions are important in controlling metabolic reaction sequences since small changes in their substrate concentration can lead to large changes in their rate over some regions of the curve.

There are enzyme-catalysed reactions which generate either hyperbolic or sigmoidal curves on plotting v against [S] depending upon the presence or absence of molecules other than the substrates. These molecules (called heterotropic effectors) bind to sites (allosteric sites) on the enzyme which are distinct from the active or catalytic sites where the substrate is transformed. Some allosteric or heterotropic effectors increase enzyme activity at a given substrate concentration whilst others decrease it. Examples of this type of behaviour are given in Figure 1-38. At a constant substrate concentration equal to S' (see Figure 1.37) the rate of reaction increases with increasing concentrations of allosteric activator and the curve becomes increasingly hyperbolic. Increasing con-

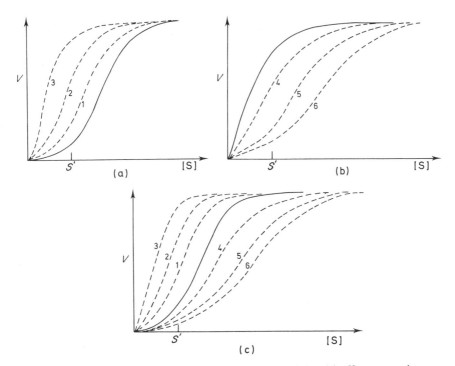

Figure 1.38 Examples of the effects of heterotropic (allosteric) effectors on the activity of enzymes. (a) Enzyme activated by effector. (b) Enzyme inhibited by effector. (c) Enzyme activated by one effector and inhibited by another. ——— no effector present; – – – – 1, 2, 3 increasing concentrations of +ve (activating) effector; – – – – 4, 5, 6 increasing concentrations of −ve (inhibiting) effector

centrations of allosteric inhibitor diminish the rate by making the curve more sigmoidal. Some allosteric effectors 'lock' the enzyme in its 'normal conformation' and so diminish the ability of the opposing effector to induce the conformational change necessary to yield a change in reaction rate at a given concentration of substrate.

Enzymes whose rate can be controlled by heterotropic effectors are of great importance in the control of metabolism since their activities can be altered at constant substrate concentration by molecules whose concentrations may vary as a result of other changes occurring in the cells.

Inhibition of Enzyme Action

The Michaelis–Menten concept of enzyme substrate interaction can also be used in descriptions of the inhibition of enzymes. Enzymes are often very sensitive to the composition of their immediate environment as the descriptions of the effects of temperature and pH have indicated. The substances now to be considered as inhibitors are those which, because of their chemical nature, alter the enzyme so that it can no longer work at its optimum rate. There are two main kinds of inhibitors. The first kind, the irreversible, non-specific inhibitors, exert their influence in a general way, for example by denaturing the enzyme protein; the salts of heavy metals or trichloroacetic acid are in this group. The second kind, the reversible specific inhibitors, can be subdivided into two types, the non-competitive and the competitive. Non-competitive inhibitors react reversibly with some sites on the enzyme so that V_{max} is altered but K_m is not. Non-competitive inhibitors cannot be overcome by increasing the concentration of substrate. Competitive inhibitors increase the value of K_m but do not affect V_{max}; in this case the inhibition can be overcome by increasing the concentration of substrate molecules.

A striking example of competitive inhibition is the action of malonate on succinic dehydrogenase. Succinic dehydrogenase catalyses the conversion of succinic acid to fumaric acid. The malonic acid molecule has the same general shape as

```
   COOH                                        COOH
    |                H       COOH               |
   CH₂                  \   /                  CH₂
    |          →         C                      |
   CH₂                   ‖                     COOH
    |                    C
   COOH            HOOC/   \H

succinic acid           fumaric acid        malonic acid
```

that of succinic acid but since it does not have adjacent CH_2 groups it cannot be dehydrogenated in the same way as succinic acid.

Competitive and non-competitive inhibition can be distinguished by plotting the reciprocal of the rate ($1/v$) against that of the substrate concentration as shown in Figure 1.39.

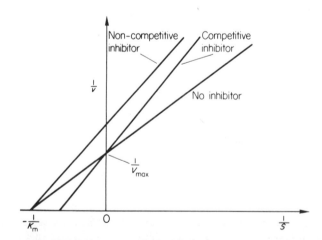

Figure 1.39 Effects of inhibitors on the kinetics of one substrate enzymes; v, initial velocity; [S] the substrate concentration; K_m, the Michaelis constant; V_{max} the maximum velocity

The Mechanism of Enzyme Action

At present the relationship between catalytic activity and three-dimensional structure is understood for only a small number of hydrolytic enzymes of relatively low molecular weight.

Lysozyme, which hydrolyses polysaccharides forming bacterial cell walls, was the first enzyme to have its three-dimensional structure solved by X-ray analysis. The most striking feature of its structure is the deep cleft which runs across the middle of the molecule and into which the polysaccharide chain of the substrate fits (Figure 1.40). The sugar residues A, B and C of the substrate are bound to the enzyme by six hydrogen bonds and a large number of non-polar interactions. As the substrate is bound, part of the enzyme to the left of the cleft tilts inwards so that the cleft closes slightly. Such 'induced' changes in enzyme structure prior to catalysis are thought to occur in the course of many, but not necessarily all, enzyme-catalysed reactions. (Similar changes of conformation induced in other enzymes by heterotropic effectors might alter their ability to bind substrate and so form the basis of allosteric control mechanisms (see p. 86).) Residue D can only

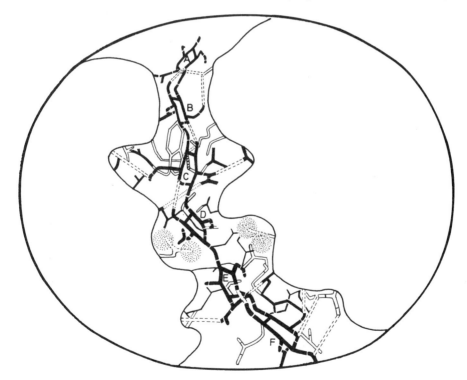

Figure 1.40 The three-dimensional structure of lysozyme showing the substrate located in the cleft that runs through the middle of the molecule. Sugar residues of the polysaccharide substrate are indicated by A, B, C, D, E and F. Bonds denoted ======= are hydrogen bonds linking substrates and enzyme or intramolecular hydrogen bonds within the substrate. The four spheres represent the oxygen atoms of the carboxyl residues of glutamate 35 and aspartate 52 at the active site

be accommodated in the cleft if it is twisted out of its chair conformation (see p. 22) so that carbons 1, 2 and 5 and the ring oxygen are coplanar; the substrate thus takes up a strained conformation on binding to the enzyme. As will be seen later, this is of significance in the subsequent hydrolytic reaction. The bond hydrolysed is that linking carbon 1 of residue D with oxygen in the glycosidic link. Two amino acid side chains, aspartic acid (52) and glutamic acid (35) are situated close to this bond. Residue 52 is in a polar environment and is present in its ionized form as aspartate: residue 35, in an apolar region, is in its protonated form as glutamic acid. The mechanism of hydrolysis is thought to be as follows (Figure 1.41). The COOH group of residue 35 donates its proton to the oxygen atom involved in the glycosidic link between carbohydrate residues D and E, thus breaking the bond and turning carbon 1 of residue D into a carbonium ion (C^+). This carbonium ion is stabilized by the carboxylate anion of residue 52

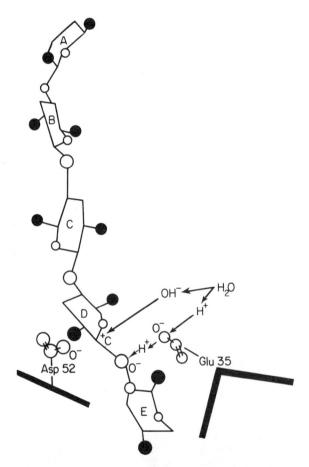

Figure 1.41 Proposed mechanism for the hydrolytic cleavage of a glycosidic link by lysozyme. See text for details

until it can combine with an OH⁻ ion derived from water. The consumption of an OH⁻ ion leaves a proton which can replace that lost from residue 35. The products of hydrolysis then fall away from the enzyme to complete the reaction. The distortion of residue D referred to above aids the progress of the reaction by forcing the sugar residue into the conformation adopted by sugar carbonium ions, thereby facilitating carbonium ion formation later in the catalytic process.

This brief summary of the mode of action of lysozyme illustrates several features of enzyme action that are probably of general significance.

(1) Prior to catalysis an enzyme–substrate complex is formed in which enzyme and substrate are linked by weak, non-covalent, bonds.

(2) Substrate binding induces conformational changes in both enzyme and substrate.

(3) The binding site(s) is distinct from the catalytic site.

(4) The substrate is bound so as to bring it into contact with reactive groups in the enzyme able to catalyse its conversion into products.

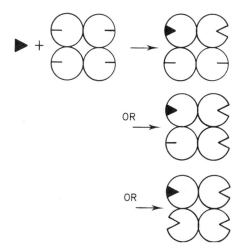

Figure 1.42 Substrate activation in an oligomeric enzyme (a positive homotropic effect) by a mechanism not involving the conservation of symmetry. ▶ Substrate, ⊖ protomer (subunit) undergoing an induced conformation change to ⟩ on binding substrate. This conformational change can be transmitted to some or all of the other protomers in the oligomer and thereby increase the ease with which they bind substrate. According to this model, heterotropic inhibitors would stabilize the ⊖ form of the enzyme whilst heterotropic activators would promote the change to the ⟩ state on being bound at an allosteric site (allosteric sites are not shown in the diagram)

Homotropic Effects and Changes in Conformation

In some of those enzymes consisting of a number of catalytically active subunits (protomers) linked together to form an oligomer, the binding of substrate to one of the protomers can facilitate substrate binding (and hence catalysis) by the remaining protomers constituting the oligomer. Such behaviour is often thought to be responsible for the appearance of sigmoidal curves in plots of $V \times$ [S] of the type discussed on p. 80, and for the well-known form of the dissociation curve of oxyhaemoglobin (see Figure 7.2). Several attempts have been made to account for positive homotropic effects in terms of changes in molecular conformation. One of these suggests that when substrate binds to any of the protomers in an oligomer, it induces a conformational change which is transmitted to all or some of the other protomers via the weak interactions that hold the assembly together and thereby increases their affinity for substrate molecules (Figure 1.42). Another model is more restrictive and requires that the oligomers are composed of symmetrically arranged protomers and that the oligomers exist in two forms of R and T which are in equilibrium with each other. Only the protomers of those oligomers in the R-state are supposed to bind substrate. If substrate binds to a protomer of an oligomer in the R-state, the equilibrium T $\rightleftharpoons$ R is displaced towards R so that the number of oligomers in the active, R-state and the rate of catalysis are increased (see Figure 1.43).

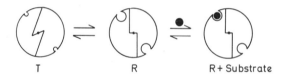

T   R   R + Substrate

Figure 1.43 The symmetry model for positive homotropic effects in enzymic catalysis ● = substrate. According to this model the oligomers can exist in two states, T and R, of which only those in the R-state can bind and transform substrate. The model also requires that the oligomers must always retain their symmetry, consequently the conformation of their constituent monomers must change *simultaneously* on passing from the T (inactive) state to the R (active) state. The binding of substrate to R-state oligomers both stabilizes them and displaces the T $\rightleftharpoons$ R equilibrium to the right, hence increasing the availability of catalytically active sites. Heterotropic activators also bind to R-state oligomers but at some special, allosteric site instead of the active site and increase the availability of active sites by stabilizing the R-state oligomers and displacing the equilibrium to the right. Heterotropic inhibitors diminish the availability of active sites by stabilizing the T-state and displacing the equilibrium to the left. Allosteric sites are not shown in the diagram

Heterotropic Effectors and Changes in Conformation

Heterotropic effectors could activate or inhibit oligomeric enzymes by stabilizing R- or T-states in those instances where symmetry is conserved (see Figure 1.43) or they could induce changes in enzyme conformation favouring or counteracting substrate binding in those systems which are less dependent on symmetry conservation (Figure 1.42).

CHAPTER 7

The Quantitative Man

Biochemistry has now passed from the state of a descriptive science to a stage where quantitative problems are becoming more important. Thus a biochemist is always interested in the following things about a metabolic sequence.

(1) The description of the enzymes and chemical changes that comprise the metabolic sequence.
(2) The rate at which material can be transformed by the sequence.
(3) The amounts of material that are utilized by the sequence in the living organism.
(4) The nature of the control mechanisms which adjust the amounts of material utilized by the sequence.

Apart from (1) the considerations are all of a quantitative nature. Table 1.4 gives some basic data on the 'standard' 70 kg man. Much of this data about the weight of tissues comes from comparatively few studies involving total dissection of corpses and weighing of the parts and applying various calculations. The three tissues, skeletal muscle, adipose tissue and skin which comprise some 60 per cent of the total body weight are particularly difficult to measure accurately and a range of weights is given. The mean of the range has been used for any calculations. Quantitatively the five major tissues are skeletal muscle, adipose tissue, bone, skin and the gastrointestinal tract. If the weight of blood is added the total is 86 per cent of the body weight. However, it is not the sheer mass of tissue which determines its quantitative contribution to metabolic activity. A better correlation is the protein content of the tissue because, as we shall see later, the activity of a tissue is determined by its enzyme content and all enzymes are protein in nature. However, the protein content of the tissues is not a complete guide since not all the protein is specifically concerned with metabolism; for example, the haemoglobin of the blood has the function of carrying oxygen but does not enter into other metabolic sequences and much of the protein of bones and skin is the inert supporting protein collagen.

The substances used by the tissue enzymes are transported to the tissues by the bloodstream and, as shown in Table 1.4, the blood flow varies to some tissues according to the circumstances, being greatly increased to the muscle and skin during exercise. In the first case the increased blood flow is to increase the supply of substrates, including oxygen, and to remove products of metabolism, including heat. By contrast, the increased flow to the skin is mainly for the dissipation of heat—to prevent the engine overheating! The consumption of oxygen by the tissues (Table 1.4) gives some idea of their relative capacities for utilizing substrates by oxidative processes. But, as we shall see in the next section, much of the muscles' utilization of the chemical fuel brought to it is by degradation not directly involving oxygen. The complexity of different tissues carrying out different amounts of chemical transformation of different sorts, and moreover varying these parameters according to the body's need of the moment, may look

Table 1.4 Tissues of the '70' kg man

Tissue	Wet weight (kg)	Protein content (kg)	Blood flow to tissue (ml/min) at rest	Blood flow to tissue (ml/min) with light exercise	Resting oxygen consumption (ml/min)
Skeletal muscle	20^a–30	4–6	1200	4500	70
Adipose tissue	9^a–13	0.14–0.2			
Skin	4.3^a–9.0^b	0.65–1.1	500	1500	5
Gastrointestinal tract	2.6^a	0.32	1400	1100	58
Blood	5.5^a	0.77	—	—	
Liver	1.7^a	0.38	1500^d		75
Brain	1.5^a	0.15	750	750	46
Heart	0.3^a	0.04	250^e	350^e	27
Kidneys	0.3^a	0.05	1100	900	16
Lung	1.0^a	0.15^c	5000	10,000	12
Bone (total)	10^a	2.2			
Red marrow	1.25^a	0.37			
White marrow	1.75^a				
Spleen	0.15^a	0.02			
Lymphoid tissue	0.7^a	0.1			
Miscellaneous glands, etc.	0.5^a	0.07			

[a] Extensive review of the data on skeletal muscle and other tissue weights by Mr J. W. Legge of Melbourne University, Dept. of Biochemistry, strongly suggests that the usual value of 30 kg for muscle is too high. Other values indicated are recalculated in the light of Mr Legge's work.
[b] The higher value includes subcutaneous tissues (fat).
[c] Calculated as 15 per cent of the wet weight.
[d] 25–30 per cent is arterial flow.
[e] Coronary circulation.

too difficult to analyse but, as we shall see, there is an underlying simplicity and universality of the chemical mechanisms the body uses to meet its needs.

The tissues which contribute the largest fraction to the body weight are the different forms of muscle; smooth muscle (for example in the intestine), cardiac muscle (found in the heart) and the majority which is skeletal muscle. The next section will therefore examine in some detail the way muscle transforms the blood chemicals brought to it so that it can use the energy of degradation to bring about muscular contractions to meet the total body's needs.

References

The Biosphere. *Scientific American*, 1970, September.
The Origin of Life: Molecules and Natural Selection. Orgel, L. E. Chapman Hall Ltd., London, 1973.
The Origin of Life. Bernal, J. D. and Synge, A. Oxford Biology Readers. Oxford University Press, London, 1972.
Bioenergetics. Lehninger, A. L. W. A. Benjamin Inc., Menlo Park, California, U.S.A., 2nd edn., 1971.
A Biologists Physical Chemistry. Morris, J. G. Arnold, London, 2nd edn., 1974.
The Structure and Action of Proteins. Dickenson, R. E. and Geis, I. Harper Row, New York, 1969.
Macromolecules: Structure and Function. Wold, F. Prentice Hall International Inc., London, 1971.
Acquisition of Three Dimensional Structure of Proteins. Wetlaufer, D. B. and Ristow, S. *Annual Review of Biochemistry*, **42**, p. 135, 1973.
The Structure of Cell Membranes. Fox C. F. *Scientific American*, 1972, February, p. 30.
The Control of Biochemical Reactions. Changeux J. P. *Scientific American*, 1965, April, p. 36.

SECTION 2

Muscle

CHAPTER 8

Muscle Contraction and Relaxation

This chapter is concerned with the molecular mechanisms that bring about the shortening and subsequent relaxation of muscles.

Skeletal muscles are composed of bundles of fibres surrounded by, and terminating in, connective tissue which is continuous with the inextensible,

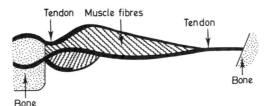

Figure 2.1 Longitudinal section of the semitendinosus muscle of the rabbit. In this fusiform muscle sheets of connective tissue extend over the surface from tendon to tendon and provide anchorage sites for the attachment of short fibres. The fibres do not run the whole length of the muscle. Connective tissue also surrounds each fibre

collagenous, tendons attached to bones. Individual fibres are cylindrical in form with diameters between about 10 μ and 100 μ but with lengths of a few millimetres to many centimetres. Each fibre is bounded by the sarcolemma or cell membrane immediately beneath which are situated many nuclei; the fibre is thus a multinucleate syncytium. The bulk of the fibre is filled with longitudinally

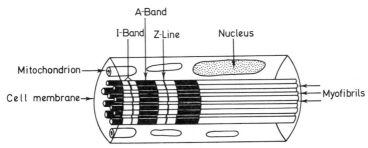

Figure 2.2 Diagram to show the arrangements of myofibrils within a myofibre

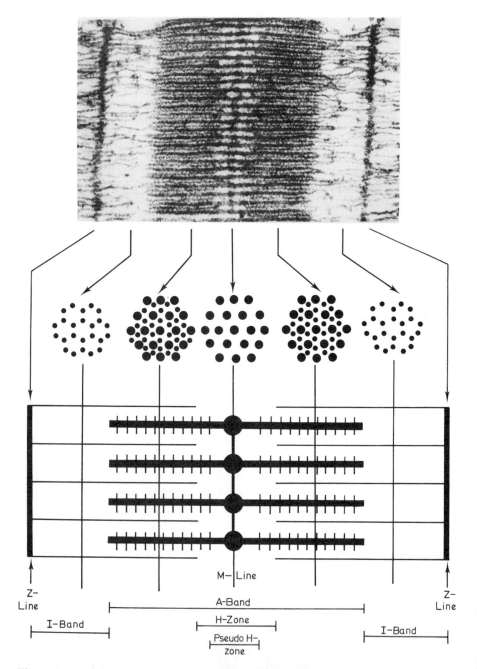

Figure 2.3 High-resolution electron micrograph showing details of filament structure and the interpretation of banding in terms of structure of overlapping thin and thick filaments. Notice that the M-line can be seen to be due to interconnections between thick filaments, the pseudo H-zone due to the absence of cross-bridges, the H-zone due to the absence of thin filaments, the I-band due to the absence of thick filaments, and the

arranged myofibrils or contractile elements which account for 60–70 per cent of the cell mass and have diameters of 1 to 1·5 μ. Each myofibril is enveloped by the precisely arranged tubules of the sarcoplasmic reticulum and transverse tubule system. Depending upon the type of muscle, variable numbers of mitochondria are wedged between the invested myofibrils and just beneath the sarcolemma.

When viewed under polarizing or phase-contrast microscopes, fresh myofibres appear to be striated in both the longitudinal and transverse directions. The longitudinal striations result from the parallel arrangement of myofibrils within the myofibre, whilst the transverse striations are caused by alterations in optical density along the length of each myofibril. In any fibre, the axial striations of the individual myofibrils are usually in register so that the whole fibre takes on a regular pattern of transverse striations (Figure 2.2). The sarcomere or repeating unit of the myofibril, which extends from one Z-line to the next, is composed of overlapping arrays of thick and thin filaments. Since the sarcomere is the basic contractile unit of a myofibril, it is necessary to examine its structure in some detail. Each sarcomere (Figure 2.3) contains a hexagonal array of thick filaments aligned parallel with the long axis of the myofibril and defining the limits of the A-band. This array of filaments overlaps with an array of thin filaments attached to, and ordered by, each of the two Z-lines. In the region of overlap between the thick and thin filaments, cross-bridges extend from the thick filaments towards the thin; these bridges are responsible for generating the relative sliding motion that draws the arrays of thin filaments deeper into the array of thick filaments to produce an overall shortening of the sarcomere. The thick filaments are held in an ordered array by fibres termed M-bridges that link their centres (Figure 2.4).

The Thick Filaments

The thick filaments are multimolecular aggregates of the protein myosin; structures closely resembling them are formed spontaneously when the KCl concentration of a solution containing myosin monomers is reduced from 0·6 M to 0·2 M. In the electron microscope, molecules of myosin appear as rods, about 1300 Å long and 15 to 20 Å wide, terminated by two globular heads some 200 Å long and 40 to 50 Å wide; thus the molecules have an overall length of approximately 1500 Å. Treating myosin briefly with trypsin (a proteolytic enzyme) breaks each molecule into two fragments termed Heavy (H-) and Light (L-) meromyosin. L-Meromyosin forms insoluble, rod-shaped, aggregates at physiological ionic strengths whilst H-meromyosin is soluble and carries both the ATP-ase activity and the actin combining site of myosin. L-Meromyosin is derived from the 'tail' of myosin whilst H-meromyosin comprises part of the 'tail' and the two 'heads' of myosin. Further proteolysis of H-meromyosin releases the two heads from the tail section. The globular heads carry the ATP-ase and actin

---

darkest part of the A-band due to the presence of thick filaments, thin filaments and cross-bridges. The Z-line is due to interconnections between these filaments. The organization of the thick and thin filaments in transverse section is also shown. (Courtesy of H. E. Huxley)

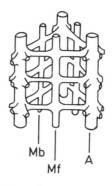

Figure 2.4 Diagram of the central region thick filaments (A) showing the M fibrils (Mf) and M bridges (Mb) responsible for holding the thick filaments in a regular hexagonal array

binding sites whilst the tail part is an inert soluble rod about 400 Å long. Other measurements indicate that L-meromyosin and Subfraction 2 (the tail portion of H-meromyosin) contain a very high proportion of their polypeptide in the α-helical form. It would appear that the protease-sensitive regions of myosin are not helical but constitute regions of flexibility. Finally, studies with dissociating agents such as urea show that myosin is composed of two large (molecular weight

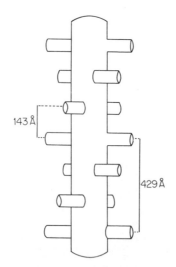

Figure 2.5 Thick filament showing arrangement of myosin cross-bridges

200,000) and four small (molecular weights about 20,000) polypeptides, the small units being confined to the globular heads (Subfraction 1) whilst the large chains extend throughout the molecule.

In forming the thick filaments, the L-meromyosin segments of myosin aggregate together to form the main axis of the filament whilst the H-meromyosin segments project from the surface about a 6/2 screw axis of symmetry (Figures 2.5 and 2.6). The central part of the filament is devoid of projections as it is formed by the 'tail-to-tail' aggregation of myosin molecules. The helical arrangement of projections at one end of a thick filament is the mirror image of that at the other end. This difference in polarity between opposite ends

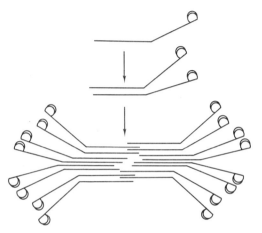

Figure 2.6 Diagram of the way in which thick filaments are thought to be formed by the aggregation of myosin monomers

of the filaments is required to effect the movement, from opposite ends of the sarcomere, of the two arrays of thin filaments into the array of thick filaments. The thin filaments on opposite sides of a Z-line also have opposing polarities for the same reason. The Z-line is a molecular device for holding the thin filaments in a regular array and for inverting their polarity in adjacent I-bands.

The Thin Filaments

Basically the thin filaments consist of two chains formed from globular molecules of the protein actin (M.W. 45,000, diameter 55 Å) wrapped around each other to form an open, double-stranded, helix (Figure 2.7). Lying head-to-tail along the grooves of the helix are molecules of tropomyosin (Figure 2.8) (M.W. 54,000, length 400 Å) to each of which is attached a molecule of troponin (M.W. 80,000) consisting of three subunits. When bound to tropomyosin, one of these subunits (TN-I) inhibits the actin-induced stimulation of Mg^{2+}-ATP hydrolysis catalysed by myosin. Another subunit (TN-C) overcomes this inhibition on binding Ca^{2+} ions and the third (TN/T) binds TN-I and TN-C to

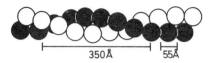

Figure 2.7 Part of a filament of fibrous actin (F-Actin) formed from two chains of globular actin monomers

tropomyosin which interacts with actin. Tropomyosin alone does not confer a Ca^{2+} dependence on Mg^{2+}-ATP hydrolysis by myosin in the presence of actin. The troponin/tropomyosin system must be regarded as a regulatory subunit of actomyosin and calcium ions as heterotropic effectors (see p. 87).

Figure 2.8 Model showing possible arrangement of troponin and tropomyosin along the thin actin filament

Actomyosin and Myofibrils

Pure myosin and actin interact to form a complex called actomyosin which hydrolyses Mg^{2+}-ATP very much more rapidly than myosin alone. Myosin hydrolyses ATP on binding ATP but the products of hydrolysis, ADP and Pi, leave the enzyme only slowly after their formation:

$$\text{myosin} + \text{ATP} \longrightarrow [\text{myosin}^* \text{ATP}] \longrightarrow [\text{myosin}^* \text{ADP} + \text{Pi}]$$
$$\downarrow \text{slow}$$
$$\text{myosin} + \text{ADP} + \text{Pi}$$

As a result the ATP-ase activity of myosin is inhibited by its products. Actin, by binding to myosin, seems to greatly facilitate the release of the hydrolysis products:

$$[\text{actin}] + [\text{myosin}^* \text{ADP} + \text{Pi}] \longrightarrow [\text{myosin}^* \text{ADP} + \text{Pi} + \text{actin}]$$
$$\downarrow$$
$$\text{myosin} + \text{ADP} + \text{Pi} + \text{actin}$$

On binding ATP, myosin undergoes a conformational change (indicated by *) and only returns to its ground state when ADP and Pi are released.

Fibres of pure actomyosin can be prepared which shorten when they hydrolyse Mg^{2+}-ATP but are easily stretched in the presence of ATP if hydrolysis is prevented by agents such as EDTA which chelate Mg^{2+} or those such as methylmaleimide which react with —SH groups at the active site. On the other hand, in the absence of ATP such fibres become inextensible. In contrast to the behaviour of pure actomyosin, isolated myofibrils (prepared by treating muscle fibres with glycerol) hydrolyse Mg-ATP and shorten only if Ca^{2+} ions are present. This calcium sensitivity is conferred upon the myofibrilar system by its complement of troponin and tropomyosin. In the absence of ATP myofibrils are inextensible, but when ATP is present (but cannot be split) they are readily stretched; this aspect of their behaviour thus mirrors that of fibres composed of pure actin and myosin.

Excitation—Contraction Coupling

In resting muscles the concentration of free Ca^{2+} ions in the sarcoplasm pervading the myofilaments is reduced to a very low level (about 10^{-7} M) by the activity of the sarcoplasmic reticulum. The cisternae of the reticulum are able to concentrate Ca^{2+} ions from the sarcoplasm by means of a membrane-bound enzyme system able to transport Ca^{2+} ions across the membrane against a concentration gradient. The energy required to effect this accumulation of Ca^{2+} ions is provided by coupling the process to the hydrolysis of Mg^{2+}-ATP. As can be seen in Figure 2.9 the sarcoplasmic reticulum investing each sarcomere consists of two terminal cisternae linked by longitudinal tubules and a fenestrated collar. At the level of each Z-line, two terminal cisternae abut onto a transverse tubule formed by an invagination of the sarcolemma. The two cisternae and intervening section of transverse tubule are referred to as a triad.*

When a muscle fibre is stimulated via its motor neurone, a wave of electrical depolarization termed an action potential (see p. 335) spreads longitudinally along the sarcolemma and is conveyed inwards at the level of each Z-line (or A/I band junction) by the transverse tubules. The transverse spread of depolarization seems to be coupled electrotonically to the membranes of the sarcoplasmic reticulum by means of the tight contacts between the membranes of the transverse tubules and those of the terminal cisternae. This electrical activity increases the permeability of the membranes of the sarcoplasmic reticulum to Ca^{2+} ions which consequently flood out into the cytoplasm to activate the troponin-inhibited contractile apparatus. The successive activation of consecutive sarcomeres ensures that one sarcomere begins to shorten before the next with the result that a wave of contraction passes along the muscle. If all the sarcomeres were activated simultaneously, shortening would be impossible as Z-bands would be pulled equally in opposite directions. When the muscle fibre ceases to be depolarized, the permeability of the membranes of the sarcoplasmic reticulum to calcim falls to the resting value and, as the pump begins to retrieve Ca^{2+} ions

---

* In some muscles the triads are situated at the level of the junction between the A- and I-bands.

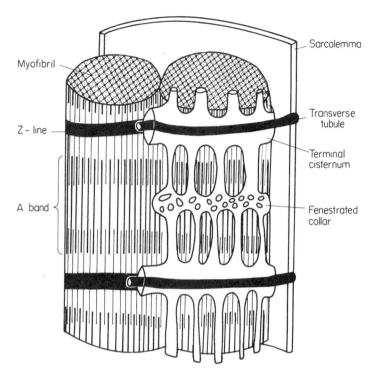

Figure 2.9 Diagram showing the relationship of the transverse tubules and sarcoplasmic reticulum to the myofibrils

from the sarcoplasm, the contractile apparatus returns to its inhibited relaxed state.

The Contractile Apparatus at Rest, Contracting and in rigor mortis

At rest, Mg^{2+}-ATP is bound to the Subfraction 1 region of the myosin crossbridges and probably hydrolysed to ADP and Pi, but these products have only a small tendency to leave the enzyme surface. The rate of product release cannot be speeded up by actin because tropomyosin molecules lying in the grooves of the thin filaments prevent the ATP-primed myosin 'heads' gaining access to the actin subunits. Since bridge formation cannot occur under these conditions, there is no linkage between the thick and thin filaments, thus the sarcomeres and hence the muscle is readily extensible and relaxed. When the muscle is activated and Ca^{2+} ions flood out into the cytoplasm, the TN-C subunit of troponin binds calcium and induces a conformational change in the TN-I subunit which is transmitted, presumably via the TN-T unit, to tropomyosin which moves deeper into the groove to expose the myosin binding site on the actin molecules. Myosin heads are then able to interact with actin to complete the hydrolysis of ATP (i.e. release of ADP and Pi). In so doing it seems probable that the angle of attachment of the myosin head to actin changes (Figure 2.10) putting the bridge (Subfraction 2) under tension and so generating a relative sliding motion between thick and thin filaments. The actin/myosin link is broken by the myosin head binding another molecule of ATP from the sarcoplasm. Once this is hydrolysed and myosin has undergone a conformational change it is ready to reestablish a link with a new actin partner. In this way a series of contractile cycles can be carried out with the consequent shortening of each sarcomere until the muscle ceases to be activated and calcium is pumped back into the sarcoplasmic reticulum. As troponin loses its calcium, tropomyosin shifts its position to hinder the access of ATP primed myosin heads to the actin molecules. Hence, links between thick and thin filaments cease to be made, ATP hydrolysis falls back to the low resting level, and the muscle returns to its relaxed, resting, state. In this state, myosin is already primed with ATP so that only the release of calcium is required to reestablish contraction.

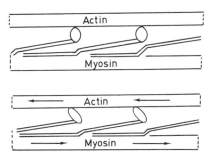

Figure 2.10 Diagram to illustrate the way in which a change in the angle of attachment of the head of the cross-bridge to actin could result in a relative sliding motion

Shortly after death muscles become inextensible, a condition called *rigor mortis*. The cause of the rigor seems to be that as the ATP concentration falls, Subunit 1 regions of myosin become denuded of ATP and in that condition are able to bind to actin (despite the presence of a Ca^{2+} free troponin/tropomyosin system) to form stable actomyosin complexes (rigor complexes). Presumably the change in the shape of ATP-free myosin differs from that of ATP-loaded myosin. However, it is also possible that Ca^{2+} ions are released from the sarcoplasmic reticulum as the ATP concentration falls and that they expose the actin molecules to myosin by binding to troponin.

The Provision of ATP for Muscle Contraction

Although the discussion of the function of ATP makes it clear that a contracting muscle must have a continuous supply of this compound, the amount of ATP in muscle tissue is sufficient for about two seconds' work only. It follows that the supply of ATP must be continually replenished. To anticipate the final picture, this replenishment comes largely from the energy made available during the breakdown of glucose and of fats but as the demands on the energy supply in muscle are unpredictable and often very large it follows that an immediately accessible store of energy must be available to buffer an immediate stress. The substance which acts as the buffer store in mammalian muscle is phosphocreatine (creatine phosphate). This compound is able to transfer a phosphate group to ADP in the presence of the enzyme creatine phosphokinase thus

$$\begin{array}{c} \text{OH} \\ | \quad \text{H} \\ \text{HO—P—N} \\ \| \quad\quad\quad \text{C=NH} \quad + \quad \text{ADP} \\ \text{O} \quad / \\ \text{CH}_3\text{—N—CH}_2\text{—COOH} \end{array} \rightleftharpoons \begin{array}{c} \text{H}_2\text{N} \\ \quad\quad\quad \text{C=NH} \quad + \quad \text{ATP} \\ / \\ \text{CH}_3\text{—N—CH}_2\text{—COOH} \end{array}$$

creatine phosphate   creatine

Creatine phosphate is one of a class of compounds called guanadino phosphates. Creatine is formed from the amino acids glycine, arginine and methionine in the liver. The guanadino group is transferred from arginine to glycine, producing ornithine and glycocyamine which receives a methyl group from methionine to form creatine.

$$\begin{array}{c} \text{NH}_2 \\ | \\ \text{C=NH} \\ | \\ \text{NH} \\ | \\ (\text{CH}_2)_3 \\ | \\ \text{NH}_2\text{—CH—COOH} \\ \text{arginine} \end{array} + \begin{array}{c} \text{NH}_2 \\ | \\ \text{CH}_2 \\ | \\ \text{COOH} \\ \text{glycine} \end{array} \longrightarrow \begin{array}{c} \text{NH}_2 \\ | \\ (\text{CH}_2)_3 \\ | \\ \text{NH}_2\text{—CH—COOH} \\ \text{ornithine} \end{array} + \begin{array}{c} \text{NH}_2 \\ | \\ \text{C=NH} \\ | \\ \text{NH} \\ | \\ \text{CH}_2 \\ | \\ \text{COOH} \\ \text{glycocyamine} \end{array}$$

$$\begin{array}{c} \text{NH}_2 \\ | \\ \text{C=NH} \\ | \\ \text{NH} \\ | \\ \text{CH}_2 \\ | \\ \text{COOH} \\ \text{glycocyamine} \end{array} + \begin{array}{c} \text{S—CH}_3 \\ | \\ (\text{CH}_2)_3 \\ | \\ \text{NH}_2\text{—CH—COOH} \\ \text{methionine} \end{array} \longrightarrow \begin{array}{c} \text{NH}_2 \\ | \\ \text{C=NH} \\ | \\ \text{N—CH}_3 \\ | \\ \text{CH}_2 \\ | \\ \text{COOH} \\ \text{creatine} \end{array} + \begin{array}{c} \text{SH} \\ | \\ (\text{CH}_2)_2 \\ | \\ \text{NH}_2\text{—CH—COOH} \\ \text{homocysteine} \end{array}$$

As the free energy of hydrolysis of the phosphate group of creatine phosphate is about 11,000 cal/mole, this compound is a high-energy phosphate. The phosphate group can therefore be transferred from creatine phosphate to ADP to form ATP. Phosphorylation of creatine is favoured at relatively high pH (the optimum being about 9) whereas the maximum rate of ATP formation from creatine phosphate is obtained at pH 6–7. The pH of the muscle itself varies and, when it is working hard, lactic acid accumulates producing a pH change which is favourable in ensuring that free creatine does not compete so well for the terminal high-energy phosphate of ATP.

CHAPTER 9

Glycolysis

Throughout the living kingdom the main source of energy under anaerobic conditions is the breakdown of carbohydrate. In mammalian muscle, the chief store of energy is the polysaccharide glycogen; in addition, there is a small mobile store of glucose. Both of these compounds yield energy on being degraded to lactate. With glucose, the energy released is about 47 kcal/mole; each equivalent 6-carbon unit from glycogen releases 52 kcal. The series of reactions bringing about this degradation in living things is called glycolysis.

The Structure of Glycogen

Figure 2.11 illustrates the structure of glycogen; each circle represents a glucose equivalent. In each molecule, there is only one glucose unit with carbon 1 free to act as a reducing agent; this forms the 'reducing end' of the molecule. The

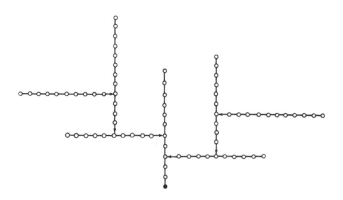

Figure 2.11 Diagram of glycogen molecule: ●, glucose residue carrying free reducing group; ○, glucose residues; ↑ , α-1,6 bonds

presence of two different chemical links in the structure is apparent from Figure 2.12; thus linear arrays of glucose molecules joined by 1–4 linkages are connected at branching points formed by 1–6 linkages. The frequency of branching is such that the average number of glucose residues between the branches is about 5.

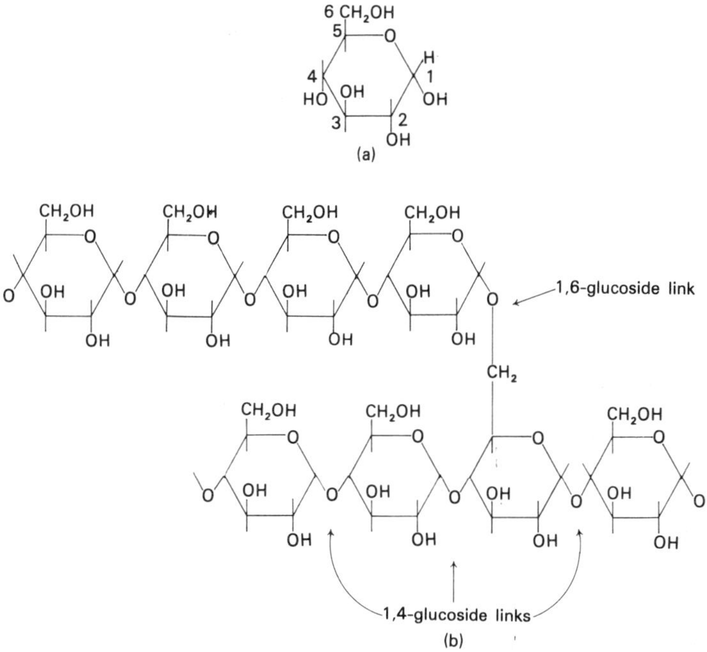

Figure 2.12 Glucosidic links in glycogen (b). The glucose molecule is represented by (a). The numbers refer to the C atoms; all C valencies with unspecified groups are occupied with H

The Degradation of Glycogen to Glucose-6-phosphate

The first step in the breakdown of glycogen involves the enzyme phosphorylase which catalyses the reaction

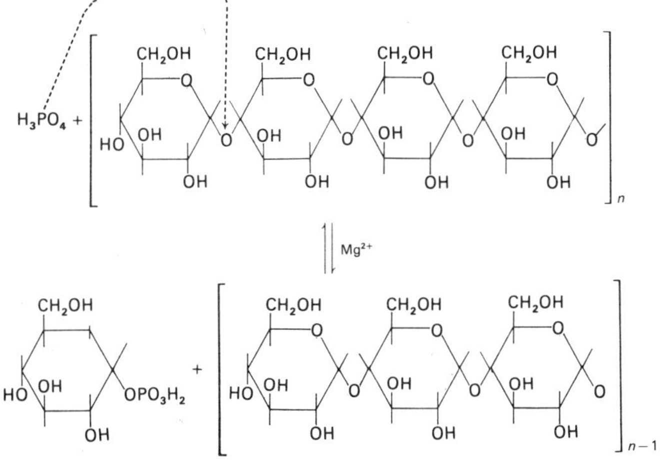

α-D-glucose-1-phosphate

Here phosphate behaves in the same way as does water in an hydrolytic reaction; the terminal 1–4 bond of the glycogen molecule is broken with the phosphoric acid providing the phosphate attached to carbon 1 of the released glucose-1-phosphate. The equilibrium constant at pH 7 is 0·3. The reaction is readily reversible and its direction is determined by the ratio of the concentrations of glucose-1-phosphate to phosphoric acid. The concentration of glycogen is not involved since every terminal glucose removed leaves an exactly equivalent terminal residue.

This action of phosphorylase will prune the outer branches of the glycogen molecule residue by residue until the fourth glucose residue before an α-1,6 linkage (at a branch point) is reached. The three residues before that carrying the α-1,6 linkage are then removed to another part of the molecule by a transglucosylase. In this way the glucose residue engaged in the α-1,6 linkage is exposed and can be liberated as free glucose by the debranching enzyme, amylo-1,6-glucosidase. This liberates the last glucose residue in each branch as free glucose, not as glucose-1-phosphate. Something like 8 per cent of the total glucose residues in glycogen are liberated as free glucose in this way.

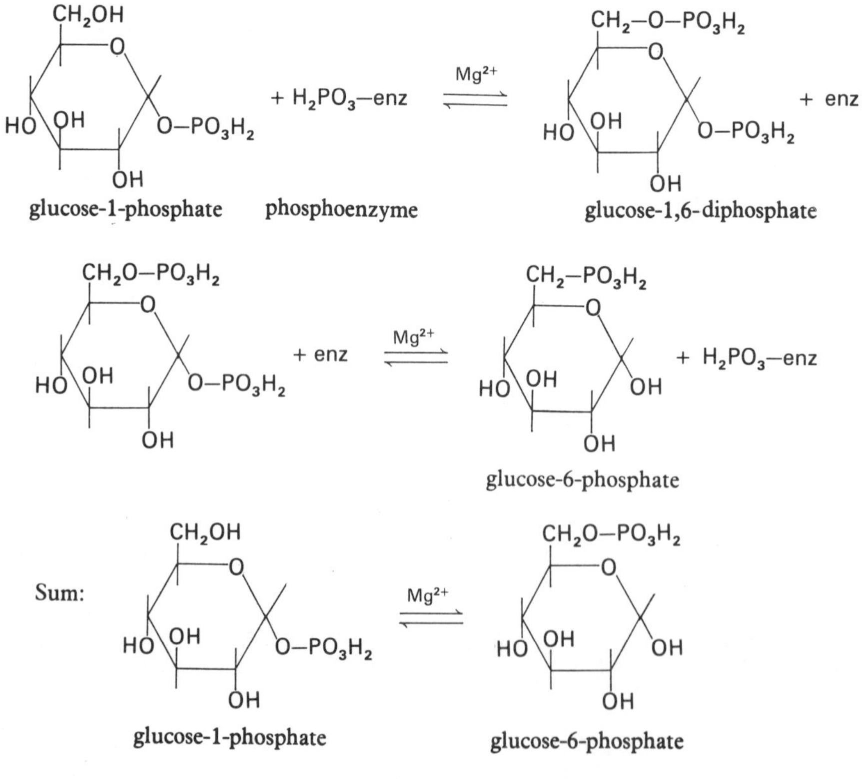

The glucose-1-phosphate is isomerized to glucose-6-phosphate by the enzyme phosphoglucomutase, which is present in muscle in high concentration (about 2 per cent of the water soluble protein). Phosphoglucomutase requires glucose-1,6-diphosphate as a cofactor. The enzyme carries a transferable phosphate attached

glucose-1-phosphate phosphoenzyme glucose-1,6-diphosphate

glucose-6-phosphate

glucose-1-phosphate glucose-6-phosphate

to a serine residue. This phosphate can be donated to either glucose-1-phosphate or glucose-6-phosphate to produce glucose-1,6-diphosphate. At equilibrium at pH 7, there is about 19 times as much glucose-6-phosphate as glucose-1-phosphate (see p. 41).

Interrelations between Glucose and Glucose-6-phosphate

Although glucose is the transportable form of carbohydrate it is important to realize that glucose itself cannot take part in any important metabolic conversion in the cell. It must be phosphorylated before it can enter into pathways of synthesis or degradation. For each molecule of glucose phosphorylated, one mole of ATP is required to provide the phosphate for attachment to carbon 6. The energy relationship may be illustrated as follows. The hydrolysis of phosphate from glucose-6-phosphate yields 3 kcal ($\Delta G^{\circ\prime} = -3$ kcal/mole). The free energy of hydrolysis of the terminal phosphate of ATP is about -7 kcal/mole. Therefore, the phosphorylation of glucose by ATP is strongly exergonic, to the extent of about -4 kcal/mole; consequently, the equilibrium constant is large ($K = 10^3$). Thus, the animal must 'pay' 4 kcal in terms of energy loss for each mole of glucose-6-phosphate formed from glucose; when glucose-6-phosphate is hydrolysed, with an energy loss of 3 kcal/mole, it gains, on the other hand, a substance which can pass through cell membranes that are impermeable to glucose-6-phosphate.

The formation of glucose-6-phosphate from glucose and ATP is catalysed by the enzyme hexokinase:

$$\text{glucose} + \text{ATP} \underset{}{\overset{Mg^{2+}}{\rightleftharpoons}} \text{glucose-6-phosphate} + \text{ADP}$$

'Hex' refers to the six-carbon sugars suitable as substrates and 'kinase' refers to the process of transfer of phosphate from ATP. In the hexokinase reaction, a phosphoryl group is transferred from ATP to glucose.

In common with many other phosphokinases, hexokinase requires magnesium. It is a relatively unspecific enzyme, phosphorylating many hexose sugars other than glucose; it has a very low K_m (37 μM) for glucose and is inhibited by the product of its own reaction, glucose-6-phosphate. There is another enzyme, glucokinase (present in the liver), catalysing the same reaction; but this is specific for glucose, has a much higher K_m (15 mM) and is not influenced by glucose-6-phosphate. The concentration of glucose in the blood is about 5 mM and, although the concentration in the cell is somewhat variable it is likely that the hexokinase is always saturated with glucose and is therefore working at its maximum rate. The characteristic properties of glucokinase are such, however, that it would seem to be ideally suited for removing any excess glucose that may be present by converting it to glucose-6-phosphate.

The Conversion of Glucose to Triose Phosphate

Glucose-6-phosphate formed either from glycogen or, with the expenditure of a molecule of ATP, from glucose, is now converted into other sugar phosphates. The next step is isomerization to fructose-6-phosphate by the enzyme phosphohexoseisomerase (or 'isomerase'). No cofactor is required and the equilibrium constant is about 0·5, so that no great energy changes are involved.

glucose-6-phosphate ⇌ fructose-6-phosphate

The succeeding conversion of fructose-6-phosphate to fructose-1,6-diphosphate is again one that has to be 'paid for' in terms of expenditure of ATP. This reaction is catalysed by the enzyme phosphofructokinase and requires magnesium. Again, as in the formation of glucose-6-phosphate, the high energy

fructose-6-phosphate + ATP $\xrightarrow{Mg^{2+}}$ fructose-1,6-diphosphate

of the ATP is used to phosphorylate the sugar hydroxyl on carbon 1; the ester formed has a low energy of hydrolysis. Thus the reaction is accompanied by a liberation of free energy ($\Delta G^{\circ\prime} = -4\cdot 2$ kcal/mole) and is not readily reversible. The enzyme is influenced by a number of important metabolites and is a key enzyme in controlling the rate of glycolysis; thus it is inhibited by excess ATP and this inhibition is overcome by ADP, AMP, phosphate and fructose-6-phosphate.

The diphosphorylated sugar is next cut in two by the enzyme aldolase. This reaction produces two triose phosphates, glyceraldehyde-3-phosphate and dihydroxyacetone phosphate. (In the equation, note that the reactive form of the hexose diphosphate is represented as a straight chain.) The enzyme reaction is freely reversible—indeed, the position of equilibrium favours the formation of the hexose diphosphate—and the enzyme is called 'aldolase' because in forming fructose-1,6-diphosphate it catalyses an aldol condensation.

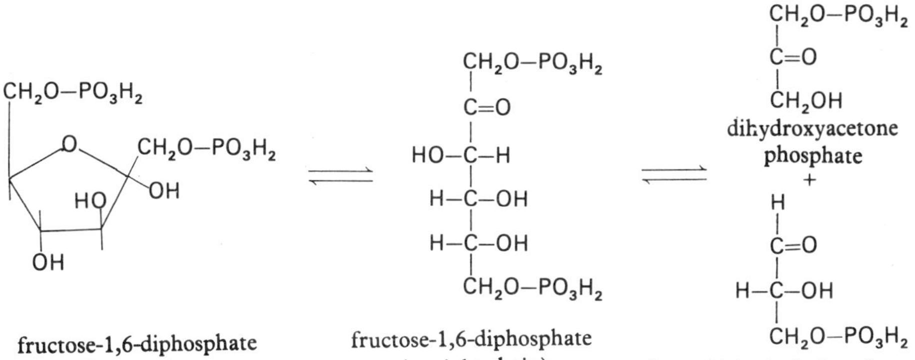

$$\text{fructose-1,6-diphosphate (furanose)} \rightleftharpoons \text{fructose-1,6-diphosphate (straight chain)} \rightleftharpoons \text{dihydroxyacetone phosphate} + \text{glyceraldehyde-3-phosphate}$$

The mixture of triose phosphates that accumulates is equilibrated by an enzyme triosephosphate isomerase, the reaction being similar to that carried out by the phosphohexose isomerase. The enzyme is very active and although there is 22 times as much of the ketose as the aldose present at equilibrium, the rate of

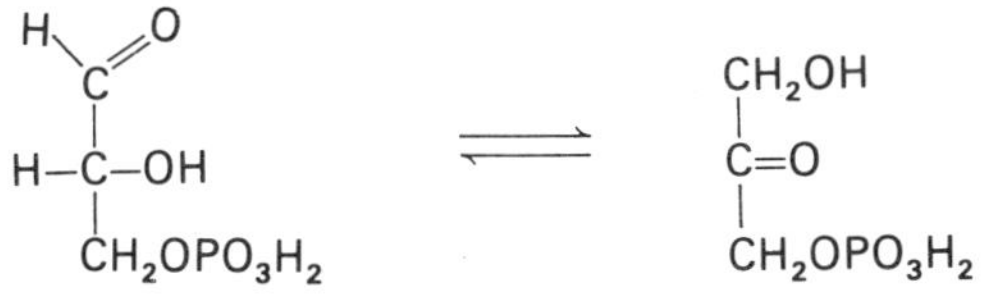

$$\text{glyceraldehyde-3-phosphate} \rightleftharpoons \text{dihydroxyacetone phosphate}$$

equilibration is so fast that even when glyceraldehyde phosphate is removed by the subsequent reaction at a very high rate, the two trioses remain virtually in equilibrium.

The Oxidation of Triose Phosphate and the Reduction of NAD^+

All the reactions considered so far have been either rearrangements of molecules or phosphorylations, but the next step involves an oxidoreduction and the coenzyme nicotinamide adenine dinucleotide (NAD^+), Figure 2.13. This dependence of an enzyme reaction on the availability of another reactant called a coenzyme is common in living things. The coenzyme may be thought of as a chemical 'handle', for carrying a reactive chemical group from one molecule to

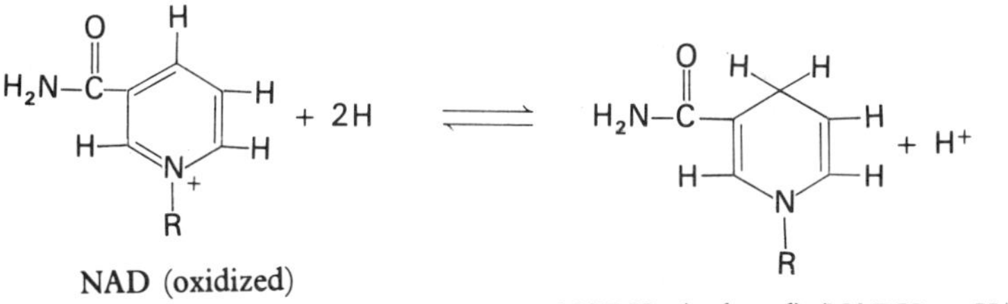

$$\text{NAD (oxidized)} (NAD^+) + 2H \rightleftharpoons \text{NADH}_2 \text{ (reduced)} (NADH + H^+) + H^+$$

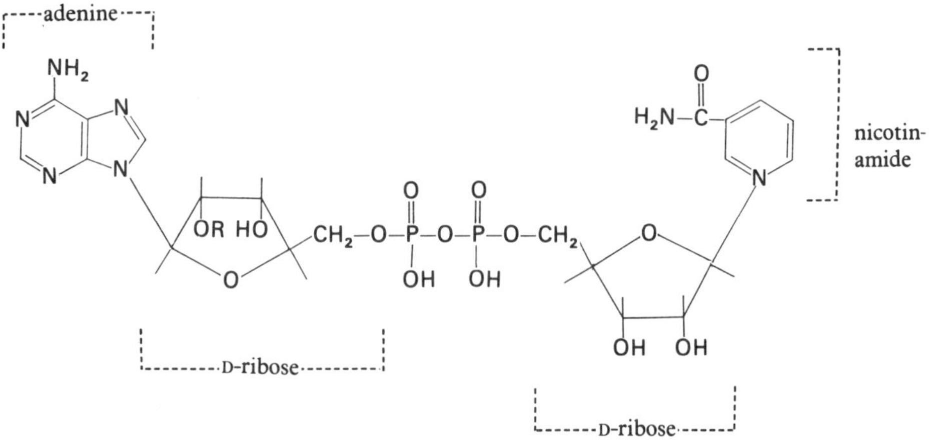

Figure 2.13 The nicotinamide coenzymes (oxidized form). NAD^+, $R = H$; $NADP^+$, $R = PO_3H_2$. Nicotinamide is a derivative of the vitamin niacin (nicotinic acid)

another in metabolic transformations. NAD^+ is a 'chemical handle' for the transport of 'reducing power' around the cell for utilization at another point. The reducing power is carried in a form equivalent to a pair of hydrogens.

The oxidation of glyceraldehyde-3-phosphate, catalysed by glyceraldehyde-phosphate dehydrogenease ('triosephosphate dehydrogenase') results in the reduction of NAD^+. The reaction involves the condensation of phosphoric acid

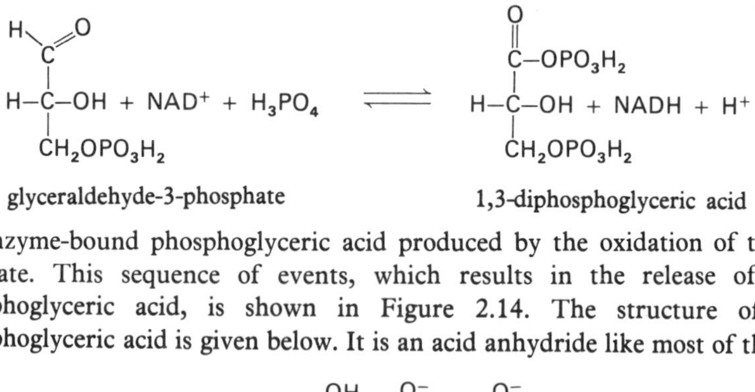

glyceraldehyde-3-phosphate ⇌ 1,3-diphosphoglyceric acid

with enzyme-bound phosphoglyceric acid produced by the oxidation of triose-phosphate. This sequence of events, which results in the release of 1,3-diphosphoglyceric acid, is shown in Figure 2.14. The structure of the diphosphoglyceric acid is given below. It is an acid anhydride like most of the so-

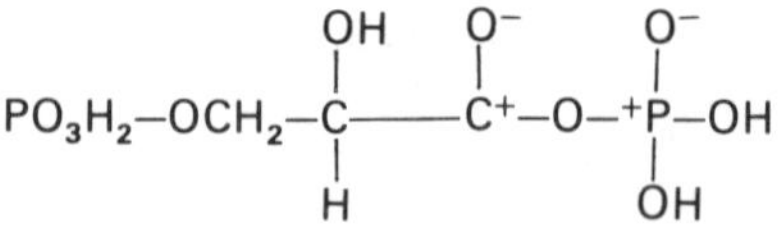

1,3-diphosphoglyceric acid (mixed anhydride)

called high-energy compounds; the proximity of the positive charges on the carbon and phosphorus atoms leads to a pronounced thermodynamic instability, which manifests itself in a large heat of hydrolysis (see Table 1.2). Thus the formation of this compound represents a conservation of some of the energy made available during the oxidation of the triose, in the form of a high-energy phosphate compound.

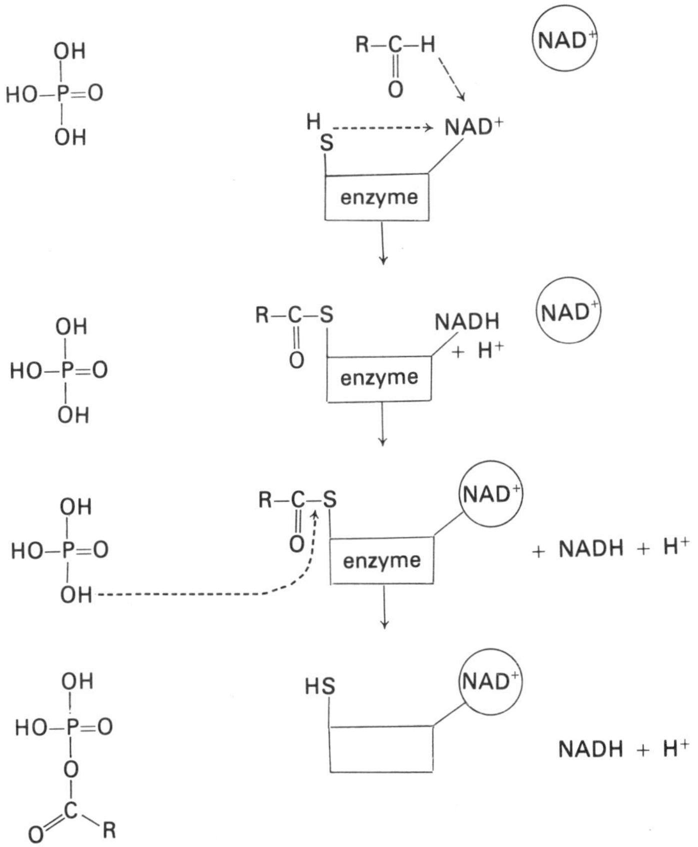

Figure 2.14 Mechanism of action of glyceraldehyde-3-phosphate dehydrogenase

Phosphoglyceraldehyde dehydrogenase is somewhat unusual in that the reduced coenzyme (NADH + H⁺) must be replaced by a molecule of NAD⁺ before the phosphorylation step can occur (see Figure 2.14); this figure also illustrates the essential role of a sulphydryl group in the enzyme during catalysis. The sulphydryl group is the point at which iodoacetate inhibits glycolysis.

The Utilization of 1,3-Diphosphoglyceric Acid

Thus, phosphoglyceraldehyde dehydrogenase produces two compounds: NADH + H⁺, whose utilization is considered later, and 1,3-diphosphoglyceric acid, which is dephosphorylated in the next step of the glycolytic sequence, catalysed by phosphoglyceric acid kinase. The free energy change in this reaction is about −4 kcal/mole, i.e. the equilibrium is far to the right. As the $\Delta G^{\circ\prime}$ for the hydrolysis of the terminal phosphate of ATP is about −7 cal/mole, the $\Delta G^{\circ\prime}$ for the hydrolysis of the acyl phosphate group of 1,3-diphosphoglyceric acid is (−7 +

$$\underset{\text{1,3-diphosphoglyceric acid}}{\begin{array}{c} \text{O} \\ \| \\ \text{C}-\text{OPO}_3\text{H}_2 \\ | \\ \text{H}-\text{C}-\text{OH} \\ | \\ \text{CH}_2\text{OPO}_3\text{H}_2 \end{array}} + \text{ADP} \underset{}{\overset{\text{Mg}^{2+}}{\rightleftharpoons}} \underset{\text{3-phosphoglyceric acid}}{\begin{array}{c} \text{HO} \diagdown \text{C} \diagup \text{O} \\ | \\ \text{H}-\text{C}-\text{OH} \\ | \\ \text{CH}_2\text{OPO}_3\text{H}_2 \end{array}} + \text{ATP}$$

$-4 \cdot 7$) kcal/mole or about -12 kcal/mole. Thus the organism 'pays the price' of losing about one-third of the available energy in the diphosphoglyceric acid for the privilege of ensuring that ADP is phosphorylated quantitatively to ATP, a price similar to that 'paid' for ensuring that glucose was phosphorylated quantitatively to glucose-6-phosphate. Reactions of this kind, in which a phosphorylated metabolite acts as the donor of a phosphoryl group for the synthesis of a nucleoside triphosphate, are called 'substrate level' phosphorylations, to distinguish them from the phosphorylations linked to the respiratory chain (see p. 139).

The Formation and Utilization of Phosphoenolpyruvic Acid

3-Phosphoglyceric acid produced by phosphoglyceric kinase is isomerized to 2-phosphoglyceric acid by the enzyme phosphoglyceromutase. Since the

$$\underset{\text{3-phosphoglyceric acid}}{\begin{array}{c} \text{HO} \diagdown \text{C} \diagup \text{O} \\ | \\ \text{H}-\text{C}-\text{OH} \\ | \\ \text{CH}_2\text{OPO}_3\text{H}_2 \end{array}} \overset{\text{Mg}^{2+}}{\rightleftharpoons} \underset{\text{2-phosphoglyceric acid}}{\begin{array}{c} \text{HO} \diagdown \text{C} \diagup \text{O} \\ | \\ \text{H}-\text{C}-\text{OPO}_3\text{H}_2 \\ | \\ \text{CH}_2\text{OH} \end{array}}$$

equilibrium constant is about $0 \cdot 17$, no great energy change is involved; the reaction is an almost precise analogy to the formation of glucose-6-phosphate from glucose-1-phosphate and, as might be predicted from this analogy, requires 2,3-diphosphoglyceric acid as a cofactor.

The next enzyme in the series is called enolase; this removes water from 2-phosphoglyceric acid to give phosphoenolpyruvic acid. This latter compound is the phosphorylated derivative of the enol form of pyruvic acid, a form which is relatively unstable compared with the keto form. Although the equilibrium cons-

$$\underset{\text{2-phosphoglyceric acid}}{\begin{array}{c} \text{HO} \diagdown \text{C} \diagup \text{O} \\ | \\ \text{H}-\text{C}-\text{OPO}_3\text{H}_2 \\ | \\ \text{CH}_2\text{OH} \end{array}} \overset{\text{Mg}^{2+}}{\rightleftharpoons} \underset{\text{phosphoenolpyruvic acid}}{\begin{array}{c} \text{HO} \diagdown \text{C} \diagup \text{O} \\ | \\ \text{C}-\text{OPO}_3\text{H}_2 \\ \| \\ \text{CH}_2 \end{array}} + \text{H}_2\text{O}$$

tant for the enolase reaction is only 3 and the reaction is freely reversible, involving little energy change, the rearrangement of the molecule has redistributed the internal energy so that the phosphate may now be considered as a high-energy group. This is apparent from the different heats of hydrolysis of 2-phosphoglyceric acid (-3 kcal/mole) and phosphoenolpyruvic acid (-13 kcal/mole).

Enolase requires magnesium and is inhibited by fluoride in the presence of inorganic phosphate due to the removal of magnesium as a complex magnesium fluorophosphate.

The final step in the glycolytic pathway is one in which the high-energy phosphate of phosphoenolpyruvic acid is used to form ATP and which requires the enzyme pyruvic kinase. The equilibrium constant of $2 \cdot 6 \times 10^4$ indicates that

$$\begin{array}{c} \text{HO} \diagdown \text{C} \diagup \text{O} \\ | \\ \text{C}-\text{O}-\text{PO}_3\text{H}_2 + \text{ADP} \\ | \\ \text{CH}_2 \end{array} \quad \underset{\longleftarrow}{\overset{Mg^{2+},\ K^+}{\longrightarrow}} \quad \begin{array}{c} \text{HO} \diagdown \text{C} \diagup \text{O} \\ | \\ \text{C}=\text{O} + \text{ATP} \\ | \\ \text{CH}_3 \end{array}$$

phosphoenolpyruvic acid   pyruvic acid

phosphopyruvate and ADP are converted almost quantitatively to pyruvate and ATP. Again, the penalty of a loss of about 6 kcal/mole of the available energy is paid to ensure that the yield of ATP is high. This reaction concludes the glycolytic sequence proper, but the problem of the regeneration of NAD^+ from the $NADH + H^+$ formed by the action of phosphoglyceraldehyde dehydrogenase must still be considered.

The Regeneration of NAD^+ and the Formation of Lactic Acid

The necessity for regenerating NAD^+ efficiently if glycolysis is to continue is obvious since the concentration of the coenzyme is strictly limited, being about 1 mM in the tissue water. The muscle of the 70 kg man, which amounts to some 25 kg of wet tissue, has about $17 \cdot 5$ kg of water (70 per cent of the wet weight). This water will contain $17 \cdot 5$ mmole of nicotinamide nucleotide, which will have a total 'hydrogen capacity' equivalent to that released from 9 mmole of glycolysed glucose. Some 50 kcal are obtained from each mole of glucose degraded to pyruvate, i.e. 50 cal/mmole. Therefore, something like 450 cal can be obtained from the amount of glucose required to saturate all the NAD^+ of the muscle with hydrogen. If the man is undertaking moderate exercise (for example, running), he may be using about 10 kcal/min, or 167 cal/sec and, therefore, without reoxidation, his NAD^+ store will suffice for only about 3 sec of exercise. As most 70 kg men can exercise for considerably longer than 3 sec, it follows that the hydrogen must be removed continuously from $NADH + H^+$ to allow glycolytic energy production to continue. One solution to this problem is to oxidize the hydrogen with atmospheric oxygen to give water—the process of respiration. The second solution, however, permits an anaerobic regeneration of NAD^+ by the enzyme lactic dehydrogenase.

Lactic dehydrogenase transfers the hydrogen of NADH + H$^+$ to pyruvic acid to form lactic acid and NAD$^+$. The equilibrium is in favour of lactate production

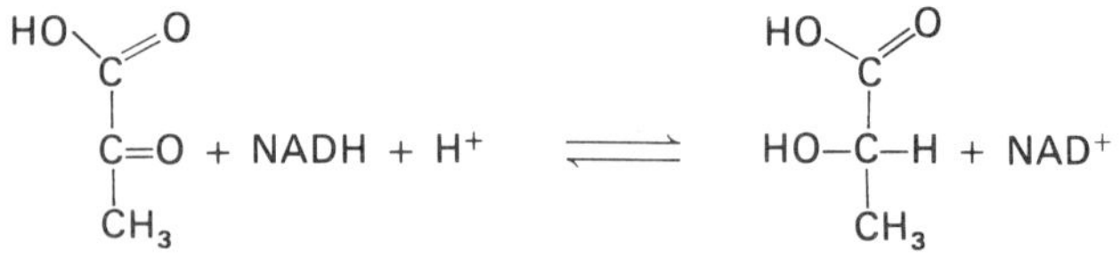

$$\begin{array}{c} HO\diagdown{}_{C}\diagup{}^{O} \\ | \\ C{=}O \\ | \\ CH_3 \end{array} + NADH + H^+ \rightleftharpoons \begin{array}{c} HO\diagdown{}_{C}\diagup{}^{O} \\ | \\ HO{-}C{-}H \\ | \\ CH_3 \end{array} + NAD^+$$

pyruvate   L-(+)-lactic acid

($\Delta G^{\circ\prime} = -6$ kcal/mole) and therefore the enzyme is efficient in regenerating NAD$^+$. The cyclic nature of the utilization of the coenzyme during the conversion of glucose to lactic acid, as shown in Figure 2.15, makes it apparent that the availability of NAD$^+$ may limit the rate of the complete pathway.

Thus, the result of glycolysis is the production of two molecules of lactic acid for every glucose unit degraded (see Figure 2.15), the conservation as ATP of some of the energy made available and dissipation of some of the energy as heat. Moreover, the availability of ADP will limit the rate of the glycolytic process, since neither 1,3-diphosphoglyceric acid nor phosphoenolpyruvate can be broken down unless ADP is available to accept the high-energy phosphates. Thus there is a strict 'coupling' of the utilization of ATP with its production in the glycolytic path.

The Energy Balance of Glycolysis

The energy balance from the glycolysis of either glycogen or glucose can be expressed in terms of ATP molecules used or gained, or in terms of the total energy lost as heat or conserved as chemical energy. The values for the balance with ATP are as follows:

Glycolysis from glycogen

6 carbon unit

ATP used	*ATP gained*
1 to phosphorylate fructose-6-P	2 from triose phosphate
	2 from phosphopyruvate
Net gain	3 ATP per glucose unit

Glycolysis from glucose

ATP used	*ATP gained*
1 to phosphorylate glucose to glucose-6-phosphate	2 from triose phosphate
1 to phosphorylate fructose-6-P	2 from phosphopyruvate
Net gain	2 ATP per glucose unit

Thus either two or three ATP molecules are synthesized as a net gain, depending on the substrate used.

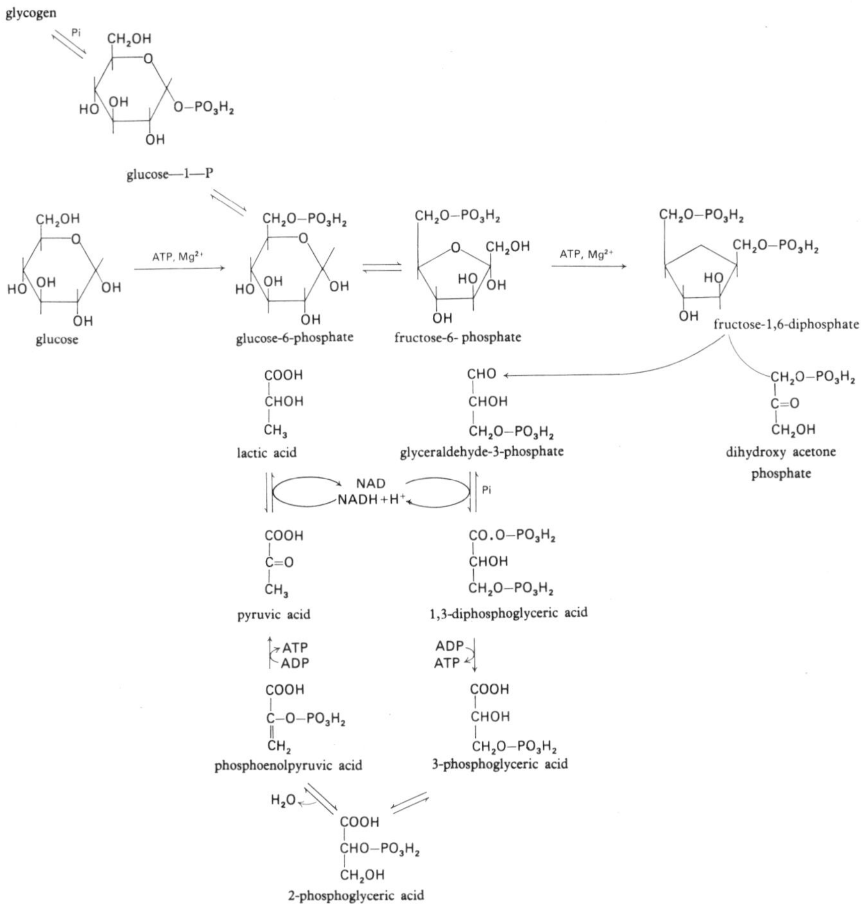

Figure 2.15 The glycolytic pathway

The energy balance sheet is given in Table 2.1. Assuming that the 'value' of each ATP molecule is 11·3 kcal (at the concentrations of reactants defined in Table 2.1) and that the net gain of ATP is 3/mole of glucose degraded, the efficiency of the process of energy conservation from anaerobic glycolysis is 33·9/56·7 = 60 per cent.

Glycolytic Enzymes in Muscular Dystrophy

One disease of muscle in which there is a considerable disturbance of the enzymes of the glycolytic pathway coupled with a breakdown of the tissue, is muscular dystrophy. This is a progressive wasting disease and, as shown in Table 2.2, there is an appreciable loss of enzymes from the muscle, sufficient to cause a marked drop in the glycolytic rate of the tissue. Whether this loss precedes

Table 2.1 Energy balance sheet from glycogen to lactate (Data from Burton, K. and Krebs, H. A., *Biochem. J.*, **54**, 105, 1958)

Free energy fed into the system for the conversion of 1 mole glucose to lactate and for the hydrolysis of 1 mole of ATP.

ΔG for formation of lactic acid	56,700 cal
+ energy of terminal phosphate of one ATP	11,300 cal
Total	68,000 cal
Expenditure at phosphorylation of fructose-6-phosphate	−4300 cal
Expenditure in formation of ATP from 1,3-diphosphoglyceric acid	−9500 cal
Expenditure in formation of ATP from phosphopyruvate	−10,000 cal
Expenditure in formation of lactate from pyruvate	−12,000 cal
Energy change in all other steps	+13,000 cal
Total known expenditure	−22,800 cal
Energy saved as ATP (4 × 11·3)	45,200 cal approx.

ΔG values are for 0·2 atm O_2, 0·05 atm CO_2, pH 7 and 10 mM concentration of all reactants other than glycogen. 'Expenditure' refers to energy loss as heat.

muscular degeneration, or vice versa, is not known but there is an accompanying rise in the amounts of the enzymes in the blood plasma. Aldolase, in particular, is an enzyme which apparently passes through cell membranes comparatively easily and the rise of aldolase activity in the plasma may be used as a method of diagnosis of the disease.

Universality of the Glycolytic Pathway in Living Matter

Although the discussion of the process of glycolysis has centred around its role in muscular tissue, the same pattern of changes is exhibited by an enormous variety of cells, animal, plant and microbial. Many cells capable of extensive glycolysis however, while preserving the same general pattern, regenerate NAD^+ by reducing a compound other than pyruvate; noteworthy examples are the production of ethanol from acetaldehyde by yeasts and of α-glycerophosphate from dihydroxyacetone phosphate in the flight muscle of some insects.

Table 2.2 Enzyme content of muscle

	Phosphorylase	Phosphoglucomutase	Aldolase	Total rate of glycolysis
Normal	23	11	19	2·9
Dystrophic	8·6	4·4	6·3	0·8

The values for the enzyme activity are expressed as millimoles of glucose equivalents transformed/g of protein/hr.

Disposal of Lactic Acid

The process of glycolysis described in this chapter results in an accumulation of lactic acid and a loss of glycogen and glucose. Since lactic acid is not excreted in the urine and does not accumulate continuously in the blood during exercise, there must be processes disposing of some of the lactate formed. (There is, of course, some rise in blood lactate after exercise, from a resting level of about 19 mg/100 ml of blood to about 80 mg/100 ml.) In the intact animal much of the muscle lactate passes into the blood and thence to the liver where it is resynthesized to glucose and glycogen. The remaining lactate is oxidized to carbon dioxide and water.

Organs differ greatly in their ability to oxidize lactate. The venous blood of skeletal muscle under all conditions contains more lactate than the arterial blood; thus, skeletal muscle always makes a net contribution to blood lactate. By contrast, the venous blood of the heart contains less lactate than the arterial blood; thus the heart is a net consumer of lactate. The reason for this difference can be understood in terms of the different functions of the two sets of muscles. Skeletal muscle undertakes only intermittent activity, but for a time this activity may be beyond the capacity of the blood stream to supply oxygen and substrate. Therefore, provision must be made for a large-scale production of energy from anaerobic pathways. On the other hand, the heart works continuously and must rely on aerobic mechanisms to make the most efficient use of the energy in the oxidizable substrates.

Energy Yields from Carbohydrate in Respiration and Glycolysis

For each mole of hexose glycolysed to lactic acid, only about 45 to 50 kcal become available to the cell, whereas the total free-energy change during the complete oxidation of a mole of glucose to carbon dioxide and water is about 700 kcal. To obtain this same amount of energy anaerobically, the amount of substrate decomposed would need to be about 15 times greater. Only in exceptional circumstances can tissues afford to consume carbohydrate in this way and the great majority of cells possessing both aerobic and anaerobic mechanisms for energy transformation rely on respiration rather than glycolysis for the production of most of their ATP.

However, under special conditions, anaerobic reactions are of advantage even to those cells of animal tissue which are generally dependent on respiration. The substrate for glycolysis can be stored in the tissues, in the form of glycogen, in relatively large quantities (in some muscles, up to several per cent of the wet weight). The additional substrate required for respiration, molecular oxygen, can be stored only in very limited quantities (bound to myoglobin) and respiration is dependent therefore on a continuous supply of oxygen from the circulating blood. Hence, when there is a need for the rapid provision of energy (as, for example, in skeletal or cardiac muscle making an acute effort) this energy can be supplied more effectively for short periods by glycolysis. This may be illustrated by the following calculations for the energy supply available to a muscle suddenly cut off from the circulation.

Muscle may contain over 1 per cent of available carbohydrate, most of it as glycogen. However, if the circulation ceases, the increasing acidity due to lactic accumulation will prevent the complete glycolysis of this carbohydrate. Assuming that for each 1 g wet weight of muscle, 5 mg of glycogen are broken down (i.e. about half the total amount present) then 5000/180 μmole of glucose units (M.W. 180 g) are degraded, i.e. 27·8 μmole. The free-energy yield for the fermentation of one glucose unit of glycogen to lactate is 52 kcal/mole, or 0·052 cal/μmole. Therefore, for the 27·8 μmole degraded by 1 g of tissue the total yield of energy would be about 1·4 cal.

Under aerobic conditions (but with no influx of oxygen from the blood) the oxidation would be limited by the total oxygen stored in the tissue. For each gram of muscle there will be

5 μl of dissolved oxygen
40 μl of oxygen combined with myoglobin (this corresponds to 10^{-4} g of myoglobin iron per g muscle)
10 μl of oxygen combined with haemoglobin (this assumes that 5 per cent of the tissue consists of arterial blood)

Thus, the total oxygen available from all sources in 55 μl/g tissue, i.e. 2·45 μmole. For each mole of oxygen used when glucose is the substrate, there is a yield of 114·3 kcal or 0·1143 per μmole of oxygen. Therefore, the 2·45 μmole of oxygen available will be sufficient to release only 0·28 cal from glucose oxidation, or about one-fifth of that available from the anaerobic source.

CHAPTER 10

The Citric Acid Cycle

The complete degradation of glucose to carbon dioxide involves the oxidative disposal of the lactic acid (or pyruvate) formed during glycolysis. As heart muscle is an efficient oxidizer of lactic acid, the process will be considered in relation to this organ, whose structure is shown in Figure 2.16. The most striking difference between this tissue and skeletal muscle (Figure 2.17) is that the latter is composed very largely of the contractile proteins and relatively few mitochondria, whereas in the heart there are rows of rounded mitochondrial structures between the muscle fibres. These mitochondria are responsible for the large oxygen uptake of the heart and its utilization of lactate.

Besides the enzymes responsible for catalysing the reaction with oxygen, mitochondria contain the integrated complex of enzymes responsible for the process called the citric acid cycle which achieves the final oxidative breakdown of lactate and pyruvate (the products of glycolysis) and of other components (fatty acids and amino acids) capable of being converted to intermediates of this cycle. The mitochondrion is thus a striking example of the segregation of a series of complex enzyme sequences in a special cell compartment or organelle. Sometimes the continuation of a metabolic pathway may require the passage of intermediates from one cell compartment to another, as, for example, the movement of the end products of glycolysis (which occurs outside the mitochondrion) into the mitochondria for oxidative degradation.

The Oxidation of Lactate to Pyruvate

The first step in the disposal of lactate is a reversal of the last glycolytic reaction catalysed by lactic dehydrogenase. This enzyme is found mainly outside the intracellular organelles in what is called the 'soluble phase' of the cell. What this means exactly in the muscle cell is far from clear—it is not known where the enzymes of glycolysis are located in respect to the muscle fibres, though they are certainly extramitochondrial.

As the equilibrium for the reaction

$$\text{lactate} + \text{NAD}^+ \rightarrow \text{pyruvate} + \text{NADH} + \text{H}^+$$

favours the production of lactate ($\Delta G^{\circ\prime}$ is 6000 cal/mole), the reaction can be reversed only if the concentrations of the products are kept low. In muscle this may be achieved by the rapid consumption of both NADH + H$^+$ and pyruvate by the mitochondria. The hydrogen from NADH + H$^+$ is finally oxidized to water,

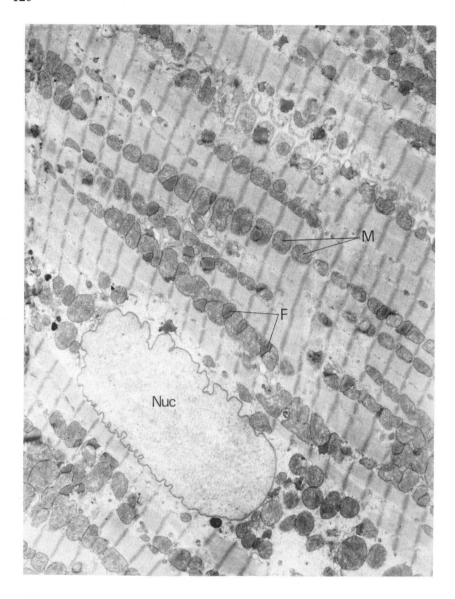

Figure 2.16 An electron micrograph of heart muscle. The abundance of mitochondria (M) and their dense arrays of cristae reflect the predominantly aerobic metabolism of this tissue. Note the location of mitochondria close to the region of overlap between the thin and thick myofilaments. The close association of fat droplets (F) with the mitochondria illustrates the ability of the heart to use fatty acids as a major metabolic fuel. By kind permission of Mr J. H. Kugler

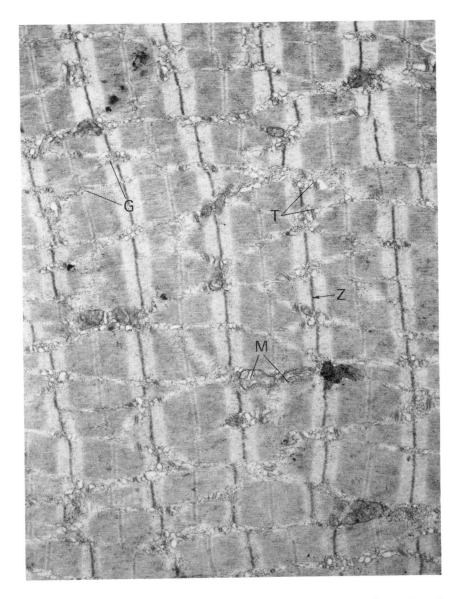

Figure 2.17 An electron micrograph of skeletal muscle. The small number of mitochondria (M) in this field reflects the lower oxidative capacity of this tissue compared with heart muscle (Figure 2.16). All the mitochondria are located near the Z-bands (Z) in the general area of the triad system (T) where ATP is required to pump Ca^{2+} ions into the terminal cisternae of the sarcoplasmic reticulum in order to bring about relaxation. Glycogen granules (G) located close to the contractile apparatus are able to provide a ready source of ATP predominantly via glycolysis. By kind permission of Mr J. H. Kugler

the pyruvate to water and carbon dioxide. For the moment we will consider only the oxidative breakdown of the carbon skeleton of pyruvate by the citric acid cycle.

The Citric Acid Cycle

The postulation of this cycle (also known as the Krebs' cycle) in 1937 by Krebs and Johnson was based on measurements of oxygen uptake by preparations of minced pigeon breast muscle, incubated with various oxidizable substrates. They observed that the oxygen uptake of the muscle preparation was increased by adding citrate and that the magnitude of the increase was greater than could be accounted for by the complete oxidation of the added substrate, i.e. its addition catalysed the oxidative degradation of some other substances in the preparation.

Reaction 1a (pyruvate dehydrogenase)

$$CH_3-\overset{O}{\underset{}{C}}-COOH + TPP \longrightarrow CH_3-\overset{H}{\underset{H}{\overset{|}{\underset{|}{C}}}}-TPP + CO_2$$
$$\text{where the middle C bears OH}$$

pyruvic acid · thiamine pyrophosphate · α-hydroxyethyl thiamine pyrophosphate

Reaction 1b (pyruvate dehydrogenase)

$$CH_3-\underset{H}{\overset{OH}{\underset{|}{C}}}-TPP + \begin{matrix}S-\\|\\S-\end{matrix}\!\!\Big\rangle \rightleftharpoons CH_3-\overset{O}{\underset{}{C}}-S\!\!\Big\rangle\ HS\!\!\Big\rangle + TPP$$

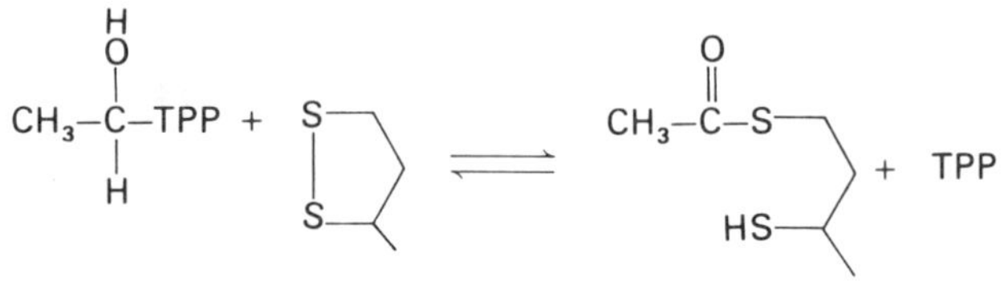

oxidized lipoic acid · acetylhydrolipoic acid

Reaction 2 (lipoyl acetyltransferase)

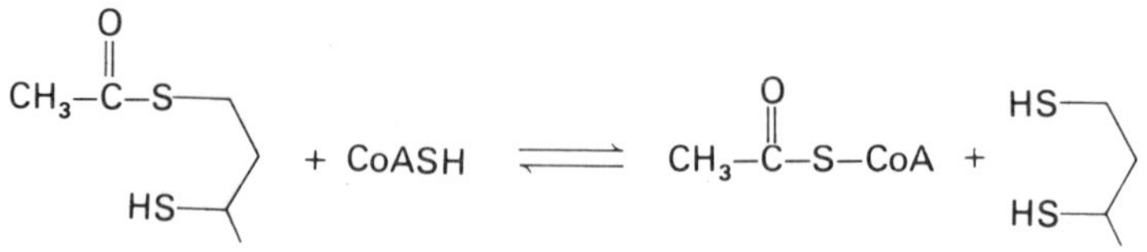

$$CH_3-\overset{O}{\underset{}{C}}-S\!\!\Big\rangle\ HS\!\!\Big\rangle + CoASH \rightleftharpoons CH_3-\overset{O}{\underset{}{C}}-S-CoA + HS\!\!\Big\rangle\ HS\!\!\Big\rangle$$

coenzyme A · acetyl-CoA · dihydrolipoic acid

Reaction 3 (lipoyl dehydrogenase)

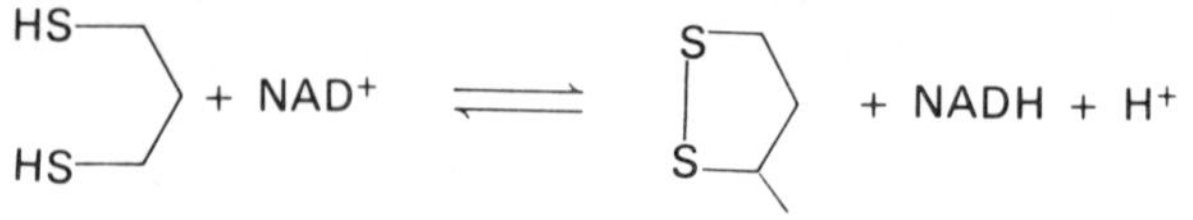

$$HS\!\!\Big\rangle\ HS\!\!\Big\rangle + NAD^+ \rightleftharpoons S\!\!\Big\rangle\ S\!\!\Big\rangle + NADH + H^+$$

Further experiments suggested that the other substances were derived from carbohydrate. Similar catalytic stimulations of oxygen uptake were found with added succinate, fumarate, malate and oxaloacetate, and it seemed that the oxidation of citrate itself involved the following compounds in order: α-oxoglutaric, succinic, fumaric, malic and oxaloacetic acids. Oxaloacetic acid could be converted into citric acid by combination with some unknown compound, thus completing a cyclic series of changes. This cycle was found in a great number of animal tissues. The compound required for the reaction with oxaloacetate to form citrate was shown subsequently to be an activated acetic acid, called acetyl coenzyme A. This compound is formed from (a) pyruvate by an energy-yielding process, (b) long-chain fatty acids by a complex series of reactions also with a net yield of energy, (c) free acetate by a reaction in which ATP must be consumed (see also p. 230) and (d) amino acids and certain other compounds (see Figure 2.22).

Formation of Acetyl-CoA from Pyruvate

This reaction, catalysed by a complex of enzymes, is a three-step reaction; the individual proteins are aggregated in a multienzyme complex with a molecular weight of millions (see Figure 1-35). Each of these steps requires the presence not only of the enzyme and its substrate but also of the appropriate coenzymes.

Reaction 1 is the oxidative decarboxylation of pyruvic acid catalysed by pyruvic dehydrogenase. We may picture this reaction as occurring in two stages. Firstly the pyruvate reacts with the coenzyme thiamine pyrophosphate (TPP or cocarboxylase—see Figure 2.18—derived from vitamin B_1). The thiazole ring of TPP forms addition complexes with keto compounds, which react with the ion produced by the dissociation of the hydrogen on the carbon lying between the nitrogen and sulphur of the thiazole ring. CO_2 is released from this complex forming hydroxyethyl thiamine pyrophosphate (HETP). In mammals the second stage is the reaction of HETP with the coenzyme lipoic acid in the enzyme complex; the lipoic acid is anchored to the protein of the pyruvate dehydrogenase through the epsilon amino group of a lysine residue thus

←—lipoic acid residue—→ ←—lysine residue of —→
                                                protein

Lipoic acid contains two sulphur atoms existing in the disulphide form or reduced to sulphydryl groups (when it may be considered as a derivative of *n*-

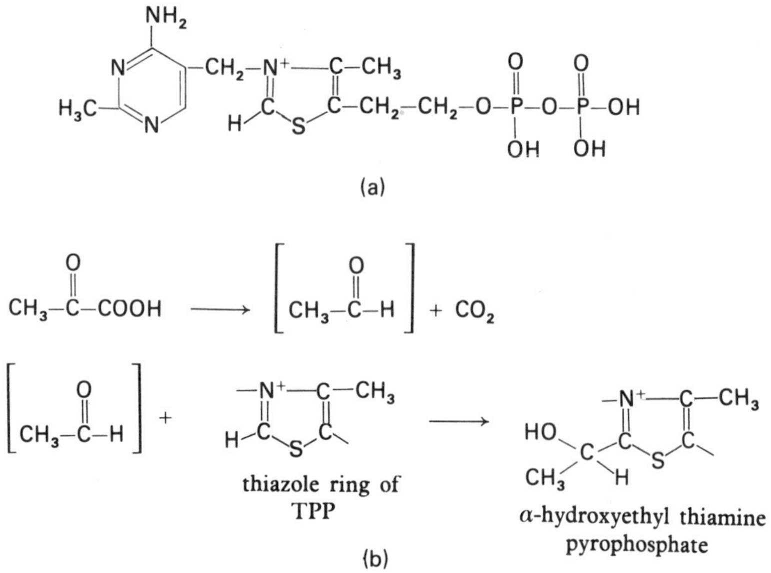

Figure 2.18 (a) Thiamine pyrophosphate (TPP). During the decarboxylation of α-oxo acids the thiazole ring acts as an acceptor for the decarboxylation product; the compound formed may be regarded as an adduct of the TPP and an aldehyde, e.g. with pyruvic acid the reaction may be represented as shown in (b). Thiamine is also known as vitamin B_1

octanoic acid, hence its other name thioctic acid. Arsenite inhibits pyruvate oxidation by combining with the two sulphydryl groups of reduced lipoic acid thus

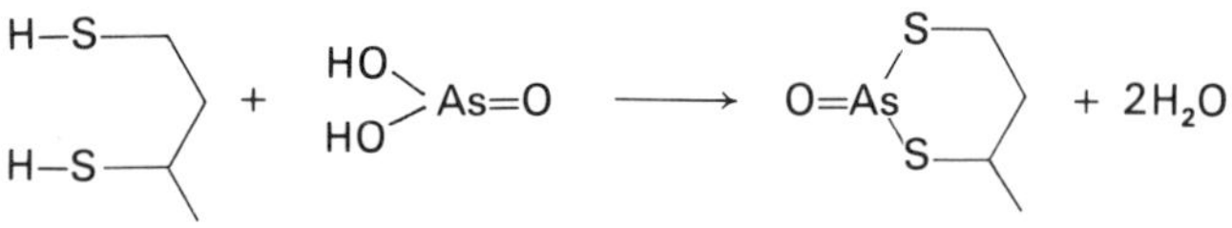

The exact details of the reaction of lipoic acid with HETP are not known, but it is supposed that the hydroxyethyl group ('active acetaldehyde') reacts with the oxidized lipoic acid and that rearrangement of the resulting compound forms the acetyl derivative of reduced lipoic acid. This is the stage in which the decarbox-

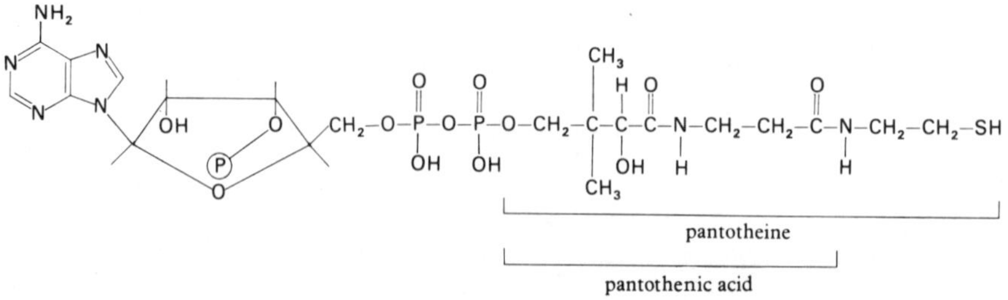

Figure 2.19 Coenzyme A (CoASH). Pantothenic acid is one of the B-group of vitamins. Ⓟ = PO_3H_2

ylation product of pyruvate (acetaldehyde) is converted to the oxidation level of acetic acid (as the acetyl group) and which reforms free thiamine pyrophosphate.

The second reaction involves the enzyme lipoyl acetyltransferase and is the transfer of the acetyl group to coenzyme A (CoASH). The group involved in the coenzyme is the sulphydryl group of the pantotheine residue (Figure 2.19). The products of the reaction are acetyl-CoA and reduced lipoic acid. The third reaction catalysed by lipoyldehydrogenase reoxidizes the lipoic acid and produces $NADH + H^+$.

Thus the overall products of the reaction are acetyl coenzyme A, $NADH + H^+$ and carbon dioxide; as the decarboxylation is virtually irreversible, the entire reaction sequence is also irreversible.

The Reaction of the Citric Acid Cycle

The first step of this cycle, the condensation of acetyl coenzyme A with oxaloacetate to form citrate, is catalysed by the 'condensing enzyme' or citrate synthetase. The thioester linkage between the acetyl group and the sulphydryl group

oxaloacetic acid

$$\begin{array}{c} COOH \\ | \\ HOOC-CH_2-C=O \\ | \\ H \\ | \\ H-C-H \\ | \\ C=O \\ | \\ S-CoA \end{array} \longleftrightarrow \begin{array}{c} COOH \\ d^+| \ d^- \quad H_2O \\ HOOC-CH_2-C-O \\ \\ H \\ H-C^- \quad H^+ \\ | \\ C=O \\ | \\ S-CoA \end{array} \rightleftharpoons \begin{array}{c} COOH \\ | \\ HOOC-CH_2-C-OH \\ | \\ H-C-H \\ | \\ C=O \\ | \\ OH \end{array} + CoASH$$

acetyl-CoA   citric acid

of the coenzyme is a high-energy bond (it may be thought of as a mixed acid anhydride); thus the equilibrium position of the reaction which forms citrate is far to the right ($\Delta G^{\circ\prime}$ is -7500 cal/mole).

Citrate is equilibrated with two other acids, isocitrate and *cis*-aconitate by the enzyme aconitase. This reaction probably involves the formation of an enzyme-bound common intermediate, so that the change from citrate to isocitrate does not necessarily pass through *cis*-aconitate. Aconitase requires iron which is apparently vital for the formation of the common intermediate. Formally, the reaction may be written as if water were removed from citric acid to produce *cis*-aconitic acid which can then be rehydrated to produce isocitric acid.

Isocitric acid is oxidized and decarboxylated by isocitric dehydrogenase to α-oxoglutarate. There are two forms of this enzyme, one of which transfers hydrogen to NAD^+, the other to the closely related coenzyme $NADP^+$ (see Figure 2.13) which functions in exactly the same way as NAD^+.

$$\underset{\text{citric acid}}{\begin{array}{c} CH_2-COOH \\ | \\ HO-C-COOH \\ | \\ CH_2COOH \end{array}} \rightleftharpoons \left[\begin{array}{c} \text{enzyme-bound} \\ \text{intermediate} \end{array} \right] \rightleftharpoons \underset{\text{isocitric acid}}{\begin{array}{c} H \\ | \\ HO-C-COOH \\ | \\ H-C-COOH \\ | \\ CH_2COOH \end{array}}$$

$$+H_2O \updownarrow -H_2O$$

$$\underset{\text{cis-aconitic acid}}{\begin{array}{c} CH-COOH \\ \| \\ C-COOH \\ | \\ CH_2COOH \end{array}}$$

The NADH + H$^+$ generated by this enzyme contributes directly to the mitochondrial formation of ATP (see p. 135). The function of NADPH + H$^+$

$$\begin{array}{c} H \\ | \\ HO-C-COOH \\ | \\ H-C-COOH \\ | \\ H-C-COOH \\ | \\ H \end{array} + \begin{array}{c} NAD^+ \\ (NADP^+) \end{array} \longrightarrow \begin{array}{c} O=C-COOH \\ | \\ H-C-H \\ | \\ H-C-COOH \\ | \\ H \end{array} + CO_2 + \begin{array}{c} NADH + H^+ \\ (NADPH) \end{array}$$

produced by the mitochondrion is not clear; it may represent a reserve of reducing power which can be

(i) transferred to NAD$^+$ (and thence to ATP generation) by a sequence of enzyme reactions—for example,

NADPH + H$^+$ + α-oxoglutarate + NH$_4^+$ $\longrightarrow$ glutamate + NADP$^+$

glutamate + NAD$^+$ $\longrightarrow$ α-oxoglutarate + NH$_4^+$ + NADH + H$^+$

Sum: NADPH + H$^+$ + NAD$^+$ $\longrightarrow$ NADP$^+$ + NADH + H$^+$

or

(ii) used for mitochondrial elongation of fatty acid chains (see p. 253) or for desaturating saturated fatty acids (for example, the conversion of stearate to oleate).

α-Oxoglutaric acid is converted to succinyl coenzyme A by a complex series of reactions analogous to that oxidizing pyruvate to acetyl coenzyme A. The same cofactors are required and coenzyme A again acts as the recipient of the acyl group formed, and the reaction is inhibited by arsenite. The products are NADH + H$^+$, carbon dioxide and succinyl coenzyme A, which is a high-energy compound (it is a mixed acid anhydride) that can be used to drive the formation of ATP from ADP and inorganic phosphate. This synthesis (a substrate-level phosphorylation) occurs in two steps. Firstly, succinic thiokinase forms GTP

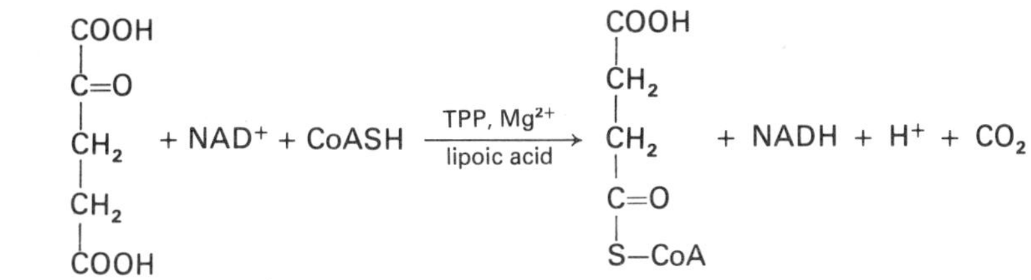

$$\begin{array}{l}\text{COOH}\\|\\\text{C=O}\\|\\\text{CH}_2\\|\\\text{CH}_2\\|\\\text{COOH}\end{array} + \text{NAD}^+ + \text{CoASH} \xrightarrow{\text{TPP, Mg}^{2+}}_{\text{lipoic acid}} \begin{array}{l}\text{COOH}\\|\\\text{CH}_2\\|\\\text{CH}_2\\|\\\text{C=O}\\|\\\text{S-CoA}\end{array} + \text{NADH} + \text{H}^+ + \text{CO}_2$$

α-oxoglutaric acid   succinyl-CoA

(guanosine triphosphate, which is a nucleoside triphosphate like ATP though containing the base guanine instead of adenine; see Figure 2.20) at the expense of

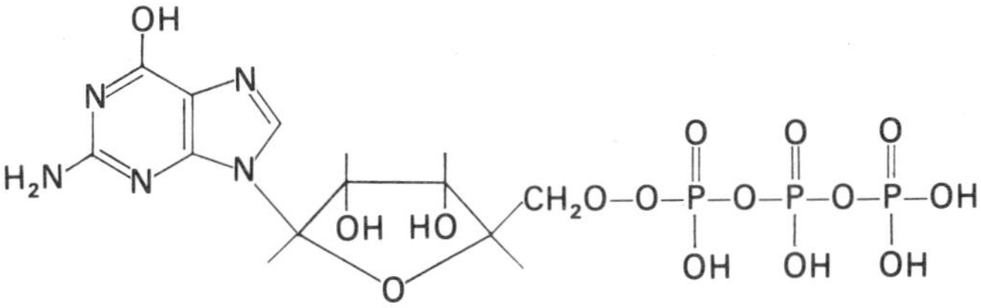

Figure 2.20 Guanosine triphosphate

the succinyl-CoA. The reaction is readily reversible (K is 3·7). Secondly, GTP can transfer its terminal phosphate to ADP, forming ATP and releasing GDP for further participation in the thiokinase reaction. The enzyme required is nucleoside diphosphokinase. Thus the products of this entire series of reactions

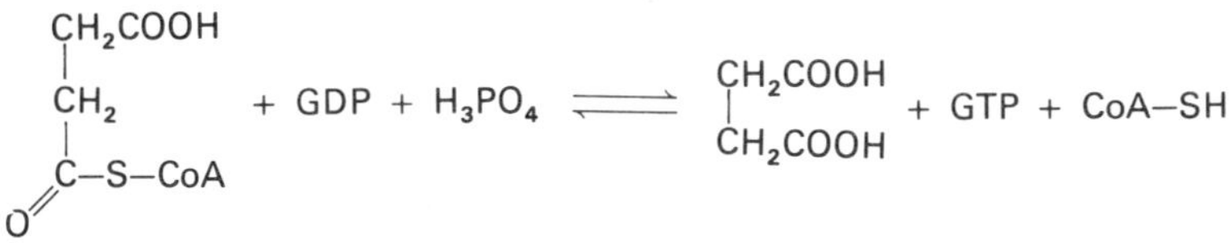

$$\begin{array}{l}\text{CH}_2\text{COOH}\\|\\\text{CH}_2\\|\\\underset{\text{O}}{\text{C}}\text{-S-CoA}\end{array} + \text{GDP} + \text{H}_3\text{PO}_4 \rightleftharpoons \begin{array}{l}\text{CH}_2\text{COOH}\\|\\\text{CH}_2\text{COOH}\end{array} + \text{GTP} + \text{CoA-SH}$$

succinyl-CoA   succinic acid

are succinate, ATP, CO_2 and NADH + H$^+$. The 6-carbon acid formed at the beginning of the cycle has been degraded to a 4-carbon acid by two decarboxylations while one molecule of ATP and two of NADH + H$^+$ have been accumulated. The oxidative decarboxylation of α-oxoglutarate is virtually irreversible and ensures that the cycle as a whole operates in one direction only.

Succinic acid is oxidized to fumaric acid by succinic dehydrogenase which forms part of the inner membrane of the mitochondrion. The hydrogen acceptor for this enzyme is FAD (Figure 2.21) which is firmly attached to the enzyme structure (it is a prosthetic group). Enzymes like succinic dehydrogenase are called flavoproteins and the coenzyme is called a flavin. The necessity for dehydrogenases with flavin as hydrogen acceptors arises from the fact that the nicotinamide nucleotides ($E^{\circ\prime}$ of $-0\cdot32$ V) are unsuitable as coenzymes for cer-

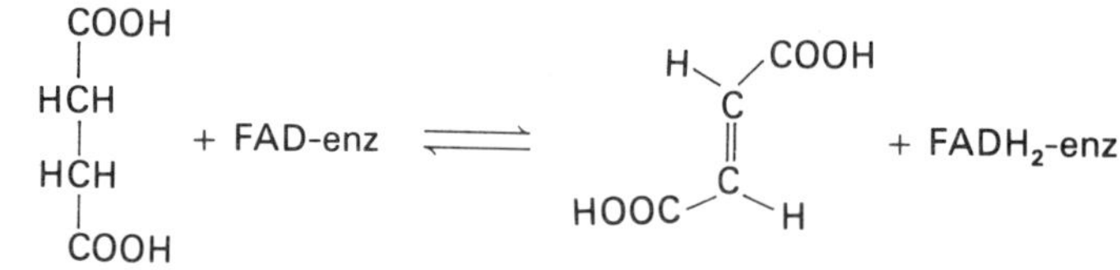

$$\begin{array}{c}\text{COOH}\\|\\\text{HCH}\\|\\\text{HCH}\\|\\\text{COOH}\end{array} + \text{FAD-enz} \rightleftharpoons \begin{array}{c}\text{H}\quad\text{COOH}\\\diagdown\;\diagup\\\text{C}\\\|\\\text{C}\\\diagup\;\diagdown\\\text{HOOC}\quad\text{H}\end{array} + \text{FADH}_2\text{-enz}$$

succinic acid   fumaric acid

tain oxidations, for example that of succinate, i.e. for reactions involving substrates which do not have hydrogen atoms of sufficient reducing power for transfer to NAD^+.

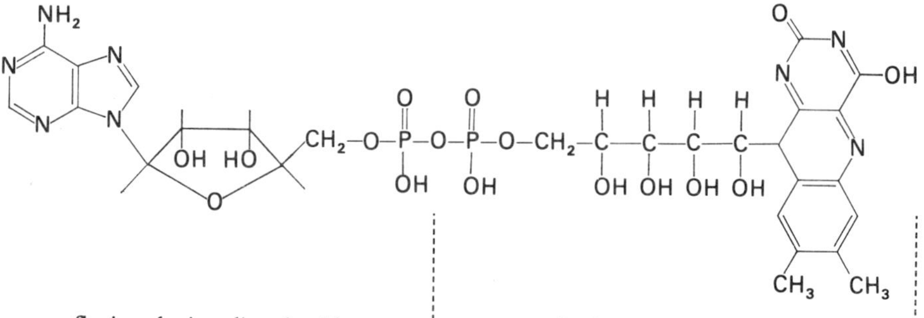

[Structure of flavin adenine dinucleotide showing adenine, ribose, pyrophosphate, ribitol and isoalloxazine moieties; flavin mononucleotide portion bracketed]

flavin adenine dinucleotide   flavin mononucleotide

[Oxidized flavin + 2H ⇌ reduced flavin structures]

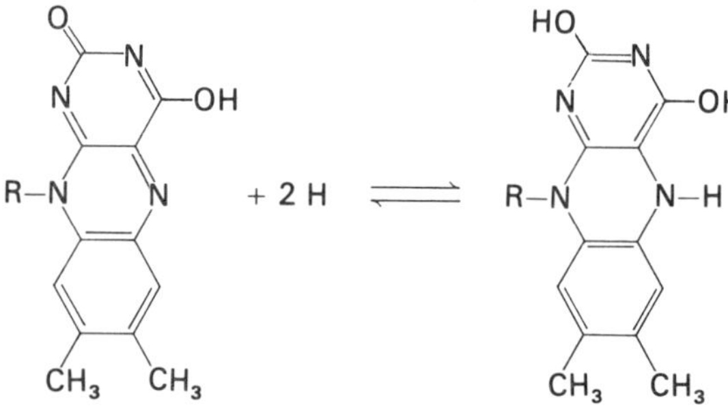

oxidized flavin   reduced flavin

Figure 2.21 Flavin nucleotides: flavin adenine dinucleotide (FAD) and flavin mononucleotide (FMN). FMN is the phosphorylated derivative of the vitamin riboflavin (vitamin B$_2$)

Fumarate is hydrated by the enzyme fumarase to form L-malic acid. The equilibrium constant is 4·5 so that the reaction is readily reversible.

Malic acid is converted to oxaloacetate by malic dehydrogenase, NAD^+ being reduced. At neutral pH, the equilibrium lies very much to the left ($K = 1\cdot3 \times 10^{-5}$) and, in addition, the reduction of oxaloacetate to malate is a very rapid reaction so that oxaloacetate added to tissues disappears in a few seconds.

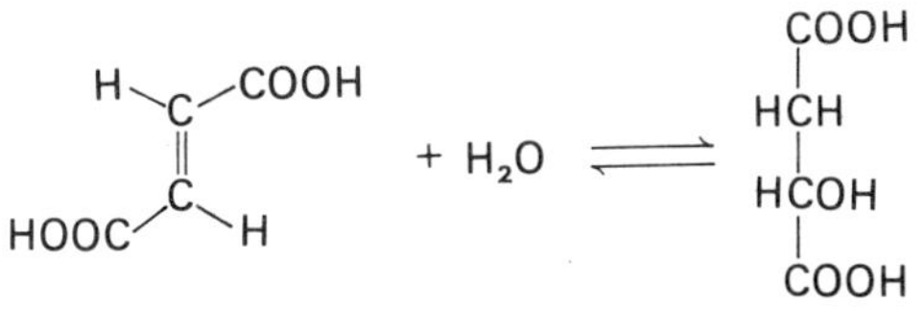

$$\underset{\text{fumaric acid}}{\underset{HOOC}{\overset{H}{\subset}}\overset{COOH}{\underset{H}{\overset{\|}{C}}}} + H_2O \rightleftharpoons \underset{\text{L-malic acid}}{\begin{matrix} COOH \\ | \\ HCH \\ | \\ HCOH \\ | \\ COOH \end{matrix}}$$

Although oxaloacetate is required for the constant operation of the citric acid cycle it is also a potent inhibitor of succinic dehydrogenase and thus any accumulation of oxaloacetate would act as an automatic cut-out, preventing its own formation through the operation of the cycle.

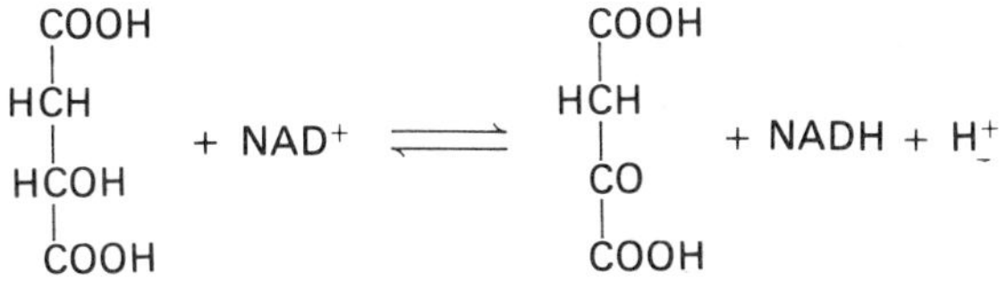

$$\underset{\text{L-malic acid}}{\begin{matrix} COOH \\ | \\ HCH \\ | \\ HCOH \\ | \\ COOH \end{matrix}} + NAD^+ \rightleftharpoons \underset{\text{oxalacetic acid}}{\begin{matrix} COOH \\ | \\ HCH \\ | \\ CO \\ | \\ COOH \end{matrix}} + NADH + \underline{H}^+$$

Subcellular Localization of the Enzymes of the Cycle

Succinic dehydrogenase, the α-oxoglutaric dehydrogenase and pyruvic dehydrogenase complex, the citrate synthetase and NAD^+-dependent isocitric dehydrogenase are present only in the mitochondria. Aconitase, $NADP^+$-dependent isocitric dehydrogenase, fumarase and malic dehydrogenase are found both in the soluble part of the cell and the mitochondria.

The Rate of Operation of the Cycle

It is possible to calculate the approximate rate of operation of the citric acid cycle from the following considerations. Suppose that a man at rest after exercise is oxidizing lactic acid; two-thirds of his oxygen consumption will come from the citric acid cycle and one-third from preliminary reactions. Table 2.3 shows the hydrogen equivalents of the oxygen used.

The oxygen consumption at rest is about 300 ml per minute, i.e. 13·4 mmole of oxygen/minute. The amount required by the citric acid cycle (2/3 of 13·4) is 8·8 mmole or 17·6 m.atoms of oxygen, which is equivalent to 35·2 m.atoms of hydrogen oxidized to water; thus 35·2/8 = 4·4 mmole of acetyl-CoA pass through the cycle/minute, or 264 mg of acetate/minute.

To calculate the frequency of operation of this cycle, we must consider in addition the concentration of oxaloacetate which limits the number of acetyl-CoA molecules that can be accepted at any one time. Usually this concentration is about 1 μm, i.e. 1 litre of tissue contains 1 μmole of oxaloacetate. If we subtract from the weight of the 70 kg man the weight of the blood and bone (neither of

Table 2.3

Reaction		Number of hydrogens
lactate	⟶ pyruvate	2
pyruvate	⟶ acetyl-S-CoA	2
isocitrate	⟶ α-oxoglutarate	2
α-oxoglutarate	⟶ succinate	2
succinate	⟶ malate	2
malate	⟶ oxaloacetate	2
	Total hydrogens in cycle	8
	Total hydrogens before cycle	4

which contain oxaloacetate available for the citric acid cycle) the mass of tissue remaining is about 55 kg, with a density of 1·1. Thus the volume of tissue containing oxaloacetate is $55 \div 1·1 = 50$ litre, and there will be 50 μmoles of oxaloacetate available at any one time for condensation with acetyl coenzyme A. Therefore, as 4·4 mmoles (= 4400 μmoles) of acetyl coenzyme A are consumed/minute, and they pass through the cycle 50 μmoles at a time, the cycle must turn 4400/50 times/minute, i.e. about 90 times per minute.

The Cycle as a Common Pathway for the Degradation of Carbohydrate, Fatty Acids and Amino Acids to CO_2

Besides oxidizing acetate derived from the pyruvate formed in glycolysis, the citric acid cycle is also involved in the conversion of fatty acids and the carbon skeletons of many amino acids to CO_2 (Figure 2.22). Fatty acids may enter the cycle after being broken down to acetyl-CoA units (see p. 230). Many amino acids, after removal of their amino group by transamination or deamination (see p. 214) can enter the cycle. For example, deamination of alanine, glutamic and aspartic acids produces pyruvate, α-oxoglutarate and oxaloacetate respectively.

Summary of the Functions of the Citric Acid Cycle

The functions of the citric acid cycle may be summarized as follows. Its operation, which begins with acetic and oxaloacetic acids, produces a series of organic acids (some of which are used as building blocks for providing essential cell constituents) reduced nicotinamide and flavin coenzymes and GTP. These compounds are formed at the expense of the conversion of one molecule of acetic acid to carbon dioxide so that the net transformation during one turn of the cycle may be represented as follows:

$CH_3COOH + 2 H_2O + 3 NAD^+ + FAD\text{-enz} + GDP + Pi \longrightarrow$

$2 CO_2 + 3 NADH + H^+ + FADH_2\text{-enz} + GTP$

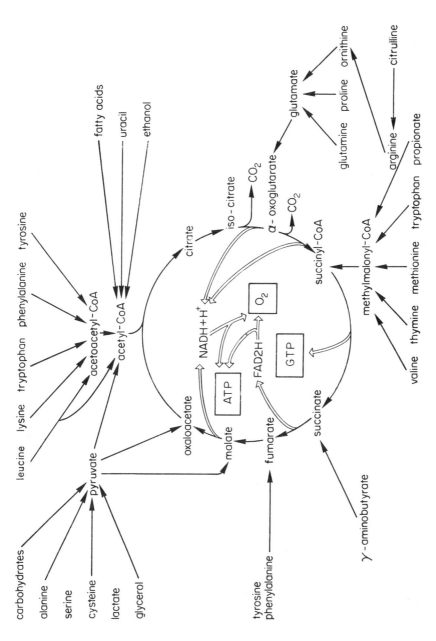

Figure 2.22 The role of the tricarboxylic acid cycle in catabolism

Table 2.4 Energy transformation during the degradation of glucose

Substrate	Energy level of substrate (kcal)	Energy lost from substrate (kcal)	Energy conserved as reduced coenzymes (kcal)	ATP (kcal)
glucose	700	—	—	—
2 pyruvate	550	150	100	40
2 acetate (+2CO_2)	440	110	100	0
6CO_2	0	440	300 (NADH + H$^+$)	20
			80 (reduced flavin)	
Totals		700	580	60

The values[a] are calculated on the following assumptions:

glucose + 6CO_2 ⟶ 6CO_2 + 6H_2O; $\Delta G^{\circ\prime} = -700$ kcal/mole
NADH + H$^+$ + O ⟶ NAD$^+$ + H_2O; $\Delta G^{\circ\prime} = -50$ kcal/mole
reduced-flavoprotein + O ⟶ flavoprotein + H_2O; $\Delta G^{\circ\prime} = -40$ kcal/mole
ATP + H_2O ⟶ ADP + phosphate $\Delta G^{\circ\prime} = -10$ kcal/mole

P/O ratios for the oxidation of NADH + H$^+$ and reduced flavoprotein are 3 and 2 respectively.

[a] All values have been expressed to the nearest 10 kcal to simplify the calculations.

Much of the energy made available during the degradation of acetate (and indeed during the complete degradation of glucose) is retained in these reduced coenzymes. This is apparent from the values in Table 2.4.

These reduced coenzymes are utilized for the production of ATP by the process of oxidative phosphorylation. In this process, the hydrogens removed from the substrates of the citric acid cycle are finally oxidized to water by a series of components comprising the respiratory chain; the energy made available is used to drive the synthesis of ATP from ADP and inorganic phosphate.

Mitochondrial Permeability to Metabolites

We have seen that the citric acid cycle can convert to CO_2 compounds derived from the degradation of a number of different fuels. As many of these degradative pathways begin in the extramitochondrial compartment of the cell, their products must penetrate the two membranes of the mitochondrion before further catabolism can occur. The outer membrane is probably permeable to all but high molecular weight compounds, but the inner one has a much more restricted permeability. Water, oxygen, carbon dioxide and ammonia all appear to pass readily through it. Monocarboxylic acids also appear to diffuse across the membrane readily, accounting for the passive movement of such compounds as pyruvate, acetoacetate and short-chain fatty acids. Permeability to many other important metabolites is much more restricted, and requires the operation of specific transport systems.

Phosphate Transport

A carrier system (sensitive to poisoning by compounds reacting with SH groups), apparently promotes the movement of phosphate into the mitochondrial matrix.

Dicarboxylate and Tricarboxylate Anions

Neither of these classes penetrate mitochondria readily; rather, a number of carriers for various anions has been identified as follows:

(a) For malate, succinate and perhaps oxaloacetate; its operation requires a simultaneous countermovement of phosphate.

(b) For citrate and isocitrate; penetration requires countermovement of an anion such as malate.

(c) For α-oxoglutarate; penetration requires countermovement of either malate or succinate.

These systems are interlocked functionally through a series of carrier-mediated exchanges thus:

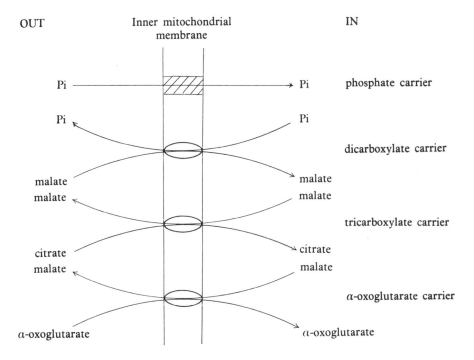

Amino Acid Transport

Glutamate is apparently transported by a specific mitochondrial carrier.

Fatty Acids

In addition to diffusion through the membrane, there appears to be a specific transport mechanism for fatty acids in the inner mitochondrial membrane (see p. 230).

Nucleotides

The nicotinamide nucleotides of the mitochondria and the cytosol do not mix to any significant extent, but the adenine nucleotides, ADP and ATP, do move from one compartment to the other. This ensures that the relatively vigorous generation of ATP in the mitochondria provides ATP for the rest of the cell, and that the ADP formed can be returned to the mitochondrial matrix for phosphorylation.

The transport mechanism is highly specific for ADP and ATP; it provides a 1:1 exchange of the nucleotides. Coupled mitochondria presented with equal external concentrations of ADP and ATP exchange the dinucleotide inwards about ten times faster than the trinucleotide. With uncoupled mitochondria, this difference disappears; though there is no difference in the rate of outward exchange of the two nucleotides. Therefore, maintenance of this asymmetric permeability requires an expenditure of energy, perhaps to maintain a negative membrane potential internally which will repel the highly negatively charged ATP molecule.

CHAPTER 11

The Respiratory Chain and Oxidative Phosphorylation

The respiratory chain is a series of enzymes and cofactors located in the mitochondria which catalyses a stepwise transfer of hydrogen from substrate to oxygen, reducing it to water. Information about the composition of the chain has been obtained largely from an examination of the characteristic light-absorbing properties of many of its component parts (by use of a variety of elaborate spectrophotometric methods) and by fragmentation of mitochondria and mitochondrial membranes followed by chemical and spectrophotometric analyses. While some doubt exists about the exact role of certain components, it is possible to draw up a generalized scheme for the composition of the chain (Figure 2.23).

The Components of the Respiratory Chain

All the components of this chain except NAD^+ are embedded more or less firmly in the mitochondrial membrane. There are two main points of entry for substrate hydrogen into the chain, one through flavins (for succinate and fatty acids), the other through NAD^+ (for α-oxoglutarate, fatty acids, etc.). $NADH + H^+$ is also oxidized by a flavoprotein which in heart tissue contains FMN as its prosthetic group. The structure of these flavin 'enzymes' in the mitochondrial membrane is very complex, and includes proteins containing both iron ('non-haem iron') and sulphur (from the thiol groups of cysteine residues and as an iron disulphide). Reducing equivalents entering through either of these two separate pathways are available for reducing all the remaining components which, therefore, constitute a final common pathway to oxygen. There is a third point of entry, which permits hydrogen from $NADPH + H^+$ to pass to oxygen; this is via the enzyme transhydrogenase, which is firmly bound to the framework of the mitochondria and catalyses the reaction

$$NADPH + H^+ + NAD^+ \rightleftharpoons NADP^+ + NADH + H^+$$

In the common pathway reducing equivalents pass through a quinone (ubiquinone or coenzyme Q) and then a series of cytochromes. These are iron containing proteins with different redox potentials, transferring single electrons with a corresponding change in the valency of the iron. The protein of the molecule is attached to a porphyrin group to which the iron in turn is attached. There is a family of such compounds (the cytochromes *a*, *b* and *c*) differing from one

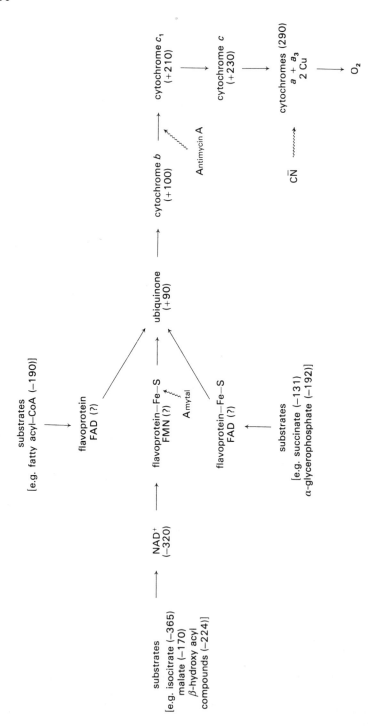

Figure 2.23 The respiratory chain. The values in parentheses are the redox potentials of the components at pH 7 in mV. The values for the flavoproteins are uncertain but are probably about 0 mV. ⤳ indicates sites of inhibition

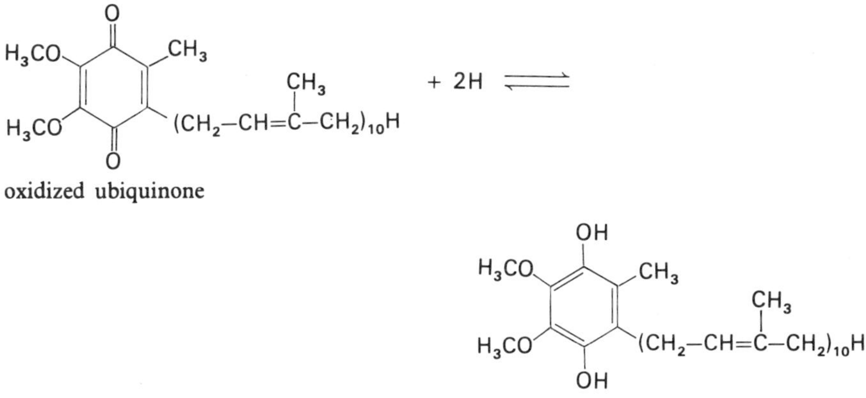

$$\underset{\text{oxidized ubiquinone}}{\begin{array}{c}H_3CO \\ H_3CO\end{array}\underset{O}{\overset{O}{\bigcirc}}\begin{array}{c}CH_3 \\ (CH_2-CH=\underset{CH_3}{\overset{|}{C}}-CH_2)_{10}H\end{array}} + 2H \rightleftharpoons$$

$$\underset{\text{reduced ubiquinone}}{\begin{array}{c}H_3CO \\ H_3CO\end{array}\underset{OH}{\overset{OH}{\bigcirc}}\begin{array}{c}CH_3 \\ (CH_2-CH=\underset{CH_3}{\overset{|}{C}}-CH_2)_{10}H\end{array}}$$

another in the mode of attachment of porphyrin to protein and in the chemical structure of the porphyrins. The porphyrin of cytochrome c (Figure 2.24) is identical to that of the haem group of haemoglobin (see p. 278) and is attached to the protein as shown in Fig. 2.25. The cytochromes of the b group have the same porphyrin structure but the mode of attachment to the protein is different; the cytochromes of the a group have a different porphyrin.

The Sequence of the Components of the Respiratory Chain

Considerable progress has been made in describing the order in which these compounds function and the sequence already given (Figure 2.23) is one generally accepted. In this scheme, there is a descending potential down which

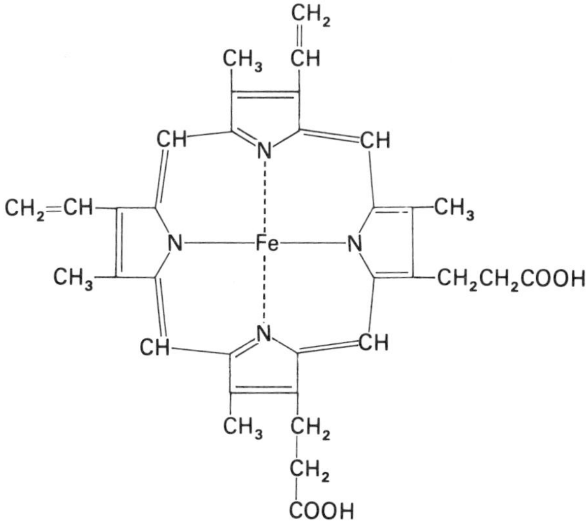

Figure 2.24 The haem prosthetic group of cytochrome c and haemoglobin

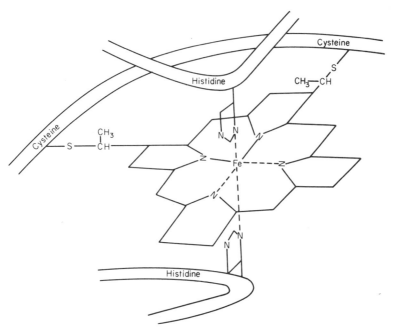

Figure 2.25 Diagrammatic representation of the binding of the prosthetic group in cytochrome c. The porphyrin ring is linked covalently to two cysteine residues in the protein; the Fe is linked to N in both the prosthetic group and the protein (the imidazole side chains of the two histidine residues)

electrons will readily flow, which indicates that such an order is probable. However, the potentials quoted have been measured on isolated redox components and it is not certain that, for example, when NAD^+ is firmly bound to its dehydrogenase during oxidation this potential remains the same. Also the potential depends on the concentrations of the oxidized and reduced forms present.

More direct evidence for the sequence has been obtained by measuring changes in the absorption spectra of the components under different conditions. Very elaborate and sensitive spectrophotometric techniques have been developed to permit the estimation of the extent of reduction and the rates of reaction of different components present in the respiratory chain. Cytochrome a_3 (also called cytochrome oxidase) is the terminal cytochrome because it is oxidized most rapidly when anaerobic mitochondria are oxygenated; cytochrome b is the first cytochrome because it is the only cytochrome reduced when substrate is added to mitochondria poisoned with antimycin a; cytochrome c precedes cytochrome a because substrate added to mitochondria extracted with salt (which preferentially removes cytochrome c) reduces cytochrome b and c_1 but not $a + a_3$; flavoprotein (and ubiquinone) are reduced by added substrate before the cytochromes. Many extensions of this approach have been used, often with inhibitors of respiratory activity. For example, if malonate is added to mitochondria oxidizing succinate, the flavin is not reduced. Amytal (a barbiturate) prevents the oxidation of

NADH + H$^+$ generated by added malate, but does not prevent the oxidation of succinate (indicating that the two flavoproteins of the chain are distinct). Cyanide, by inhibiting cytochrome a_3, permits all the carriers to be reduced by added substrate. The points of action of the inhibitors of the electron transport chain are shown in Figure 2.23.

The Functioning of the Respiratory Chain

There is little detailed information about the mechanism of the redox reactions in the chain. While it is certain that many of the enzymes reducing NAD$^+$ do so by transferring one of the hydrogen atoms from the substrate together with an electron to the coenzyme and that a similar movement of hydrogen atoms occurs with a few flavoproteins, the mechanism of redox transfers by the flavoproteins and cytochromes of the respiratory chain is not understood; nevertheless, the assumption that electrons are transferred between these components has led to the use of the term 'electron' transport chain as a synonym for respiratory chain. What is certain, however, is that maximal respiratory activity depends on the maintenance of the structure of the mitochondrial membranes containing the respiratory carriers. These membranes also contain phospholipids (about 25 per cent by mass) which are essential for the transfer of reducing equivalents.

Respiratory Chain Phosphorylation

During the transfer of hydrogen from substrate to oxygen by the respiratory chain, sufficient energy is made available to drive the synthesis of ATP from ADP and phosphate. This process, called 'respiratory chain' or 'oxidative' phosphorylation, is localized in the inner membrane of mitochondria, probably in the knob-like structures detectable by electron microscopy.

Energy Transformations in the Respiratory Chain

It appears to be a fact of life that the production of about 10–12 kcal/mole in one reaction is about the maximum that living things can handle. Reactions such as the oxidation NADH + H$^+$ or reduced flavin by oxygen which produce more than this amount of energy are therefore routed through a series of steps, which liberate the energy in smaller 'packets' with which the cell can deal. The respiratory chain is a good example of such a series and it is possible to calculate the magnitude of the release of energy at its different stages. For the synthesis of ATP, there must be sites in the chain where the redox drop is sufficient to free about 8 kcal/mole. Using the equation

$$\Delta G^{\circ\prime} = \frac{\Delta E^{\circ\prime} nF}{4\cdot 18}$$

(see p. 51) and the potentials listed in Figure 2.23, it can be seen that this occurs between NAD$^+$ and flavoprotein, cytochrome b and cytochrome c and cytochrome c and oxygen. Therefore at least three molecules of ATP could be formed during the oxidation of one molecule of NADH + H$^+$ and at least two during the oxidation of reduced flavoprotein in the respiratory chain. This conclusion is supported by quantitative analysis of oxidative phosphorylation.

The Coupling of Respiration to Phosphorylation

Carefully prepared mitochondria, incubated aerobically with substrate under conditions where all the necessary cofactors are available except ADP, have a low rate of respiration. Addition of ADP accelerates the respiratory rate; the rate of respiration increases with increasing concentration of added ADP until a maximum rate is reached. The respiration falls to a low level when almost all the ADP has been converted to ATP. Under these circumstances respiration is said to be 'coupled' to phosphorylation. The mitochondria are also said to show 'respiratory control by ADP'—that is, the rate of respiration is controlled by the rate at which ADP is made available (*in vivo* by the utilization of ATP). This simple control mechanism has the obvious advantage of diminishing respiratory activity when there is no need to synthesize ATP.

The Quantitative Relationship between Oxygen Uptake and Phosphorylation

Not only is there a qualitative relation between oxidation and phosphorylation but there is also a relation between the quantities of oxygen consumed and of ATP synthesized. This has become apparent in many experiments in which the amounts of oxygen respired and ATP produced have been measured. Most of the practical difficulties in such experiments arise because the ATP formed can be utilized for other reactions by the mitochondrial preparations, i.e. the particles have an 'ATP-ase' reaction. To avoid this difficulty, the ATP is often 'trapped' by conversion to glucose-6-phosphate (formed by added glucose and hexokinase) and the glucose-6-phosphate is estimated by a suitable specific method. An additional difficulty is that not all the oxygen uptake of such preparations is coupled to phosphorylation; there may be a built-in 'leak' of respiration. This may be allowed for by measuring the rate of oxygen uptake with a limited amount of phosphate in the absence of the 'glucose–hexokinase trap' (the unstimulated rate), then adding glucose and hexokinase to release ADP from the ATP in the system and measuring the amount of oxygen taken up in the time interval before the respiratory rate returns to its unstimulated value. When this rate is restored, all the phosphate is used up. From such experiments the ratio

$$\frac{\mu\text{mole phosphate incorporated into ATP}}{\mu\text{g atoms oxygen consumed}}$$

can be calculated; this is termed the P/O ratio. When such ratios are measured with intermediates of the citric acid cycle and related compounds the following values are obtained; for pyruvate, approaching 3; for isocitrate, 1–2; for α-oxoglutarate, about 4; for succinate, about 2; for malate, about 3. The most obvious difficulty in interpreting these values is that the addition of any of these substrates initiates the whole series of reactions of the cycle. However, if β-hydroxybutyrate, which can be oxidized only to acetoacetate by liver mitochondria, is added to a suspension of these particles P/O ratios of about 3 are obtained. Thus it seems that all substrates which transfer hydrogen to NAD^+ have a P/O ratio of the same value, about 3; this has been confirmed directly by measuring the P/O ratio with $NADH + H^+$ itself as the substrate. On the other hand, with succinate and other substrates reducing the flavoproteins, less energy is available during the oxidation of the hydrogens and the P/O ratio is only 2. The value listed above which is not explained is the P/O ratio of about 4 for the oxidation of α-oxoglutarate. However, this oxidation involves the formation of ATP from the GTP synthesis during the deacylation of succinyl coenzyme A (see p. 127), as well as the generation of $NADH + H^+$; the P/O ratio for the entire reaction is therefore 4. The addition of dinitrophenol, which prevents oxidative but not substrate-level phosphorylation, accordingly diminishes the P/O ratio for this reaction to about 1.

The Sites of Phosphorylation

The initial consideration of the redox properties of the components of the respiratory chain suggested certain possible sites for ATP synthesis, and there is direct experimental support for the conclusions. Thus the respiratory chain can be reduced at the level of cytochrome *c* by using ascorbic acid as a substrate and when the reduced cytochrome *c* is oxidized by oxygen, P/O ratios of rather less than 1 are obtained. If succinate is used as a substrate under anaerobic conditions and the reduction of cytochrome *c* (added as an acceptor of the reducing equivalents) is measured spectrophotometrically, the P/O ratio (actually P/2*e* ratio) is about 0·5; with β-hydroxybutyrate as substrate under the same conditions, the ratio rises to between 1·5 and 2. Therefore, it is generally agreed that there is one phosphorylation site between NAD^+ and flavoprotein, one between flavoprotein and cytochrome *c* and one between cytochrome *c* and oxygen (Figure 2.23).

The Dissociation of Respiratory Activity and Phosphorylation

Although the process of electron transfer and phosphorylation are normally coupled, it is possible to dissociate or 'uncouple' them so that respiration continues without phosphorylation. Uncoupling may result from a variety of physical or chemical treatments of mitochondrial preparations and only a few will be considered.

An uncoupler frequently used is 2,4-dinitrophenol, a drug popular some time ago for reducing weight, until it was found to be toxic. If administered to whole animals, it increases the basal metabolic rate and in many ways mimics hyperthyroidism. When added to mitochondrial suspensions it produces a complete loss of the ability to phosphorylate ADP and, often, an increase in the rate of oxygen consumption. The particles so treated are 'uncoupled', that is, respiration proceeds at a rate controlled only by the ability of the respiratory chain to transfer electrons and by the amount of oxidizable substrate available.

Phosphorylation can also be prevented without uncoupling. For example, if respiration is blocked in a coupled system by adding antimycin or cyanide then phosphorylation also ceases, because electron transport has ceased.

The Mechanism of Oxidative Phosphorylation

Much ingenuity and effort has been expended in attempting to describe the details of the reactions which link electron transfer in the respiratory chain to the phosphorylation of ADP to form ATP. However, no generally acceptable formulation has been proposed. Some workers believe that there are probably specific chemical intermediates (possibly proteins) which connect the two processes and refer to them as 'coupling factors' (Figure 2.26). Various lines of

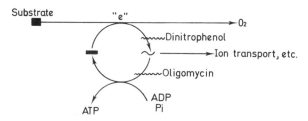

Figure 2.26 The coupling-factor mechanism for oxidative phosphorylation. Reducing equivalents 'e' passing down the respiratory chain convert a specialized component of the mitochondrial structure from a 'low-energy' state (—) to a 'high-energy' state ($\sim$). This 'high'-energy intermediate may be used to generate ATP from ADP and Pi ('oxidative phosphorylation') or for other energy-dependent reactions. The characteristic actions of the two inhibitors dinitrophenol and oligomycin have been important in developing proposals of this type

evidence point to their existence, perhaps the most compelling being experiments with oligomycin which, while it prevents ATP formation in respiring mitochondrial preparations, does not inhibit a number of other energy-dependent mitochondrial reactions. Therefore, it is argued, there must be at least one high-energy compound (the coupling factor) formed as the result of respiratory activity and capable of driving the synthesis of ATP or (in the oligomycin inhibited system) the other energy-dependent reactions. Obviously the problems in isolating and identifying any such possible intermediates from the organized and extremely complex structure of mitochondria are very great.

An alternative scheme for oxidative phosphorylation which does not require the participation of coupling factors has been proposed (Figure 2.27). In essence, it envisages that electron transport causes a vectorial movement of protons which generates an ionic gradient across the mitochondrial membrane and that this gradient provides the energy to phosphorylate ADP. In this scheme, therefore,

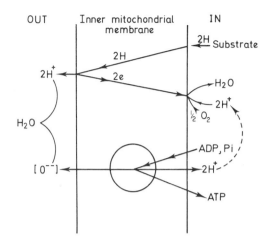

Figure 2.27 The chemiosmotic model for oxidative phosphorylation. The respiratory chain is arranged in the inner mitochondrial membrane in such a way that during its function protons are discharged to the *outside* and are collected from the inside; the membrane will thus acquire a net negative charge on the inside. Its effect is to favour the entry of ADP from the cytosol for phosphorylation. The adenine nucleotide carrier system is specifically inhibited by atractyloside and bongkrekic acid (see p. 147). The membrane-bound ATP synthetase combines ADP and Pi, eliminating water in such a way that protons are released on the *inside* and are removed on the *outside*. Thus the operation of the respiratory chain produces favourable conditions for the synthesis of ATP

the 'high-energy intermediate' of the mitochondrion is an electrochemical potential maintained across the inner membrane. In the chemiosmotic theory, dinitrophenol is thought to make the membrane freely permeable to protons and thus to uncouple phosphorylation from electron transport whilst oligomycin is considered to inhibit the ATP-synthetase.

Inhibitors of Mitochondrial Function

As we have seen, the main function of mitochondria is to oxidize foodstuffs and produce chemical energy. This requires the cooperation and interaction of three mitochondrial systems—the citric acid cycle, the respiratory chain and the phosphorylation mechanism. Each of the three systems has its own characteristic inhibitors, which have been used to elucidate the various sequences. This section discusses the most important inhibitors of each pathway.

Inhibitors of the Citric Acid Cycle

Fluoroacetate: $CH_2F.COOH$. This compound occurs naturally in some South African plants and has been used extensively as a rat poison; a closely related compound, fluoroacetamide, has also been used as a systemic insecticide in plants. However, information about their extreme toxicity has produced legislation restricting their use. When injected into rats, fluoroacetate causes large increases (up to seventy-fold) in the concentration of citrate in many tissues. α-Oxoglutarate does not accumulate, suggesting that fluoroacetate prevents the conversion of citrate to α-oxoglutarate. It has also been found that fluoroacetate could be converted to fluorocitrate by citrate synthetase, which does not distinguish between acetate and fluoroacetate, and that fluorocitrate is a potent inhibitor of aconitase. Thus fluoroacetate is poisonous because of a 'lethal synthesis', in which the body poisons itself by using its own enzymes to synthesize a toxic substance.

Arsenite:

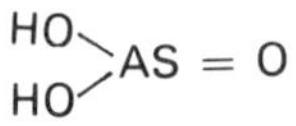

$$\begin{array}{c} HO \\ HO \end{array} As = O$$

inhibits oxidative decarboxylation of pyruvate and α-oxoglutarate by combining with the dithiols of lipoic acid (see p. 122).

Malonate:

$$HOOC.CH_2.COOH$$

(see p. 82) inhibits succinic dehydrogenase competitively and may cause accumulation of succinate if injected into the intact animal.

Parapyruvate:

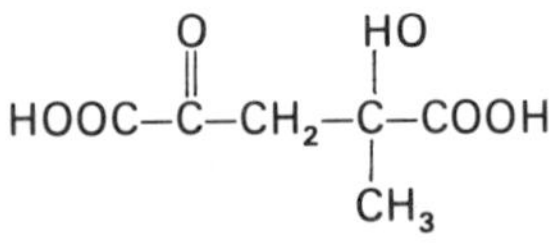

$$HOOC-\underset{\displaystyle\|}{C}-CH_2-\underset{\displaystyle CH_3}{\underset{\displaystyle |}{\overset{\displaystyle HO}{\overset{\displaystyle |}{C}}}}-COOH$$

is a fairly specific inhibitor of α-oxoglutaric dehydrogenase. It is formed by the condensation of two molecules of pyruvic acid and may be considered a structural analogue of α-oxoglutarate.

Oxaloacetate:

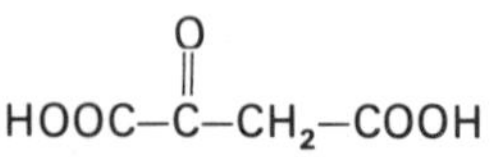

$$HOOC-\overset{\displaystyle O}{\overset{\displaystyle \|}{C}}-CH_2-COOH$$

although part of the citric acid cycle is itself a potent competitive inhibitor of succinic dehydrogenase. The inhibition amounts to about 50 per cent at a concentration of 10^{-5}–10^{-6} M. Oxaloacetate is a labile substance, breaking down readily to pyruvate and carbon dioxide, while at the same time it is essential for the functioning of the cycle. Both the equilibrium of malic dehydrogenase and the inhibition of succinic dehydrogenase ensure that oxaloacetate is conserved. Despite this, there is always some loss of oxaloacetate, which must be replaced. In muscle, the most important replenishment probably comes from the conversion of glutamate to α-oxoglutarate by glutamic dehydrogenase which is found in mitochondria. Free glutamate is one of the most common amino acids found in muscle; its concentration in isolated muscle mitochondria may be upwards of 1 mM.

Inhibitors of Electron Transport

Barbiturates (e.g. amytal) stop respiration by preventing the oxidation of $NADH + H^+$. Thus with barbiturates the oxidation of NAD^+-linked substrates is lowered whilst succinate oxidation is unaffected.

Antimycin is an antibiotic too toxic for clinical use. It is effective at a concentration of less than $0\cdot 1$ $\mu g/ml$ and interferes with the reduction of cytochrome c_1.

Inhibitor of Transphosphorylation

Oligomycin is a fungicide which, when added to a mitochondrial suspension, inhibits the respiration coupled to phosphorylation. Thus it has no effect on respiration occurring in the absence of phosphate and phosphate acceptor, but inhibits completely the stimulated respiration brought about by adding phosphate acceptor such as ADP or hexokinase and glucose. The amount of respiration left after addition of oligomycin is a measure of the tightness of coupling of the mitochondria. The inhibition of respiration is completely relieved by the addition of dinitrophenol.

Inhibitors of Oxidative Phosphorylation: Uncoupling Agents

Nitrophenols (like dinitrophenol) and *halophenols* are uncouplers. Pentachlorophenol has been extensively used in the study of oxidative phosphorylation. *Thyroxine* may be active as an uncoupler because of its ability to disorganize the mitochondrial framework, a property shared by other substances such as detergents (for example deoxycholate) which also uncouples oxidation and phosphorylation.

Dicoumarol is an uncoupler whose mode of action is not understood. It is formed naturally during the fermentation of hay and other green vegetable matter, giving rise to the haemolytic disease of cattle called sweet clover disease.

dicoumarol   vitamin K_3

In structure it is closely related to vitamin K but the effects of dicoumarol and vitamin K are antagonistic and the uncoupling due to the former may be reversed by the latter. The uncoupling effect of dicoumarol has nothing to do with its anti-bloodclotting action.

Calcium is a potent uncoupler, which is accumulated in mitochondria by an energy-requiring process. The affinity of the transport system for calcium is greater than that of oxidative phosphorylation for ADP, so that at low concentrations (less than 100 microgram atoms of calcium per gram mitochondrial protein) it uncouples by competing effectively for the mitochondrion's 'high-energy intermediate'. At higher concentrations (greater than 100 milligram atoms per gram protein) calcium produces an irreversible uncoupling with a

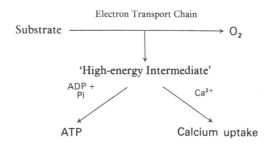

stimulated ATP-ase activity and mitochondrial swelling. The mechanism is unknown, though it has been suggested that calcium at this level stimulates a mitochondrial phospholipase which liberates sufficient fatty acid to damage the inner membranes. The uncoupling effects of calcium are enhanced by phosphate and, at least to some extent, offset by magnesium.

As it is likely that there is a strict coupling of respiration and phosphorylation it is probable that there are natural uncouplers which allow respiration to proceed slowly, even when there is no great demand for ATP. A promising candidate for this role is *carbon dioxide*, which has a marked effect in the physiological range of concentrations. It may be the accumulation of CO_2 that initiates respiration in tightly coupled mitochondria.

Arsenate uncouples by replacing phosphate in the high-energy compound which normally transfers its phosphate group to ADP to yield ATP. The arsenate analogue is readily hydrolysed by water so that energy is lost rather than conserved.

Inhibitors of Adenine Nucleotide Transport

Atractyloside is a drug obtained from the thistle *Atractylis gummifera*. In the Mediterranean basin at the beginning of the Christian era, it was used to produce abortions. Ingestion may be followed by death, after hypoglycaemic convulsions. Atractyloside is a specific inhibitor of the adenine nucleotide carrier of mitochondria, to which it binds about 100 times more firmly than ADP (see p. 134).

Bongkrekic acid. This compound was isolated in the 1930s from the defatted residue of coconut or peanuts which had caused fatal food poisoning in Javanese peasants. Subsequently, it was shown that it is produced by the bacterium *Pseudomonas cocovenenans*. Bongkrekic acid is highly toxic also producing hypoglycaemic convulsions and death, but its mode of action is quite different to that of atractyloside. It increases about a hundred-fold the affinity of the carrier for the adenine nucleotide, greatly decreasing the rate of nucleotide transport.

CHAPTER 12

Special Metabolism of Muscle Types

Red and White Muscle

Skeletal muscles vary in colour from white in the breast of chickens to deep crimson in the corresponding region of pigeons. These differences in colour are correlated with differences in physiological function. Red muscles engage in sustained activity to maintain posture or to effect prolonged movement: white muscles, on the other hand, engage in relatively short bursts of high activity. This difference in function is well illustrated in the wings of falcons which, during hovering, are extended by the prolonged contraction of red muscles whilst the rapid, controlling, movements of the pinions are effected by the intermittent activity of white muscles. In man, on account of his upright gait, postural and locomotor functions are not as clearly separated as in some species with the result that most muscles contain a mixture of red and white fibres.

Red muscles owe their colour to the presence of myoglobin (see p. 70) within the sarcoplasm of the muscle fibres. In most species the role of myoglobin is primarily to facilitate the diffusion of oxygen through the cytoplasm to the mitochondria. However, in diving mammals, such as whales, it is present at much higher concentrations and serves as a reservoir for oxygen. Red muscle also has a much richer vascular supply than white muscle so it is able to rely more heavily on oxidative metabolism. The high activity of its glycolytic enzymes allows white muscle to contract very actively for short periods under relatively anaerobic conditions by degrading its stored glycogen to lactate to provide ATP (Table 2.5). This profligate use of stored glycogen is justified by the requirement for muscular effort in excess of that which can be supported by the supply of fuel and oxygen from the circulation. Such extremes of effort may be required in nature to escape from predators or to attack prey. However, muscle stores of glycogen are inadequate for prolonged activity and must be supplemented by fuel delivery from the circulation. Red muscles have a very much higher capacity to oxidize pyruvate, ketone bodies and fatty acids than white muscles on account of their greater content of mitochondria and better blood supply. The relative activity of oxidative and glycolytic metabolism in red and white muscles is reflected in their complement of lactic dehydrogenase isoenzymes; red muscles have a higher ratio of H/M isoenzymes than white muscles (Table 2.5). The H isoenzyme characteristic of heart is inhibited by NAD^+ and pyruvate together whilst the M, or muscle, form is not. Thus tissues with a high H/M ratio would not convert pyruvate to lactate rapidly and as a result pyruvate oxidation via the citric

Table 2.5 Enzyme activities in red and white muscles. Data taken from Newsholme and Start (1973) Regulation in Metabolism, John Wiley & Sons, London, and Opie and Newsholme (1967) *Biochem. J.*, **103**, 391. Note the activity of triglyceride lipase reflects the capacity of the muscle to use triglyceride fatty acids; the carnitine palmitoyl transferase, the capacity to oxidize fatty acids and the citrate synthetase, the capacity to direct the formed acetyl-CoA into the citric acid cycle

							L-α-glycerophosphate dehydrogenase		Lactic dehydrogenase isoenzyme ratio H/M	Citrate synthetase	Triglyceride lipase	Carnitine palmitoyl transferase
Animal	Muscle	Hexokinase	Phosphorylase	Phosphofructokinase	Fructose diphosphatase		Cytoplasmic	Mitochondrial				
Rabbit	Semitendinosus (red)	1.9	8.0	8.0	0		4.6	0.2	5.3	—	0.03	0.51
	Adductor longus (white)	0.3	30.0	26.0	72		55.0	0.8	0.9	—	0.01	0.06
Rat	Heart (red)	6.1	12.0	10.0	—		6.0	0.3	—	95	—	2.2
	Quadriceps femoris (white)	1.9	50.0	47.0	—		48.0	1.2	—	—	—	—
Pigeon	Pectoral (red)	3.0	18.0	24.0	43		33.0	1.2	—	100	0.07	3.2
Domestic fowl	Pectoral (white)	1.1	83.0	105.0	211		76.0	0.6	—	—	—	—
Pheasant	Pectoral (white)	2.3	120.0	143.0	—		103.0	2.8	—	15	0.01	0.03

Enzyme activities (μmol/min/g wet tissue at 25°)

acid cycle would be favoured. A low ratio of H/M isoenzymes would, in contrast, allow lactate to accumulate the NAD^+ to be regenerated anaerobically from $NADH + H^+$.

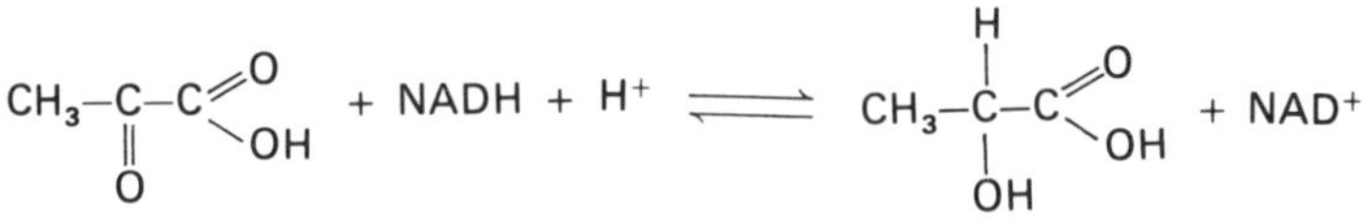

$$CH_3-\underset{\underset{O}{\|}}{C}-C\underset{OH}{\overset{O}{\diagup\!\!\!\diagdown}} + NADH + H^+ \rightleftharpoons CH_3-\underset{\underset{OH}{|}}{\overset{\overset{H}{|}}{C}}-C\underset{OH}{\overset{O}{\diagup\!\!\!\diagdown}} + NAD^+$$

In this context it is interesting that white muscles contain considerably more carnosine and anserine than red muscles. These two dipeptides

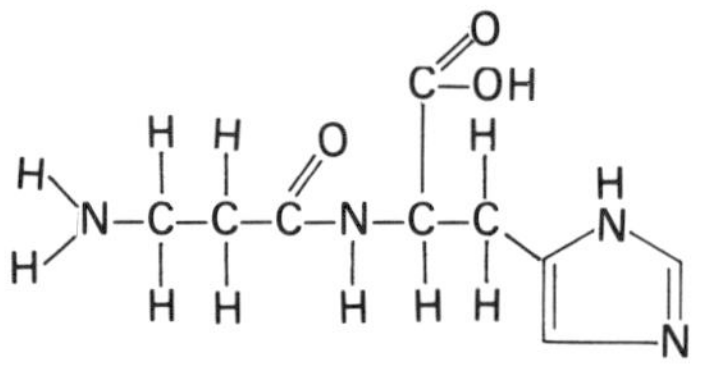

carnosine
(β-alanyl-L-histidine)

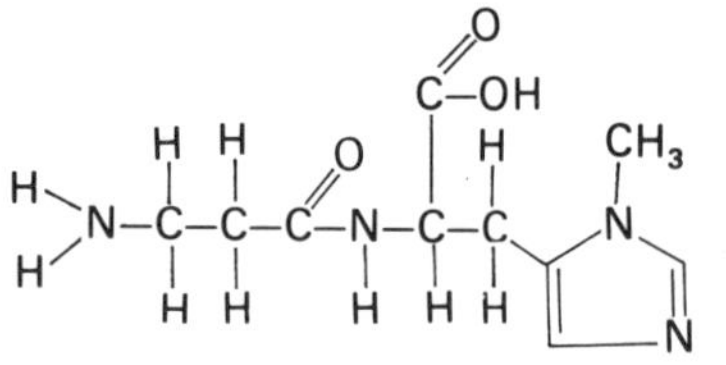

anserine (β-alanyl-methyl-L-histidine)

probably act as buffers to reduce the fall in pH that occurs as the production of lactate rises during periods of muscular effort.

A further feature that distinguishes white muscles from red is their higher content of fructose-1,6-diphosphatase and of both the cytoplasmic and mitochondrial forms of L-α-glycerophosphate dehydrogenase. The presence of these enzymes enables the muscle to reoxidize $NADH + H^+$ by reducing dihydroxy acetone phosphate in addition to pyruvate. When white muscles are activated and glycogenolysis is stimulated, $NADH + H^+$ is produced in substantial quantities by glyceraldehyde-3-phosphate dehydrogenase before pyruvate begins to accumulate. Therefore an alternative to pyruvate is required as a hydrogen sink if glycolysis is to be allowed to continue at a high rate before pyruvate production has risen. This requirement is met by dihydroxyacetone phosphate which can be reduced to L-α-glycerophosphate by the cytoplasmic L-α-glycerophosphate dehydrogenase. L-α-glycerophosphate is unable to leak out of muscle fibres because, unlike pyruvate, it is phosphorylated. It must, therefore, be removed either by oxidation or by reconversion to glycogen when the muscle returns to its resting state. The former is effected by the FMN linked L-α-glycerophosphate dehydrogenase of the mitochondria whilst the latter involves the cytoplasmic

L-α-glycerophosphate dehydrogenase and fructose-1,6-phosphatase (Figure 2.28).

White skeletal muscles also contain phosphoenolpyruvate carboxykinase whose function is somewhat obscure. It may supply oxaloacetate for the citric acid cycle or allow oxaloacetate, formed as a result of the operation of a malate shuttle (Figure 2.29), to be converted to glucose following periods of muscular activity. It is quite certain that the PEP carboxykinase does not form part of a cycle permitting lactate to be reconverted to glycogen, as muscle is deficient in pyruvic carboxylase (see p. 200).

Red and white muscles differ in their physiological responses to nerve stimulation in that red muscles are tetanized at lower frequencies of stimulation than white and can develop a greater and more prolonged tension during tetanus. On the other hand white muscles can develop much greater twitch tensions than red muscles and relax more rapidly when stimulation ceases (Figure 2.30—compare a and b). These properties correlate well with observations that the specific activity of myosin for ATP hydrolysis derived from white muscle is higher than that from red muscle and that white muscles have a better developed sarcoplasmic reticulum with a far greater capacity to accumulate Ca^{2+} ions.

Cardiac Muscle

Cardiac muscle can be regarded as a special example of red muscle. On account of its excellent blood supply and high content of mitochondria, its metabolism is highly aerobic and it uses lactate, fatty acids and ketone bodies in preference to glucose when they are available. The low ATP-ase activity of cardiac actomyosin has long been recognized and the sarcoplasmic reticulum, although present, is poorly developed. However, the control of contractility is probably similar to that described for skeletal muscle as both troponin and tropomyosin have been found in heart.

Smooth Muscle

Although actin and myosin are present in smooth muscle (Figure 2.31) their arrangement within the cells is still not well understood and the ATP-ase activity of the actomyosin formed from them is very low compared with that from skeletal muscle. Although contraction seems to be controlled by variations in the cytoplasmic concentration of calcium ions, the sarcoplasmic reticulum is very poorly developed and it is probable that the calcium ion concentration within the cell is regulated by the cell membrane. The low activity of the actomyosin and the virtual absence of reticulum correlate well with the role of smooth muscles in maintaining a tonus or in engaging in slow phasic contractions.

Mitochondria are not abundant and the energy supply seems to be largely glycolytic.

Effects of Denervation and Cross-Innervation on the Enzymology of Skeletal Muscle

Skeletal muscle fibres are normally sensitive to acetylcholine only in the region of their motor endplate and acetylcholine esterase which terminates the activity

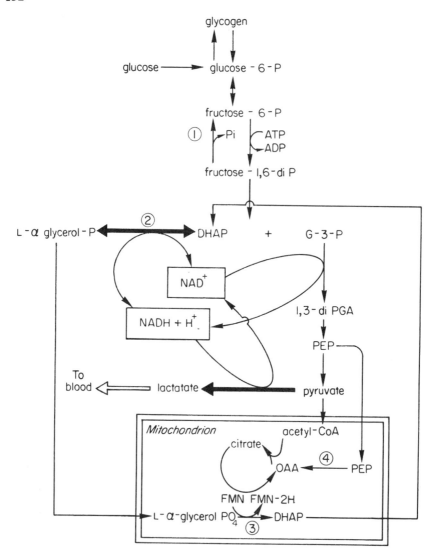

Figure 2.28 Diagram to illustrate the roles of L-α-glycerophosphate dehydrogenase, fructose 1:6 diphosphatase and phosphoenolpyruvate carboxykinase in white skeletal muscle. 1, Fructose-1:6-diphosphatase; 2, cytoplasmic L-α-glycerophosphate dehydrogenase; 3, mitochondrial L-β-glycerophosphate dehydrogenase; 4, phosphoenolpyruvate carboxykinase. DHAP = dihydroxyacetonephosphate; G3P = glyceraldehyde-3-phosphate; PEP = phosphoenolpyruvate; OAA = oxaloacetate.

Notes. In white muscle the carbon flow through mitochondrial pathways is small compared with that through the cytoplasmic pathways. Although L-α-glycero PO$_4$ can be oxidized in the mitochondria it is unlikely that a glycerophosphate cycle of the type found in insects serves to transfer large numbers of reducing equivalents from the cytoplasm to the mitochondria in mammalian white muscles. Phosphoenolpyruvate carboxykinase may serve to supply oxaloacetate for the TCA

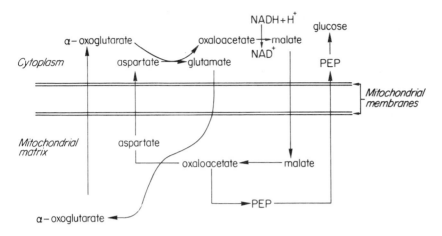

Figure 2.29 The malate–oxaloacetate shuttle. This set of reactions permits some NADH + H⁺ produced in the cytoplasm to be oxidized in the mitochondria

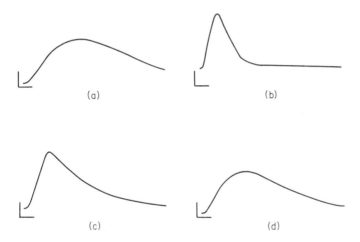

Figure 2.30 Twitch responses from soleus and flexor digitorum longus (FDL) muscles. (a) Self-reinnervated soleus. (b) Self-reinnervated FDL. (c) Cross-innervated soleus. (d) Cross-innervated FDL.

Time scale (horizontal bar = 20 msec). Tension (vertical bar = 125 g in a, b and d and 50 g in c). Derived from Buller, A. J., Mommaerts, W.H.F.M., and Seraydian, K. *J. Physiol* (1969)

---

cycle as indicated. However, it is also possible that it allows oxaloacetate involved in a malate oxaloacetate cycle (see Figure 2.29) transferring reducing equivalents to the mitochondria to be converted to glycogen following periods of contraction

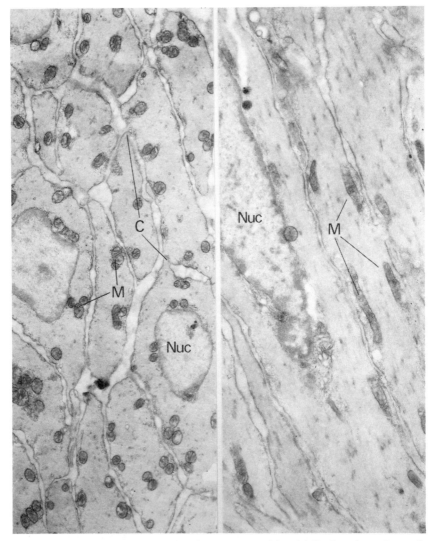

Figure 2.31 Electron micrographs of transverse and longitudinal sections of smooth muscle. Nuc = nucleus, M = mitochondria. Notice the intercellular contacts C that permit electrotonic coupling between cells. By kind permission of Mr J. H. Kugler

of acetylcholine by hydrolysis is also confined to that area. However, if the motor nerve is cut and the endings are allowed time to degenerate, acetylcholine sensitivity and acetylcholine esterase activity spread across the whole surface of the sarcolemma. Following reinnervation, acetylcholine sensitivity and esterase activity again become limited to the region of the motor endplate. It is thus clear that innervation (which is merely a specialized form of intercellular contact) can have marked effects upon the distribution of receptor sites and enzyme activity.

This type of cellular interaction is illustrated even more dramatically by experiments in which a white (twitch) muscle, such as the flexor digitorum longus, and a red (tonic) muscle, such as the soleus, were denervated and then cross-innervated (Figure 2.32). Following successful cross-innervation, the contraction characteristics of the two muscles were profoundly altered. The soleus developed some of the characteristics of a twitch muscle whilst the flexor showed behaviour typical of the soleus (see Figure 2.30). These physiological changes were paralleled

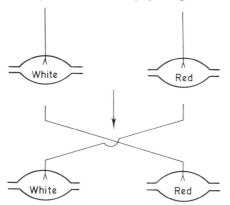

Figure 2.32 Diagram to illustrate cross-innervation experiments between red and white muscles

by remarkable biochemical adaptations. The glycogen content and glycolytic capacity of the soleus increased whilst its oxidative capacity fell. On the other hand, the metabolism of the flexor became more aerobic. The myofibrillar ATP-ase activity of the flexor fell to a value similar to that of the normal soleus but that of the cross-innervated soleus rose only slightly. Cross-innervation decreased the activation time of the soleus and increased the activity of Ca^{2+} stimulated ATP-ase in the sarcoplasmic reticulum whilst the converse changes occurred in the flexor digitorum longus. From this brief survey of cross-innervation experiments, it is clear that the metabolic and physiological typing of muscle fibres is largely dependent upon the type of innervation they receive. An intriguing but largely unanswered question is the extent to which the metabolism of neurones is determined by the type of cells that they innervate.

References

The Control of Muscle Contraction, Perry, S. V. in *Symposium XXVII of the Society for Experimental Biology on Rate Control of Biological Processes.* Cambridge University Press, 1973.

Machina Carnis, Needham, D. M. Cambridge University Press, 1971.

Molecular Control Mechanisms in Muscle Contraction. Weber, A. and Murray, J. M. *Physiological Reviews*, **53**, 1973, pp. 612–673.

SECTION 3

The Intestine

CHAPTER 13

The Digestion and Absorption of Food

Most people equate their environment with the part of the world that impinges on their skin. Biochemically this is incorrect. The chemical environment of the skin is something of great stability, changing only fractionally in its chemical composition. It is in fact the intestinal lining which bears the brunt of the changing chemical environment and responds to it by transforming, absorbing or rejecting the mixture of chemicals presented to it. No microorganisms or plants suffer the same variety of chemical insults as does the mammalian intestine, and, in the case of the human, the intestine must cope with the mixed bag of esoteric chemicals that make up the modern doctor's armoury of drugs. The epithelial cells of the intestine (perhaps because of the traumatic conditions to which they are exposed) are rapidly replaced. On average the total lining of the intestine is replaced in about three days. In weight this amounts to about 450 g, roughly equivalent to 10 g of fat and 20 g of protein daily. Besides this endogenous protein, the pancreatic juice contributes protein amounting to about 10 g daily and other intestinal secretions a further 13 g. This protein and fat is largely reabsorbed. If the intake of food is equivalent to 2500 cal it is likely that at least half of it will come from carbohydrate sources. The remainder will amount on average to about 60 g of protein and 100 g of fat. Thus the intake of protein from endogenous sources approximates to the dietary intake and in the poorer countries where the daily protein intake is less than 30 g, 60 per cent or more of the protein digested is derived from endogenous sources. Because of the large contribution of endogenous protein, the amino acid composition of the gut contents is relatively constant and direct experiments have shown that this amino acid composition is entirely unaffected by the composition of the diet. In spite of this, the proportion of protein in the diet can alter the enzymic make-up of the liver.

Besides removing substances from the lumen of the intestine, the intestine contributes substances to the lumen content. In general the substances contributed by the intestine are required to facilitate the intestine's task of absorption. For convenience the bile and the secretions of the pancreas will be considered together with the secretions of the intestine proper. Table 3.1 shows the main substances (excluding salivary amylase) discharged into the lumen by the intestinal tract.

Table 3.1 Substances secreted into intestinal lumen. The list of materials is not comprehensive but includes those which are quantitatively most important. The functions listed are the main functions

Proteins	Origin	Functions
Pepsinogen	Stomach	To yield pepsin which preferentially splits peptide links adjacent to aromatic amino acids
Amylase	Pancreas	To hydrolyse polysaccharides mainly to maltose
Trypsinogen	Pancreas	To yield trypsin for the splitting of peptide links mainly between the carboxyl group of lysine or arginine and the amino group of another amino acid
Chymotrypsinogen	Pancreas	To yield chymotrypsin for splitting peptide links mainly between carboxyl of aromatic acid and the amino group of an amino acid other than aspartate or glutamate
Maltase	Pancreas	Splits maltose to glucose
Carboxypeptidase	Pancreas	Removes amino acids from carboxyl end of a peptide
Lipase	Pancreas	Splits triglycerides in stepwise fashion to fatty acid and glycerol at rates depending on the fatty acid composition
Maltase	Intestinal mucosa	Splits maltose to glucose
Sucrase (Invertase)	Intestinal mucosa	Splits sucrose to glucose and fructose
Lactase	Intestinal mucosa	Splits lactose to galactose and glucose
Amino peptidase	Intestinal mucosa	Removes amino acids from amino end of a peptide
Dipeptidase	Intestinal mucosa	Splits dipeptides to amino acids
Phosphatase	Intestinal mucosa	Removes phosphate from organic phosphates
Lecithinase	Intestinal mucosa	Breaks down phospholipids to constituents
HCl	Gastric mucosa	Stops enzymic actions in the material ingested; converts pepsinogen to pepsin; provides favourable pH for action of pepsin; hydrolyses fructose-containing sugars
$NaHCO_3$	Pancreatic juice	Brings material discharged by the stomach into the intestine to a roughly neutral pH suitable for the action of the digestive enzymes
Bile salts	Liver	Emulsifies and solubilizes water-insoluble material

Activities of the Digestive Enzymes

Within the exocrine cells the proteolytic enzymes pepsin, trypsin and chymotrypsin are stored as the inactive forms pepsinogen, trypsinogen and chymotrypsinogen, presumably to prevent their hydrolysing the cellular protein. The activation on discharge into the gut lumen consists in removing a short peptide. With pepsinogen, the gastric HCl can remove the masking peptide and the pepsin thus liberated is also able to activate the remaining pepsinogen. With trypsin the primary activator is an enzyme called enterokinase, present in the intestinal secretion. Once again the trypsin liberated can hydrolyse the remaining trypsinogen. Chymotrypsinogen has no special activating substance but requires trypsin to remove its masking polypeptide and the chymotrypsin liberated is unable to act as an activator for the remainder of the chymotrypsinogen. The carboxypeptidase of the pancreatic juice is also secreted in an inactive form and is activated by trypsin, but, whereas the polypeptides removed from the inactive precursors of the other proteinases were comparatively small, in this case a protein of roughly two-thirds the original molecular weight is split off.

Digestion and Absorption

Before examining the digestion and absorption of the various foodstuffs, it is worth considering the quantity of food eaten. Table 3.2 shows the average composition of the British diet for an intake of 2500 kcal daily. As an example of the amount of food supplying this number of calories, one large brown loaf (800 g)

Table 3.2 Composition of British diet of 2500 kcal daily

Foodstuff	Percentage of calories from each foodstuff in diet	Weight of foodstuff (g)	Calories contributed (kcal)
Fat	35	97	875
Protein	11	70	275
Carbohydrate	54	340	1350

eaten with 110 g butter (comprising about 105 g fat, 68 g protein and 400 g carbohydrate) would provide about 2800 kcal. If 110 g of cheese is eaten as well, the bread ration could be dropped to 1½ small loaves (600 g) and the composition of the food ingested would be about 142 g fat, 79 g protein and 300 g carbohydrate. For a conveniently carried ration, a 500 g bar of milk chocolate will supply about 2700 kcal consisting of 174 g fat, 27 g protein and 240 g carbohydrate. Suppose on the other hand the meal consists of steak (170 g) chips (230 g), green vegetables or salad (110 g), biscuits and cheese (60 g of each), all washed down with 0·5 l of good beer, about 1800 kcal would be ingested with an intake of 92 g fat, 57 g protein, 143 g carbohydrate and 32 g alcohol. Thus 450 to 900 g of material may be eaten at a meal and of that about 300 g will be dry solids. The half kilogram or so of food passing into the stomach is met by the gastric juice. In the fasting state this juice is roughly equivalent to 0·1 M HCl and its volume will be about 100 ml although this would be greatly increased if the meal had been pleasurably anticipated. The protein of the meal is a very efficient buffer and the 50 or so grams eaten will mop up the acidity of the HCl with the result that the pH of the stomach contents rises to about 6. Some 300–400 ml more of gastric HCl is required to lower the pH to between 2 and 3. Direct measurements of the pH of the stomach contents during a meal, either with a conventional glass electrode or with the 'radio pill', show that initially the pH is not much below neutrality, but during the next hour or so it falls to about 3. This is still above the pH of about 2, which is usually accepted as the optimum for action of pepsin. However, it appears that the human stomach contains two types of pepsin and that both of these have two pH optima, one around 2 and the other at 3·3. The exact positions of these pH optima depend on the nature of the protein that is being digested. Thus pepsin does not require such acid conditions as hitherto believed for its efficient action.

The Secretion of Hydrochloric Acid

The gastric secretion of HCl (as H^+ and Cl^-) is a special case of the more general problem of ion transport (see p. $\overline{283}$). Because of the large differential between the H^+ concentration of the gastric juice and of the blood, the work required for transport is considerable. The oxyntic cells secreting HCl have a very high rate of respiration and appear to be one of the most efficient of chemical transducers. Consequently anaerobic conditions and uncouplers of oxidative phosphorylation are inhibitory for HCl secretion. The Cl^- apparently accompanies the H^+ passively in order to maintain electrical neutrality. Carbon dioxide under the influence of carbonic anhydrase combines with water to give bicarbonate which passes into the blood stream to compensate for its deprivation of chloride. The alkaline tide in blood and urine subsequent to a meal is supposed to be brought about by this compensatory movement of bicarbonate.

Intestinal Digestion and Absorption

The addition of the pancreatic secretion to the lumen contents besides contributing the enzymes listed in Table 3.1 adjusts the pH to the neighbourhood of 6·5. Cleavage of the proteins to amino acids then takes place by the cooperative action of the proteinases and other peptidases listed in Table 3.1. Polysaccharides such as starch and glycogen will be hydrolysed to maltose and glucose by the pancreatic amylase. The disaccharides sucrose, maltose and lactose are apparently absorbed as such into the cells of the intestinal mucosa and are then hydrolysed into their constituent monosaccharides by the appropriate enzymes.

The Absorption of Carbohydrates

The main dietary monosaccharides presented to the intestine for absorption are glucose, galactose and fructose; a smaller quantity of mannose and the pentose sugars are also utilized, depending on the dietary habits. Of these sugars, glucose and galactose are absorbed by an active process which is characteristically inhibited by phlorizin. Oxidative phosphorylation has been thought to be necessary for the active uptake of the two sugars because the process has been inhibited by uncouplers and by anaerobic conditions. Na^+ ions are also required for the intestinal transport process. A carrier molecule bearing sugar and sodium is supposed to traverse the epithelial cell, driven by the difference in Na^+ concentration between lumen and cell. The active transport of the sodium out of the cell by means of a sodium activated ATP-ase (see p. 286) maintains the sodium gradient between lumen and cell.

Fructose does not appear to be absorbed by the same carrier mechanism as glucose, but it is subject to a variable conversion to glucose in crossing the intestine. In man, up to 70 per cent of the fructose absorbed can appear as glucose in the mesenteric blood.

The Digestion and Absorption of Lipids

The breakdown of the lipids is somewhat more complicated, being influenced by the composition of the material presented for digestion. It is generally agreed that the formation of both a fine emulsion and micelles aids the digestion and later absorption of fats and their breakdown products. Bile salts and other surface active agents, such as phospholipids and monoglycerides are constituents of a mixture which promotes digestion by altering the interaction between water and the normally water-insoluble glycerides. Dietary fat is usually considered to be triglyceride, but in fact phospholipids constitute a substantial, or even the major, fraction of the ingested lipid, depending on the food eaten. For example, fish muscle has about 20 per cent of its lipid as phospholipid, animal skeletal muscle around 30 per cent, bird muscle about 40 per cent and offal about 70 per cent. The breakdown of phospholipids (see p. 30) requires the cooperation of three enzymes; phospholipase-A (from the pancreas) removes the fatty acid from the β position of the phospholipid to give a lyso compound. The second fatty acid is removed from the α position of the lipid by β-phospholipase to liberate a free fatty acid and, for example, glycerylphosphorylcholine. The latter substance is then hydrolysed by a diesterase to glycerol phosphate and choline. The breakdown of triglycerides is also a stepwise process dependent on the enzyme lipase, which appears to work effectively only at oil/water interfaces and therefore the substrate must be in sufficient concentration to form an emulsion. Lipase also has the physical property of accumulating at the oil/water interface and thus the enzyme is naturally concentrated in the most effective place. The lipase hydrolyses only the α-ester bond and the last fatty acid is removed only after isomerization to the 1 monoglyceride (see Figure 3.1).

The water solubilities of the various fatty acids presented to the intestine are very different. Whereas all the sugars and the amino acids are highly water-soluble, the fatty acids decrease in water solubility until by C_{12} they may be considered water-insoluble. The transport of fatty acids up to C_5 in chain length has been shown to be by an active process inhibited by phlorizin, dinitrophenol and anaerobic conditions. The short chain fatty acids mainly pass through the intestine into the bloodstream without forming triglycerides and it is only when they pass as the free fatty acids that the transport is active. Quantitatively these short chain fatty acids are minor components of the diet, for example your intake of acetic acid is probably dependent on the quantity of pickles you eat.

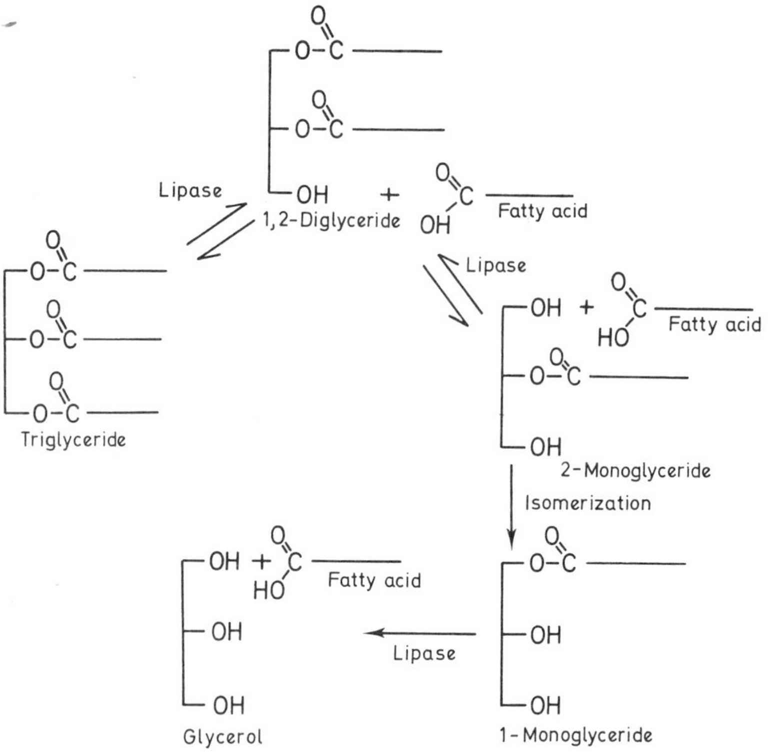

Figure 3.1 Course of hydrolysis of triglyceride by pancreatic lipase

The majority of the fatty acids are incorporated into the chylomicra and then pass into the lymph, although some of the long chain fatty acids do pass into the blood, presumably in association with albumen, and thus contribute to the free fatty acids (FFA) of the blood. Apparently a lack of conjugated bile salts may lead to a larger absorption of long chain fatty acids in the unesterified form. Some monoglycerides are also absorbed in the same way.

The absorption of fat, like that of protein, is very efficient; the amount of faecal fat is only about 5 per cent of the amount ingested. Details of the absorption of fat are still obscure, but measurements of the fat composition of the human intestinal lumen has shown that a very substantial part consists of free fatty acids. If glycerides are absorbed, they are apparently broken down to fatty acids within the intestinal mucosa. The absorption process and the constitution of the chylomicra which appear in the blood require the resynthesis of triglycerides. The glycerol split from the triglycerides during hydrolysis is not used for resynthesis, instead glucose supplies L-α-glycerophosphate as in the synthesis of triglycerides in the fat depots (see p. 252). Monoglycerides that are absorbed can also be converted to triglycerides by successive reactions with acyl CoA derivatives. The passage of resynthesized triglycerides through the cell and their collection to form chylomicra is not well understood. Some electron microscope evidence suggests that the triglycerides pass along the fine pores of the endoplasmic reticulum by a process like pinocytosis. Whatever the mechanism, the triglycerides that pass into the lymph are different from those presented to the intestinal mucosa both in the arrangements of the fatty acids in the glycerides and in the carbon skeletons of the glycerol. The formation of the chylomicra requires a final wrapping of the triglyceride package in protein and phospholipid. Without these specific syntheses the triglyceride cannot get out of the cell into the lymph. In some rare diseases where there is a failure to synthesize β-lipoprotein there is an accumulation of triglycerides in the gut wall after a fatty meal but no rise in blood triglycerides. A similar effect may be produced in experimental animals by the administration of inhibitors of protein synthesis.

The Absorption of Amino Acids

The formation of amino acids by the various proteinases has already been described (see Table 3.1). The efficiency of amino acid absorption by the intestine is very high; of the total amino acids presented to the intestine only about 5 per cent remains unabsorbed. Apart from glutamate and aspartate, the amino acids are absorbed by an oxygen-dependent active process with one amino acid competing with another more or less effectively for the transporting site. The clear demonstration of movement of the amino acids against a concentration gradient was made by the use of the everted sac technique. In this technique a segment of intestine is turned inside out and tied at either end after the lumen has been filled with a suitable fluid. Movement of material through the intestinal wall can be assessed by measuring the change in content of substances in the lumen of the sac.

Experiments on the administration of ^{15}N labelled proteins to the human show that the maximum labelling appears in the urine after about 1 to 1·5 hours suggesting a time slightly less than this for digestion and absorption together.

The Absorption of Water and Salt

The intestine transports water very efficiently from its contents to the blood stream. This may be seen from the fact that the water excreted in the faeces (100–200 ml) shows little change irrespective of the amount of liquid consumed. The intestine has to transport not only the fluid ingested but also the fluid contributed by the various secretions of the digestive system. The quantities involved are shown in Table 3.3.

Roughly half the water load of the intestine is contributed by the internal secretions. Besides water, the internal secretions contribute substantial quantities of Na^+ and Cl^- to the lumen contents, much more than comes from exogenous sources (see Table 3.3) and again there is an almost quantitative absorption of these substances. Some at least of the movement of these three substances from the intestine into the blood is against a chemical gradient and like the other active transport mechanisms of the intestine they are inhibited by anoxia, by inhibi-

Table 3.3 Contributions to water and salt contents of intestinal lumen

	Daily contribution of water to intestinal contents (litre)	Daily contribution of Na$^+$ (g)	Daily contribution of Cl$^-$ (g)
Saliva	1·5	0·8	1·5
Gastric juice	3·0	3·3	15·0
Bile	0·9	4·0	2·9
Pancreatic juice	0·7	2·1	1·7
Shed cells of intestinal lining and other secretions	0·2	0·5	0·6
Total from endogenous sources	5·3	10·7	21·7
Content of food	2·5	4·0	5·0
Content of drink	3·0		
Total exogenous sources	5·5	4·0	5·0
Total all sources	11·0	14·7	26·7
Excreted in faeces	0·2	0·12	0·09
Percentage absorbed of lumen contents	98·5	92·0	99·7

tion of oxidative phosphorylation and by phlorizin. It is probable that much of the chloride is passively transported as a counter-ion to the actively moved sodium, but there is evidence that at some sites in the intestine there is also an active transport of chloride.

Calcium Absorption

Normal adults are recommended to take 1 g of calcium in their daily diet and of this about one third is absorbed. Pregnant and nursing mothers require up to about double this amount and growing children something in between, depending on the growth rate. The absorption of calcium is by a combination of active transport and diffusion. The efficiency of the uptake process is greatly increased by vitamin D or more probably by one of its metabolites and in times of calcium shortage the vitamin D effect is larger. The intestine contains a specific calcium binding protein which increases in amount in times of calcium shortage or when vitamin D is administered.

Drug Metabolism by the Intestine

Apart from its role in preparing and absorbing nutrients, the gut is the first line of defence against chemical insult. In this role it is sometimes in conflict with our own wish to absorb various chemicals for therapeutic purposes or for personal satisfaction. As the preferred route of drug administration is oral, it is surprising that so little is known about the quantitative aspects of the intestine's abilities to transform drugs (detoxification, see p. 225) into more active or less active forms. In this context the gut behaves like a minor liver. In what follows we shall not be concerned with drug absorption but with drug metabolism although obviously this process has secondary effects on drug absorption. The chemical transformation of drugs can also be brought about by the gut flora. In general the gut can carry out all the detoxification reactions of the liver but to a much smaller degree. Selected activities of the gut are therefore presented where they may appreciably affect the overall functioning of the body.

Alcohol Dehydrogenase

As discussed on p. 471 most of the body's alcohol dehydrogenase is in the liver. The gut has about one tenth of the liver's activity and can deal with ethanol at the rate of about 1 to 1·5 g hour (roughly half of a small wine glass of light wine).

Hydrolysis Reactions

These occur with many substrates, such as acetylsalicylic acid, glucuronides and sulphates. Sometimes their activity is about 20 per cent of that of the liver's. Since drugs such as chloramphenicol and morphine are excreted as glucuronides in the bile this may result in recycling as occurs in the enterohepatic circulation of the bile salts and the pigments.

Conjugation Reactions

Glucuronide and sulphate ester formation occurs in the intestine and drugs such as sulphonamides and isoniazid are acetylated and methylated. Of particular interest, because of their widespread use, is the ability of the intestine to metabolize steroids (oestrogens etc.) and to form the glucuronides of the initial or transformed products.

In rats it has been shown that the ability to metabolize drugs varies with the diet and with the hormone balance. Presumably similar factors will influence man's ability to alter his ingested drugs. Intestinal metabolism may provide a partial reason for some drugs being apparently poorly absorbed and the recycling process already mentioned will prolong the effect of some drugs. The quantitative aspects of the intestinal drug metabolism are not clear but the possibilities must be kept in mind when considering drug metabolism.

CHAPTER 14

Protein Synthesis

It is clear from the preceding description of the function of the intestine that the synthesis of protein is an extremely active and important process in this tissue; therefore, while it is true that most tissues of the body undertake protein synthesis, this process will be discussed now.

Table 3-1 makes it clear that the intestine forms a variety of different proteins and the discussion of protein structure (p. 60) establishes that these characteristic differences arise basically from differences in the sequence in which amino acids are condensed to form the peptide chain.

A molecular account of protein synthesis must include a description of both the formation of the peptide bonds and the mechanism for determining the order of the amino acids. The process is energy-dependent and nucleic acids, both RNA (ribonucleic acid) and DNA (deoxyribonucleic acid), are essential for the synthesis. Evidence for the nuclear localization of DNA and for the movement of RNA from nucleus to cytoplasm leads to the concept that DNA, as the genetic material, directs the synthesis of specific forms of RNA which, in turn, direct the synthesis of specific proteins. This concept visualizes the attachment of particular amino acids to specific transfer RNA molecules (*t*-RNA) and their condensation in an order determined by the interaction of the *t*-RNA-amino acid complex with another species of RNA (messenger or *m*-RNA) which transfers the genetic message from the nuclear DNA to cytoplasmic sites of synthesis on the ribosomes. In some viruses the genetic message can be encoded in the form of RNA; however, the same basic principles apply in the decoding of this message.

The Structure and Biosynthesis of DNA

Much evidence points to the fact that DNA, found in greatest concentration in the nuclei of cells, is the genetic material of the cell carrying the 'hereditary plan' for cell growth. The structure of DNA isolated from many types of cells conforms to the same general pattern; it contains the purine bases, adenine and guanine, the pyrimidine bases, thymine and cytosine, the pentose, deoxyribose and phosphate. The backbone of the molecule consists of alternating sugar and phosphate units, with the bases linked to the sugar residues (Figure 3.2). X-ray studies of molecules of DNA suggests that they contain two chains which spiral around each other (Figure 3.3a). The nature of the linkage between these chains was deduced from the fact that the amounts of adenine and thymine were always

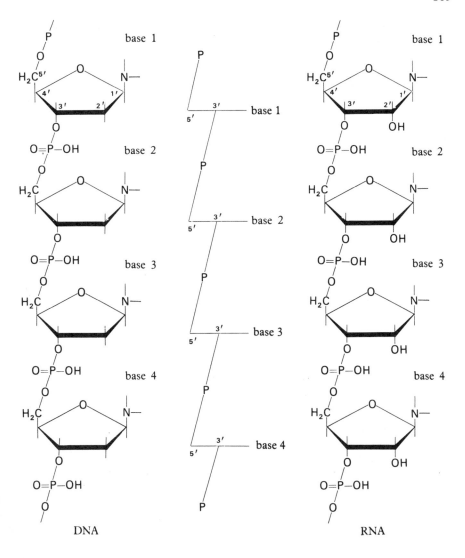

Figure 3.2 Schematic form of DNA and RNA

equal as were those of guanine and cytosine. This is accounted for by the fact that the bases are hydrogen-bonded to each other with specific pairings of adenine and thymine, and of guanine and cytosine (Figures 3.3b and 3.4). Such a structure can also account for the all-important conservation of hereditary information during cell division. In principle, it follows from the specificity of the base pairing that each strand of the DNA should be capable of directing the assembly of its complementary strand when DNA synthesis is occurring so that two new double-stranded DNA molecules will be formed of composition identical to the parent molecule and to one another (Figure 3.5).

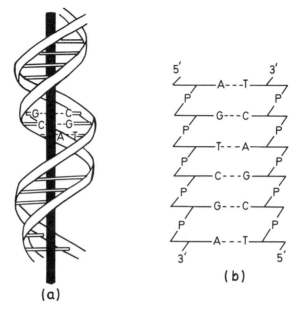

Figure 3.3 The double helix of DNA; (a) the two strands are held together by the base pairing described in Figure 3.4. In each strand the sequence of components is as shown in (b) in which the pentose residues are represented by ⤎⤏ with their hydroxyl groups on carbon five ⤎— (5′) and carbon three —⤏ (3′); P is the phosphodiester link between these hydroxyls; and the bases are adenine (A) hydrogen bonded (· · ·) to thymine (T) and guanine (G) hydrogen bonded (· · ·) to cytosine (C) (Figure 3.4). Note that in the upper and lower ends of the strands the two corresponding pentoses differ in that one has a free hydroxyl on carbon five and the other on carbon three; the strands are said to be 'anti-parallel'

The first enzyme DNA polymerase catalyses the addition of new nucleotide units to a DNA strand. Each new unit must be available as the deoxynucleoside triphosphate. In addition, the enzyme requires Mg^{2+}, a nucleic acid chain to which new units are added (the primer) and another chain to copy, i.e. to serve as a template for the alignment of successive bases by base pairing (see Figure 3.4). The new units can be added only to free hydroxyl groups at the 3′ position (Figure 3.3b). As the two chains of the DNA double helix are anti-parallel (Figure 3.3b) it is apparent that different routes must be followed for their synthesis. Replication begins with the unwinding of the original helix, as shown in Figure 3.6. In this diagram the upper chain can provide a suitable template for a new complementary strand which can grow steadily from left to right thus presenting a new free 3′ hydroxyl for further elongation. This cannot happen on the lower chain; instead short stretches of DNA, each growing from right to left, are

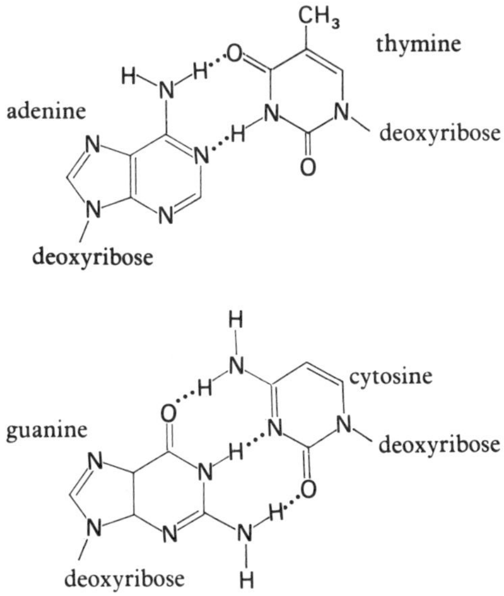

Figure 3.4 Base pairing between adenine and thymine (2 hydrogen bonds) and guanine and cytosine (3 hydrogen bonds)

formed on this template. Subsequently they are linked by a ligase to form the complete complementary chain.

The fine details of this entire process are not understood. However, it is certain that each strand is provided with a new complementary strand, in which the bases have been aligned in the correct sequence by the requirements for specific base pairing already described. Thus in each mitotic cycle, the daughter cells will receive DNA molecules identical to those of the parent cell. It was suggested that the linear sequence of bases in DNA acts as a chemical 'code', indicating to the cell the order in which amino acids should be strung together to make particular polypeptide chains and therefore particular proteins. As the DNA is largely confined to the nucleus, and as much synthesis of protein occurs in the rest of the cell, it was also clearly apparent that the information in the chemical 'code' must be transmitted to the cytoplasm in some way. An obvious candidate for the 'transmitter substance' is RNA, which occurs both in the nucleus and outside it.

DNA Repair

Cells also contain enzyme systems that can 'repair' DNA molecules damaged by, for example, radiation. Probably damage involves a break in one strand of the DNA chain (e.g. by destruction of a base) and the repair system reforms the original sequence using the appropriate deoxyribonucleotide and the complementary DNA strand as a template.

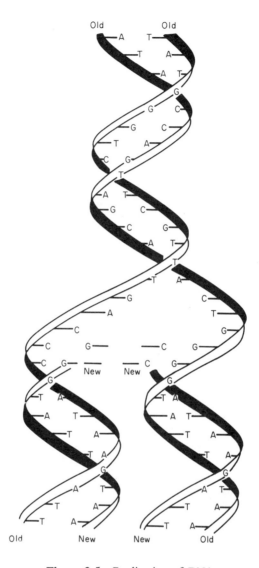

Figure 3.5 Replication of DNA

The Structure and Function of RNA

The RNA content of cells comprises a number of functionally different classes of RNA. All these classes contain the bases adenine, uracil, guanine and cytosine, bonded to chains of molecules of the sugar ribose linked through phosphate diester bonds (Figure 3.2). The bases are capable of the specific pairing reactions, adenine with uracil, guanine with cytosine (Figure 3.7).

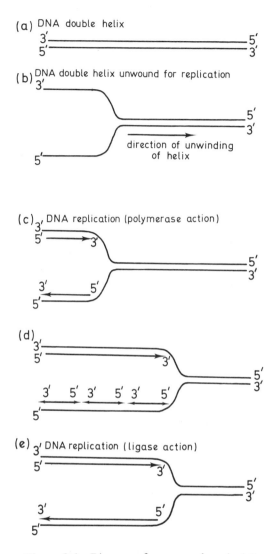

Figure 3.6 Diagram of a proposed method for DNA replication (see text for details)

Transfer RNA (t-RNA)

This is a class of RNA of relatively low molecular weight (about 25,000), each molecule containing 75–100 nucleotides. A number of *t*-RNA species is present in all cells, each one being capable of reacting with a particular amino acid during protein biosynthesis. About 40 *t*-RNA nucleotide sequences are known; all have a secondary structure stabilized by regions of base pairing, and the 'clover-leaf' configuration represented in Figure 3.8.

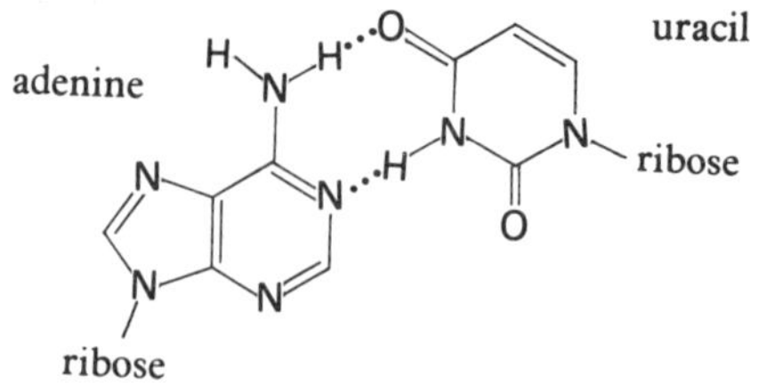

Figure 3.7 Base pairing between adenine and uracil in RNA

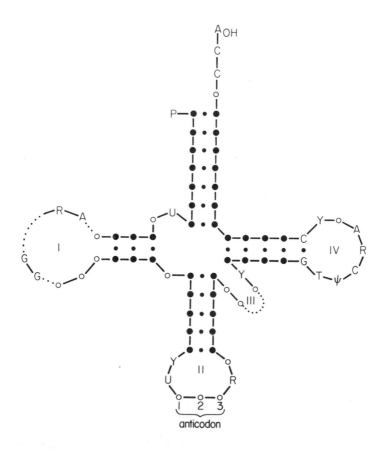

Figure 3.8 Generalized clover-leaf secondary structure of t-RNA showing those bases constant in t-RNA sequences. A, adenine; C, cytosine; G, guanine; R, purine; T, thymine; U, uracil; Y, pyrimidine; ψ, pseudouracil. ● Base hydrogen bonded structure. ○ Base not hydrogen bonded in the clover leaf

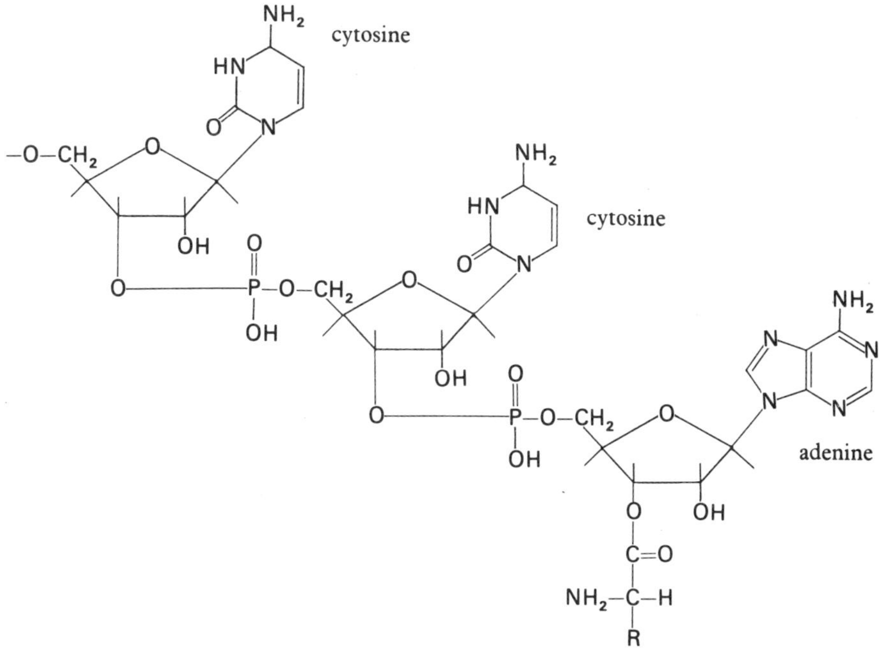

Figure 3.9 The three terminal nucleotides in a *t*-RNA molecule, with the bases adenine, cytosine, cytosine. The amino acyl group is attached to the 3' position of the ribose in the terminal adenine nucleotide

The end of the chain with which the amino acid reacts has the base sequence AMP–CMP–CMP–; attachment of the amino acid residue involves esterification of the 3' position on the ribose of the terminal adenosine (Figure 3.9). Another region of this molecule has a sequence of three bases which is specific for the particular amino acid and which is complementary to a corresponding set of three bases in a 'messenger' RNA molecule; this is the 'anticodon' (Figure 3.8) region of the *t*-RNA, which is recognized by a 'codon' of 3 bases in the messenger RNA.

Ribosomal RNA

Ribosomes contain about 60 per cent RNA (ribosomal or *r*-RNA) and 40 per cent protein. Each ribosome contains two separable subunits, one larger than the other. The two subunits come together to form functional ribosomes during protein synthesis. During this synthesis, functionally active ribosomes are distinguishable from other ribosomal material by their attachment to molecules of messenger RNA (*m*-RNA) to form polysomes.

Messenger RNA (m-RNA)

This type of RNA directs the biosynthesis of a particular protein. Most of the data concerning messenger-RNA has come from the study of bacterial systems

Table 3.4 The genetic code

5'-OH Terminal base	Middle base				3'-OH Terminal base
	U	C	A	G	
U	Phe	Ser	Tyr	Cys	U
	Phe	Ser	Tyr	Cys	C
	Leu	Ser	CTS[a]	CTS	A
	Leu	Ser	CTS	Trp	G
C	Leu	Pro	His	Arg	U
	Leu	Pro	His	Arg	C
	Leu	Pro	Gln	Arg	A
	Leu	Pro	Gln	Arg	G
A	Ile	Thr	Asn	Ser	U
	Ile	Thr	Asn	Ser	C
	Ile	Thr	Lys	Arg	A
	Met[b]	Thr	Lys	Arg	G
G	Val	Ala	Asp	Gly	U
	Val	Ala	Asp	Gly	C
	Val	Ala	Glu	Gly	A
	Val[b]	Ala	Glu	Gly	G

[a] CTS = Chain termination signals.
[b] Chain initiation.

where it appears to be relatively short-lived, being degraded by hydrolytic enzymes fairly rapidly; in higher organisms, some species of *m*-RNA are much longer-lived. The most important properties of any messenger RNA molecule are its capacity for combining with ribosomes and its possession of a base sequence complementary to some part of the DNA containing the cell's genetic information. The base triplets which form the genetic code have been established for all the common amino acids (see Table 3.4). It is striking that there are commonly a number of codons for a single amino acid, i.e. the code is 'degenerate'. Most frequently, these alternative codons differ only in the third nucleotide of the triplet. During protein synthesis, discrete sections of the information contained in the nuclear store of DNA are transcribed into specific molecules of *m*-RNA by a DNA-dependent RNA polymerase which requires Mg^{2+} and all four ribose nucleoside triphosphates. The *m*-RNA provides a programme to direct the assembly of proteins in the ribosomes. A variety of experiments has confirmed the feasibility of such a role for messenger-RNA, perhaps the most striking being those in which the synthetic messenger-RNA, polyuridylic acid, promotes the synthesis of a polypeptide of one particular amino acid, phenylalanine, when added to a cell-free system capable of protein synthesis. The attachment of ribosomes to the messenger-RNA produces clusters of ribosomes called 'polysomes' or 'polyribosomes'. It is not yet clear whether *m*-RNA leaves the nucleus to combine with ribosomes in the cytoplasm, or whether the ribosomes are programmed with *m*-RNA in the nucleus before they enter the cytoplasm. A further problem that has yet to be solved is how the decision is taken as to which part of the DNA is transcribed at any particular time. It is generally held that there are biochemical 'switches' which control the accessibility of a particular part of the DNA for transcription of its information, but the nature of these switches is poorly understood.

Polysomes

Polysomes can be separated from other RNA material by centrifuging. A number of ribosomes may attach simultaneously to a single molecule of messenger-RNA (Figure 3.10). Each separate ribosome acts as a site for protein synthesis during the period of its attachment to the messenger, which may direct the synthesis of a number of individual protein molecules simultaneously.

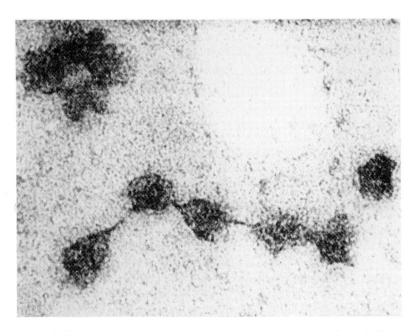

Figure 3.10 An electron micrograph of polysomes showing the individual ribosomes linked together by a fine thread of RNA. Reproduced by kind permission of Dr H. S. Slayter from *J. Mol. Biol.*, 1963

The description of messenger-RNA has indicated that there should be a detectable relationship between the base sequence of the cell's DNA and that of the messenger-RNA molecules. There is now good evidence from work with bacterial systems that only one of the DNA strands (the sense strand) acts as a primer for *m*-RNA synthesis *in vivo*, and as might be expected there is complementarity between the base sequence of the *m*-RNA and the sense strand of the DNA.

The Biosynthesis of Proteins

The Activation of Amino Acids

This requires the attachment of an amino acid to the corresponding t-RNA molecule by a transacylation process using ATP thus:

$$AA + ATP + enz \longrightarrow enz-AA-AMP + PP$$

$$enz-AA-AMP + t\text{-RNA} \longrightarrow AA-t\text{-RNA} + AMP + enz$$

(The amino acid (AA) reacts with ATP, the amino acyl–AMP product (AA–AMP) remaining bound to the enzyme surface (enz); pyrophosphate (PP) is released and is hydrolysed to Pi by pyrophosphatase.)

There are specific amino acid-activating enzymes (or amino acyl–t-RNA synthetases) for each L-amino acid found in proteins; they occur in the soluble phase of the cell. The amino acyl–AMP intermediates are bound to the surface of the enzyme throughout the course of the reaction.

The Condensation of Amino Acids: Peptide Bond Formation

This involves the cooperation of the amino acyl–t-RNA complexes, a molecule of messenger-RNA, ribosomes and soluble enzymes and protein factors. The entire process is thought to proceed as described in Figure 3.11.

(1) The m-RNA for a particular polypeptide chain is formed by the DNA-specific RNA polymerase. It contains the triplets of bases which code for particular t-RNA species and therefore for particular amino acids; for example, UAC codes for tyrosine and UUU for phenylalanine (see Table 3.4). Thus it is a portion of the RNA structure of the amino acyl–t-RNA complex that is recognized by the messenger-RNA and not the amino acid structure itself (see Figure 3.12). This was shown conclusively by experiments in which cysteine was attached to its specific t-RNA, then treated with Raney nickel. This treatment removes the —SH group of the amino acid, which is converted to alanine, i.e. the product is alanine attached to the t-RNA molecule specific for cysteine. However, in a cell-free protein synthesizing system, the complex behaves as if it still contained cysteine.

(2) m-RNA attaches to the smaller ribosomal subunits; its 5' end (the beginning of the coded message) attaches at one specific site (the 'P' or peptidyl site) while the second triplet is located at a second (the 'A' or amino acyl) site.

(3) Initiation of the synthesis requires the presence of a particular form of methionyl t-RNA coded for by AUG ('initiator' codon) in the m-RNA.

(4) This methionyl–t-RNA combines with the m-RNA triplet at the P-site in a process which requires energy (GTP), Mg^{2+} and soluble protein factors M_1 and M_2 ('initiation factors').

(5) The larger ribosomal subunit binds to form the 'initiation complex' of ribosomes, m-RNA and methionyl–t-RNA.

(6) The amino acyl–t-RNA coded for by the second m-RNA triplet at the A-

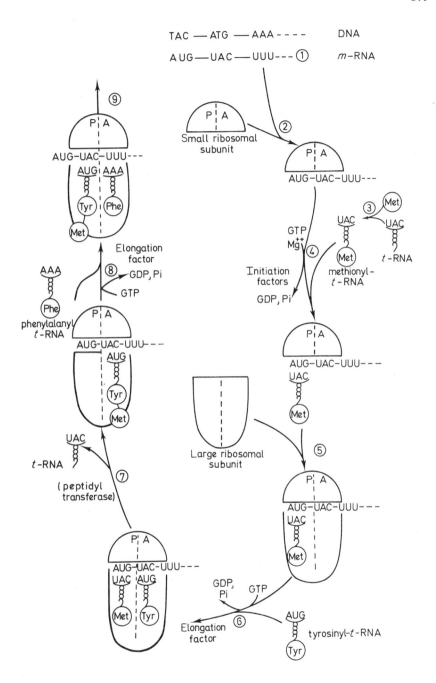

Figure 3.11 Protein synthesis

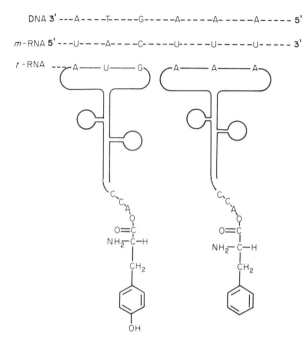

Figure 3.12 The relationship between DNA, m-RNA, t-RNA and the alignment of amino acyl groups on the m-RNA. The amino acid tyrosine has as its 'code word' the base sequence UAC; for phenylalanine it is UUU

site (for example, tyrosine) is bound in a GTP-dependent process; again, soluble protein ('elongation factor') is also required.

(7) The first peptide bond is formed under the influence of a ribosome-bound peptidyl transferase and the first t-RNA released.

(8) The dipeptidyl t-RNA at the A-site is transferred to the P-site and a relative movement of the ribosome and the m-RNA places the third triplet of m-RNA at the A-site. Again, GTP and soluble protein ('elongation factor') are required.

(9) The process is repeated with successive ribosomes each carrying a lengthening polypeptide chain, bound to the single m-RNA molecule. Eventually particular codons (the 'terminator' codons, UAG, UAA and UGA) are reached in the m-RNA and the chains are completed. The protein is released, and the two ribosomal subunits dissociate.

For at least one protein, haemoglobin, it appears that the methionine residue used for the initiation of synthesis is removed before the protein chain is completed. The released polypeptide has valine at the N-terminus, rather than methionine. The requirement for methionine incorporation to initiate protein synthesis in higher animals, and its subsequent removal from the N-terminal end, is probably a common feature of this synthetic pathway. In summary (Figure 3.11), DNA provides the fundamental 'message' describing the nature of the protein to be synthesized. This message and the units which implement the message, are carried on two separate kinds of RNA, messenger-RNA and transfer-RNA, to the sites of protein synthesis on the ribosomes, which contain their own characteristic form of RNA.

Protein Turnover

Even in the mature animal, the need for a protein synthesis continues; for example, replacement of the intestinal lining requires the formation of about 20 g of protein/day. Isotope studies have shown that most of the proteins of the body are in a constant state of flux, being continually built up and broken down. With the adult in nitrogen balance, when there is little or no change in the protein content of the body, it is possible to equate the rate of breakdown with the rate of synthesis. Table 3.5 gives an approximate idea of the quantities of protein syn-

Table 3.5 Daily synthesis of protein in the 70 kg man

	(g)
Intestinal lining	20
Intestinal secretion	23
Plasma	14
Liver	24
Skeletal muscle	6
Haemoglobin	7
Rest of body	62
Total	156

thesized in different organs of the body; in all, the turnover will amount to about 150 g/day. There are three main factors responsible for protein turnover; the replacement of damaged cells as in the intestinal mucosa and blood, the secretion of proteins by exo- and endocrine tissues and the normal intracellular degradation and resynthesis of proteins.

Although Table 3.5 lists the rates of protein synthesis on a daily basis, this does not imply that these amounts of new proteins must be ingested daily. Man can be in protein balance with as little as 10 g of dietary amino acid per day, provided that he receives the essential amino acids in the right proportions. During periods of restricted protein intake, the body's amino groups are carefully preserved, though there is still a constant synthesis and degradation of all the body's proteins.

Inhibitors of Protein Synthesis

Actinomycin D is an antibiotic, and also a potent antitumour agent. It inhibits DNA-dependent RNA synthesis by forming a complex with DNA, the site of attachment being the deoxyguanosine residues. By preventing RNA synthesis, it also inhibits protein synthesis. It is ineffective in preventing RNA-dependent RNA synthesis in viruses whose genetic potential is provided by RNA.

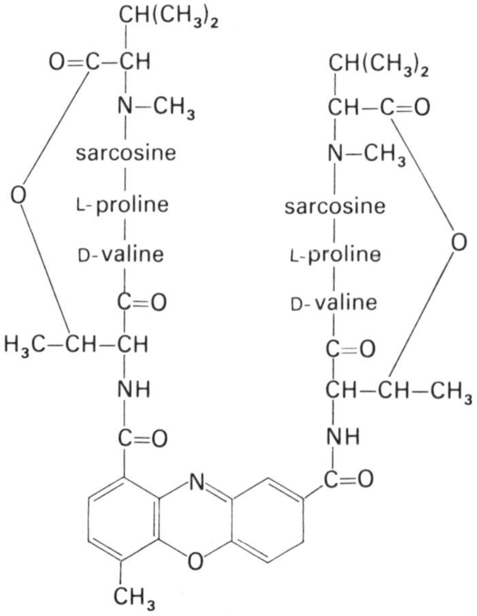

Puromycin inhibits the formation of proteins, probably by substituting for an amino acyl–*t*-RNA, whose shape it closely resembles (Figure 3.8). It differs in that the amino acyl residue of puromycin is attached to the sugar residue by an amide, rather than ester, linkage. It is likely that the inhibitory action follows the transfer of the growing peptide chain to the free-NH$_2$ group of puromycin which is bound instead of an acyl–*t*-RNA at the ribosomal site (Figure 3.11). Further transfer is then impossible because of the different mode of attachment of the resulting peptide to the sugar group and the peptidyl puromycin is released from the ribosome.

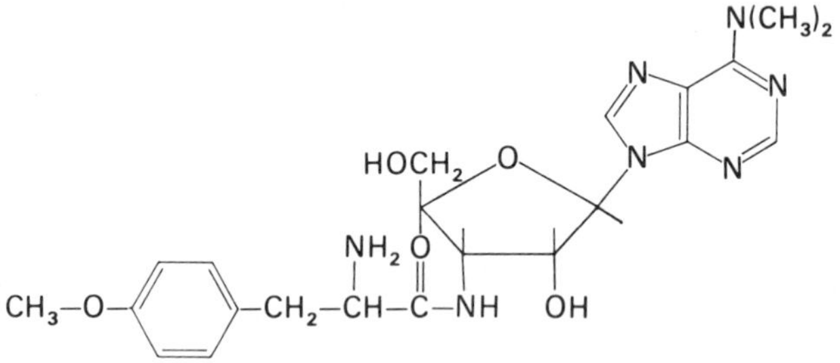

Chloramphenicol is a potent inhibitor of bacterial protein synthesis, completely preventing the growth of a wide range of species at concentrations of about

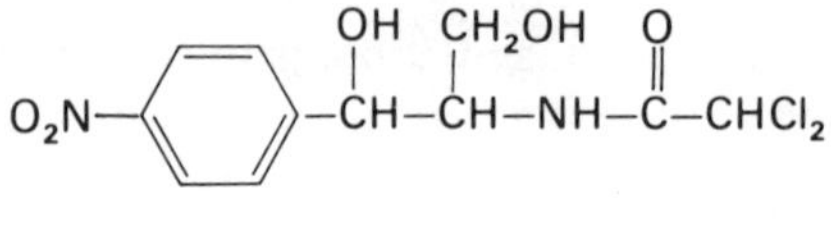

10 μg/ml. It interrupts the elongation of the protein by binding to the larger ribosomal subunit and interfering with the transfer of the growing peptide chain to the incoming amino acid.

The *tetracyclines* also inhibit the elongation of the growing peptide chain by preventing the binding of the amino acyl–*t*-RNA to the ribosomal A-site. They are effective against a wide range of bacterial species. Tetracycline R_1, $R_2 = H$; chlorotetracycline (aureomycin) $R_1 = Cl$, $R_2 = H$; oxytetracycline $R_1 = H$, $R_2 = OH$.

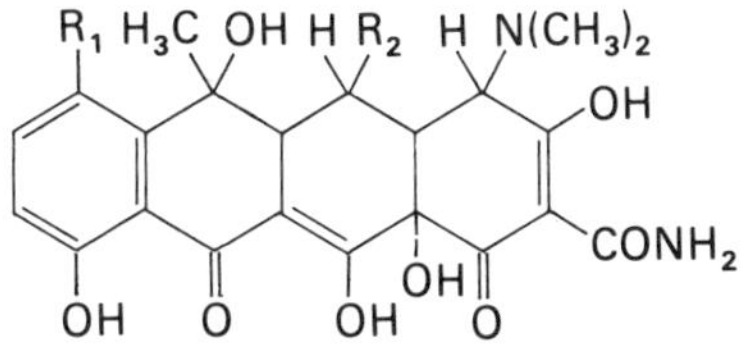

Streptomycin appears to be bound in such a way to the ribosomes of sensitive bacteria as to 'distort' the messenger RNA, so that the genetic code is 'misread'. The resulting dislocation of protein synthesis produces bacteriostasis. It also suppresses chain elongation by interfering with the binding of amino acyl–*t*-RNA to the ribosomal A-site.

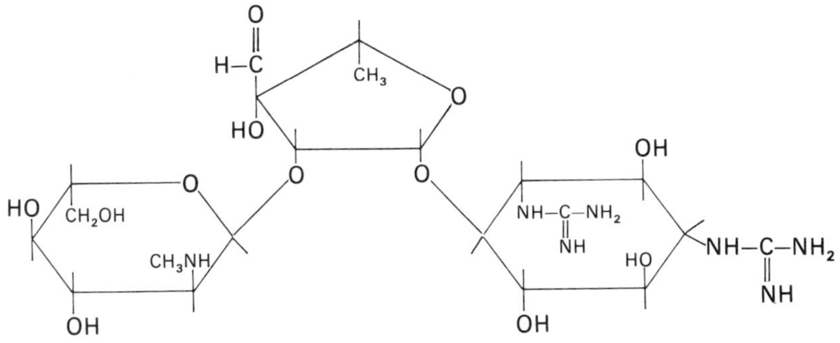

Cycloheximide interacts with the larger ribosomal subunit of higher organisms (not with those of bacteria). It inhibits the transfer of peptidyl–*t*-RNA from the A- to the P-site by suppressing the release of the deacylated *t*-RNA from the A-site (see Figure 3.11).

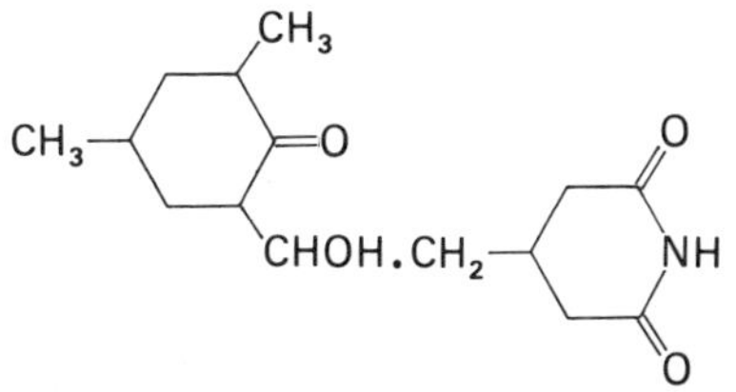

Certain *amino acid analogues* have also been used as inhibitors, e.g. ethionine (replacing methionine) and *p*-fluorophenylalanine (replacing phenylalanine and tyrosine). In some systems, however, protein synthesis continues and active enzymes containing appreciable quantities of the analogues may still be formed.

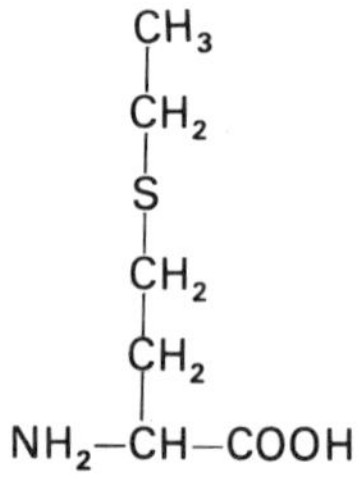

$$\begin{array}{cc}
CH_3 & \\
| & F \\
CH_2 & | \\
| & \bigcirc \\
S & | \\
| & CH_2 \\
CH_2 & | \\
| & NH_2-CH-COOH \\
CH_2 & \\
| & \\
NH_2-CH-COOH & \\
\text{ethionine} & p\text{-fluorophenylalanine}
\end{array}$$

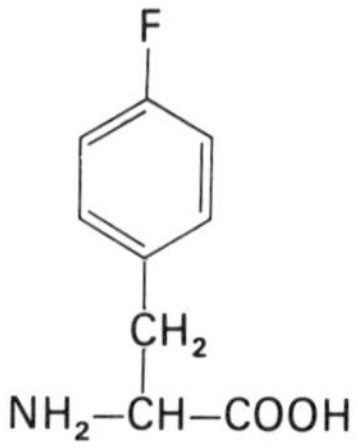

CHAPTER 15

Purine and Pyrimidine Synthesis

As the description of the structure and biosynthesis of the nucleic acids has made clear, the pyrimidine and purine bases (in the form of the appropriate nucleotides) play an extremely important part in genetic mechanisms. Moreover, the different nucleotides have characteristic roles in many other metabolic processes. Therefore, we will consider briefly the biosynthesis of these important compounds.

Biosynthesis of Pyrimidines

The pathways for pyrimidine biosynthesis are shown in Figure 3.13. The precursors are aspartate and carbamyl phosphate, the latter being derived from a reaction between glutamine, bicarbonate and ATP. Mammals have two enzyme systems capable of generating carbamyl phosphate. One, important in urea synthesis in liver, makes use of ammonium ion as the nitrogen source (see p. 220).

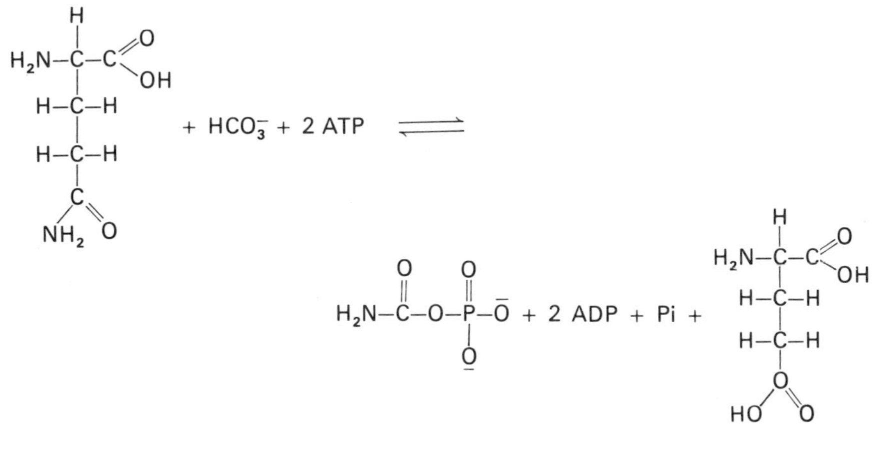

$$\text{glutamine} + HCO_3^- + 2\,ATP \rightleftharpoons H_2N-C-O-P-O^- + 2\,ADP + Pi + \text{glutamate}$$

The second, a soluble enzyme widely distributed in different tissues, utilizes glutamine directly as the nitrogen source. Tissues with a high mitotic rate (and therefore a high demand for pyrimidines) have high concentrations of this carbamyl phosphate synthetase.

The condensation of aspartate and carbamyl phosphate, the first reaction characteristic of the pyrimidine pathway, is catalysed by the enzyme aspartate

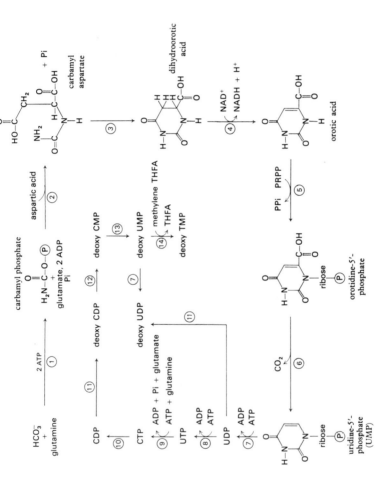

Figure 3.13 Biosynthesis of pyrimidines. The enzymes of the pathway are: 1, Carbamyl phosphate synthetase; 2, aspartate transcarbamylase; 3, dihydroorotase; 4, dihydroorotic acid dehydrogenase; 5, orotidylic acid pyrophosphorylase; 6, orotic acid decarboxylase; 7, nucleoside-monophosphate kinase; 8, nucleoside diphosphate kinase; 9, cytidine triphosphate synthetase; 10, triphosphatase; 11, ribonucleotide reductase; 12, nucleoside diphosphatase; 13, deoxycytidine monophosphate deaminase; 14, thymidylate synthetase

transcarbamylase, which in lower organisms forms part of a single enzyme complex with the carbamyl phosphate synthetase. This is probably true also in mammals, so that the carbamyl phosphate formed will be immediately available for reaction with aspartate. After the condensation, two further steps lead to the production of orotic acid. Orotic acid then reacts with phosphoribosyl pyrophosphate (PRPP), to form orotidine monophosphate. The phosphoribosyl pyrophosphate is synthesized from ribose-5-phosphate and ATP by phosphoribosyl pyrophosphate synthetase thus:

[ribose-5-phosphate structure] + ATP ⟶

ribose-5-phosphate

[phosphoribosyl pyrophosphate structure] + AMP

phosphoribosyl pyrophosphate

The conversion of orotic acid to UMP, which proceeds in two steps, also appears to be catalysed by a single protein. It is possible indeed that all the enzymes of the pathway are bound into a single multi-enzyme complex (compare fatty acid biosynthesis, p. 253).

The entire pathway is subject to feedback inhibition by UTP acting on the carbamyl phosphate synthetase (and perhaps also on the aspartate transcarbamylase which is complexed with the synthetase). In addition, it is likely that genetic control is exerted on the mammalian pathway at two points. The synthesis of aspartate transcarbamylase (and probably of carbamyl phosphate synthetase, may be inhibited (repressed) by an end-product of the reaction sequence (e.g. uridine), while the synthesis of all the other enzymes except carbamyl phosphate synthetase can be stimulated (induced) by carbamyl aspartate or dihydroorotic acid. Cellular demand for pyrimidines will act therefore initially to increase the production of carbamyl aspartate by removing the feedback inhibitor, UTP, and possibly by removing a repressor of the synthesis of the first two enzymes in the sequence. As carbamyl phosphate accumulates, the terminal enzymes in the pathway are induced. As would be expected from this, the two enzymes aspartate transcarbamylase and dihydroorotase increase in activity during active growth of mammalian tissues and probably control the overall rate of pyrimidine biosynthesis.

Once UMP is available, it can be converted to UTP by a series of reactions involving ATP as the phosphoryl donor.

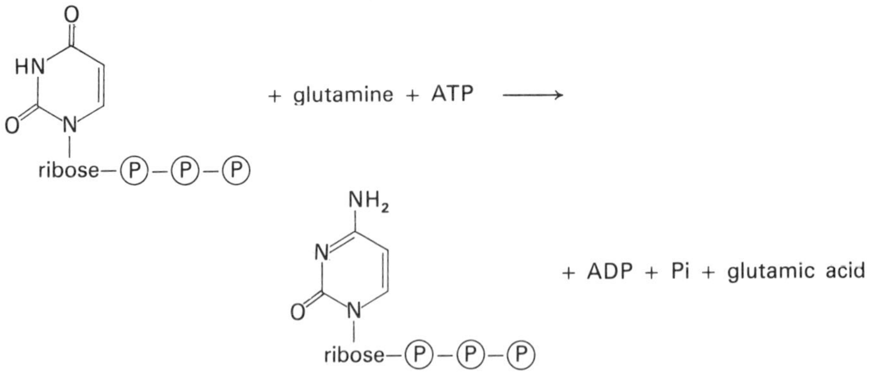

[UTP structure] + glutamine + ATP ⟶ [CTP structure] + ADP + Pi + glutamic acid

The cytidine nucleotides can be formed from uridine nucleotides by the enzyme CTP synthetase. CTP is in turn an allosteric inhibitor of aspartate transcarbamylase.

The deoxyribonucleotides deoxy CTP and deoxy TTP are formed from the cytidine and uridine nucleotides. During this transformation, the ribose residues are reduced directly to the deoxyribose form, by a reductive system which requires NADPH + H$^+$ as the ultimate source of the reducing power. The reduction also involves an —SH protein (thioredoxin) and a flavoprotein enzyme (thioredoxin reductase) thus:

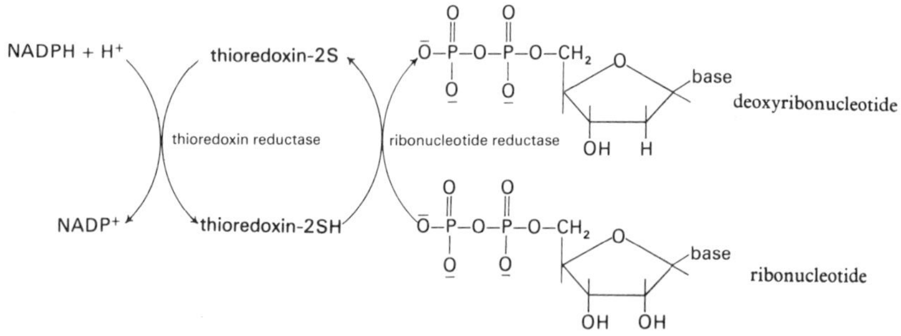

NADPH + H$^+$ ⟶ thioredoxin-2S ⟶ [ribonucleotide diphosphate] ⟶ deoxyribonucleotide
NADP$^+$ ⟵ thioredoxin-2SH ⟵ [ribonucleotide] ⟵ ribonucleotide
(thioredoxin reductase) (ribonucleotide reductase)

Hence, deoxyCDP arises directly from CDP; deoxyuridine nucleotides by deamination of deoxyCMP or reduction of UDP; and deoxythymidine nucleotides from deoxyUMP by methylation of the ring structure by the enzyme thymidine synthetase in a reaction requiring N^{5}N^{10} methylene tetrahydrofolic acid. This reaction is inhibited by the folic acid analogues aminopterin and methotrexate (amethopterin).

Synthesis of the deoxyribonucleotides is subject to feedback inhibition by products (especially deoxyATP and deoxyGTP) and to activation by the ribonucleotides (especially ATP). These interactions between potential substrates and products of the metabolic pathway are presumably designed to adjust the concentrations of the deoxyribonucleotide triphosphates to levels optimal for DNA synthesis.

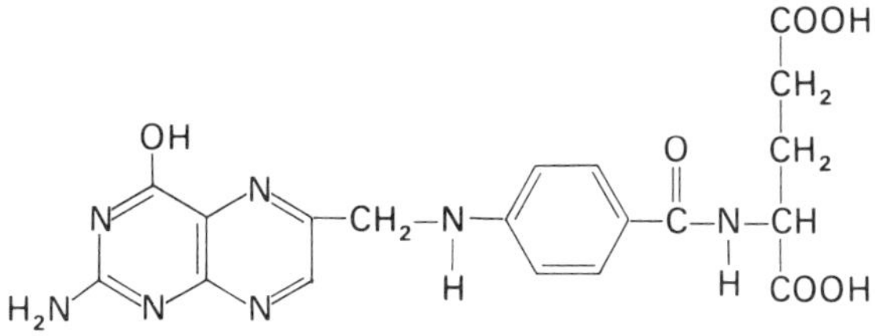

folic acid

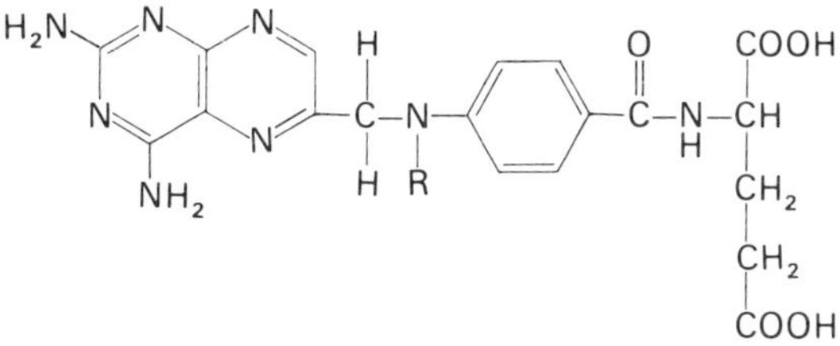

aminopterin R = H
amethopterin R = CH$_3$

Biosynthesis of Purines

The biosynthesis of the purines is outlined in Figure 3.14.

Synthesis *de novo* begins with the transfer of an amino group from glutamine to 5 phosphoribosyl pyrophosphate, to form 5 phosphoribosyl amine (PRA). Reaction of PRA with glycine yields glycinamide ribonucleotide (GAR) which is formylated with N^5,N^{10}-anhydroformyl tetrahydrofolic acid producing formyl glycinamide ribonucleotide (formyl GAR) and aminated with glutamine to give formyl glycinamidine ribonucleotide (formyl GAM). Elimination of water forms the ring compound amino-imidazole ribonucleotide (AIR) which is carboxylated (carboxy AIR). The amide of this acid (amino-imidazole carboxamide ribonucleotide, AICAR) is formed by two reactions in which the amino group from aspartate is introduced into the molecule and fumarate is freed. The ring structure is completed by the insertion of a formyl group from N^{10}-formyl THFA, and the purine nucleotide inosinic acid (inosine monophosphate, IMP) is the product.

AMP can be derived from UMP by reactions requiring aspartate as nitrogen donor and whilst GMP is formed in a two-step process involving oxidation to xanthine monophosphate (XMP) and amination by glutamine to GMP (Figure 3.15).

The corresponding di- and tri-phosphonucleotides can be produced by phosphoryl transfer catalysed by the appropriate enzymes.

As with the synthesis of the pyrimidines, that of the purines can be blocked by the folic acid antagonist, aminopterin and amethopterin (methotrexate); these compounds, which in consequence inhibit nucleic acid synthesis, have been used in the treatment of cancer. Nitrogen transfer from glutamine can also be blocked

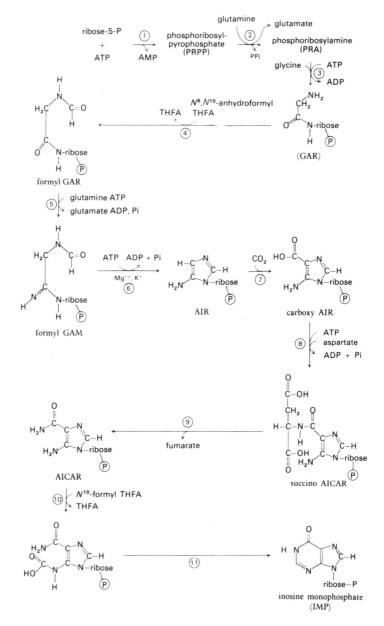

Figure 3.14 Purine biosynthesis. 1, Phosphoribose pyrophosphokinase; 2, amidophosphoribosyl transferase; 3, phosphoribosyl glycinamide synthetase; 4, phosphoribosyl glycinamide formyltransferase; 5, phosphoribosyl formylglycineamidine synthetase; 6, phosphoribosyl aminoimidazole synthetase; 7, phosphoribosyl aminoimidazole carboxylase; 8, phosphoribosyl aminoimidazole succinocarboxamide synthetase; 9, AICAR succino lyase; 10, phosphoribosyl aminoimidazolecarboxamide formyltransferase; 11, IMP cyclohydrolase

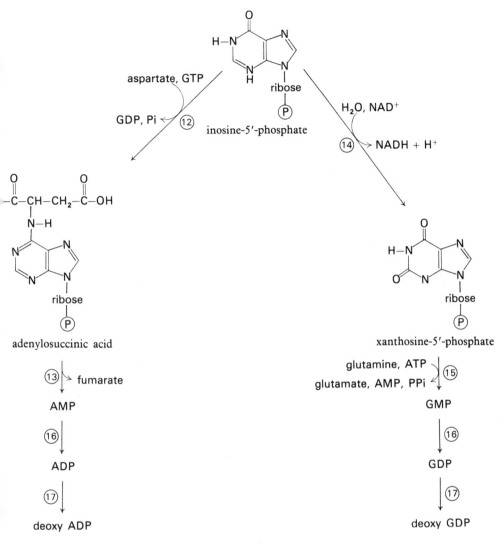

Figure 3.15 Conversion of inosine monophosphate to AMP, deoxyADP, GMP and deoxy GMP. 12, Adenylosuccinate synthetase; 13, adenylosuccinate lyase; 14, IMP dehydrogenase; 15, GMP synthetase; 16, kinase; 17, ribonucleotide reductase

specifically by azaserine and diazooxo norleucine (DON) which prevent the formation of formyl GAM from formyl GAR.

It is also possible for the free bases adenine, hypoxanthine and guanine to be combined with phosphoribosyl pyrophosphate to form the corresponding nucleotide monophosphates. These reactions are catalysed by specific transferases for adenine on the one hand and for both hypoxanthine and guanine (hypoxanthine–guanine phosphoribosyl transferase, G-PRT) on the other.

The Control of Purine Synthesis

It seems likely that an important regulation of purine biosynthesis is exerted at the first and rate-limiting step (formation of PRA), catalysed by amidophosphoribosyltransferase. This control is probably exerted by a feedback inhibition by AMP, IMP and GMP, though its effect is not thoroughly documented in animal tissues. AMP and GMP also inhibit the formation of formyl-GAR. Each nucleotide exerts a further feedback control on a specific step in its own *de novo* formation, GMP by inhibiting IMP dehydrogenase and AMP by inhibiting the formation of adenylosuccinate. Finally, AMP and GMP inhibit their respective phosphoribosyl transferases, thereby preventing synthesis from the free purine bases.

The supply of PRPP may itself be an important limiting factor for synthetic reactions. There is clinical evidence that the supply is a rate-limiting factor for purine biosynthesis in human tissues. Moreover, the synthetase reaction which forms this compound is subject to feedback inhibition by ADP and GDP as well as by the nucleoside triphosphates.

The Degradation of the Purine Bases: Gout

The breakdown of the purine nucleotides proceeds by loss of phosphate under the influence of $5'$ nucleotidases and then conversion to the free bases by the enzyme purine nucleoside phosphorylase. With this enzyme the nucleoside and inorganic phosphate react to form the purine and either deoxyribose phosphate or ribose phosphate, depending on the nature of the nucleoside. The reaction is slowest with adenosine.

Adenine and guanine are not degraded extensively but are excreted by man as uric acid or reconverted to the corresponding nucleotides as described above. The adenine ring gives rise to uric acid following deamination either of adenosine (forming inosine which is split to hypoxanthine and ribose) or of adenosine monophosphates (forming inosine monophosphate which reacts with pyrophosphate to form hypoxanthine and phosphoriboside pyrophosphate). Hypoxanthine is oxidized by xanthine oxidase firstly to xanthine (2,6-dihydroxypurine) and then by the same enzyme to uric acid. Guanine (2-amino-6-hydroxypurine) is hydrolysed by guanase, which removes the amino group and gives xanthine directly. The pathways are summarized opposite.

The enzyme xanthine oxidase is of special interest since, besides being specific for the oxidation of hypoxanthine and xanthine, it has the quite separate ability to catalyse the oxidation of a wide range of aldehydes to acids. Its presence in milk is apparently concerned with the oxidation of aldehydes. The enzyme has flavin as a prosthetic group and requires molybdenum as a cofactor; it reacts directly with oxygen, producing hydrogen peroxide.

Some individuals have a congenital defect in purine metabolism in which uric acid accumulates to a much greater extent than normal, resulting in a high level of uric acid in the blood and urine. The increase is similar to that occurring in ordinary subjects after a meal rich in purines (for example, one containing fish roes

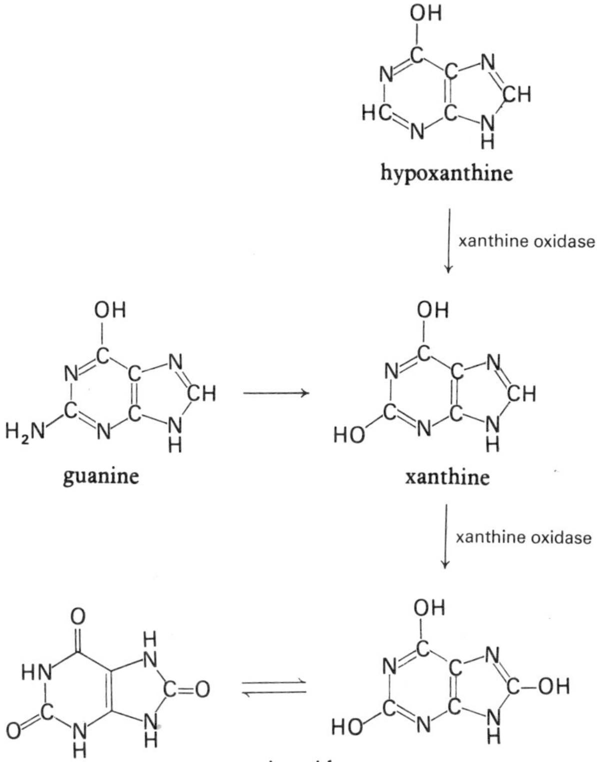

hypoxanthine

↓ xanthine oxidase

guanine → xanthine

↓ xanthine oxidase

uric acid ⇌

or sweetbreads) but in the diseased condition the level of uric acid is always high and may produce deposition of urates in the joints giving rise to gout. In some, but not all, individuals suffering from this disorder, the defect appears to be in a gene responsible for the formation of hypoxanthine–guanine phosphoribosyl transferase (i.e. it is an 'inborn error of metabolism', see p. 383). In consequence large quantities of uric acid are produced and the typical gouty arthritis develops. The defect leads to a general excessive synthesis of purines, apparently as the result of an increase in the concentrations of phosphoribosylpyrophosphate—in cultured fibroblasts from patients these levels may be up to three times normal, though the reason for this is not clear. It is also possible that in these patients GMP levels are lower than normal, thus releasing the first step in the synthetic pathway from feedback inhibition.

A recently developed and promising treatment of the condition of gout involves the use of the specific and potent inhibitor of xanthine oxidase, allopurinol. This

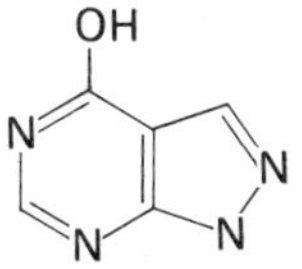

allopurinol

compound—4-hydroxypyrazole pyrimidine—can be administered safely in doses of 400–800 mg per day and greatly decreases the formation of uric acid. In addition to inhibiting xanthine oxidase, allopurinol is a substrate for phosphoribosyl transferase (G-PRT) and can be converted to the ribonucleotide form, which is an effective inhibitor of amidophosphoribosyl transferase. In consequence, it causes a general suppression of purine synthesis. Allopurinol inhibition of xanthine oxidase will increase the levels of hypoxanthine and therefore, in the presence of G-PRT, of inosine monophosphate which is a feedback inhibitor of the first step in purine synthesis. In consequence, the synthesis of purines *de novo* will be inhibited by allopurinol though the inhibition will be less effective in those patients which are deficient in G-PRT. Thus, the effectiveness of allopurinol in the treatment of gout will be variable, depending on the particular enzymic defect responsible.

References

The Biochemistry of the Nucleic Acids. Davidson, J. N. Chapman Hall, London, 7th edn., 1972.

Mechanism of Protein Synthesis and the use of Inhibitors in the Study of Protein Synthesis. Kira, A. *Progress in Molecular and Subcellular Biology,* **3**, 85–143, 1973.

Physiology of the Digestive Tract. Davenport, H. W. Year Book Medical Publishers, Chicago, 3rd edn., 1976.

SECTION 4

The Liver

CHAPTER 16

Carbohydrate Metabolism

The Glucostatic Function of the Liver

The glucostatic activity of the liver is the principle means for buffering the glucose concentration of the blood against metabolic changes. Following a meal, blood glucose rises to about 130 mg/100 ml and then falls, over a period of one to two hours, to the fasting level of between 70 and 100 mg/100 ml. Part of this fall is accounted for by the removal of glucose from circulation by the liver, which converts it first to glucose-6-phosphate using glucokinase (see p. 107) and thence to glycogen. Glucose is also converted to triglyceride by the liver and adipose tissue. Between meals the fasting level of blood glucose is maintained at the expense of the stores of glycogen in the liver. The stores of glycogen present in muscle, although totalling 120 g in a 70 kg man against only 70 g in the liver, cannot contribute directly to the glucose content of the blood because glucose-6-phosphatase is absent from muscle.

Carbohydrate Synthesis

Glycogen is synthesized from glucose originating from a number of sources. Glucose may be derived from the diet together with two other sugars, fructose and galactose, both of which can be converted into glucose. Glucose can also be synthesized from the carbon skeletons of some amino acids and from the glycerol of triglycerides. However, one of the major sources of liver glycogen and glucose is the lactate produced during the contraction of skeletal muscles and transported to the liver via the blood. Muscles, unable to use lactate in this way, synthesize their glycogen from glucose absorbed from the blood and originating in the liver. This cycle (muscle glycogen ⟶ blood lactate ⟶ liver glycogen ⟶ blood glucose ⟶ muscle glycogen) is called the Cori cycle after its discoverers.

Although the synthesis of glucose or glycogen from lactate should be possible by simply reversing the reactions of glycolysis, this is not the pathway used. Some of the component reactions have unfavourable positions of equilibrium and consequently present large energy barriers to such a simple reversal. If we consider the formation of glucose, there are three reactions that provide substantial energy barriers to the reversal of the glycolytic sequence of reactions. The formation of phosphopyruvate from pyruvate and ATP has an equilibrium constant of $2 \cdot 6 \times 10^4$ in favour of the formation of pyruvate which is equivalent to an energy barrier of 6200 cal. The formation of fructose-6-phosphate from fructose-1,6-

diphosphate also involves a large positive free energy change equal to about 4200 cal. The hexokinase reaction must be reversed as well and this requires 4200 cal since the equilibrium constant is 10^3. If glycogen, rather than glucose, is to be synthesized this last reaction is not required. It is clear from these considerations that appreciable amounts of carbohydrate cannot be synthesized by the direct reversal of glycolysis with the known concentrations of glycolytic intermediates found in the cell. There must, therefore, be some way around these energy barriers. In addition, if we look at the concentrations of ATP and ADP that are required for the three reactions just discussed, we find that reversal of both the hexokinase and phosphofructokinase reaction would require high ADP and low ATP concentrations. On the other hand, the reversal of the pyruvic kinase reaction would require low ADP and high ATP concentrations. These two opposite conditions cannot be present simultaneously in the same part of the cell. The solution to these problems is the use of different pathways for the difficult reactions and the segregation of the reactions into different cell organelles.

The conversion of lactate into phosphopyruvate is a prerequisite for its conversion to glucose. This conversion involves oxidation of the lactate to pyruvate, catalysed by lactic dehydrogenase followed by carboxylation of the pyruvate to

$$\begin{array}{c} \text{COOH} \\ | \\ \text{CHOH} \\ | \\ \text{CH}_3 \end{array} + \text{NAD} \xrightleftharpoons[\text{lactic dehydrogenase}]{} \begin{array}{c} \text{COOH} \\ | \\ \text{C}=\text{O} \\ | \\ \text{CH}_3 \end{array} + \text{NADH}_2$$

oxaloacetic acid, which is then converted to phosphoenolpyruvate. Carboxylation of pyruvate is achieved by pyruvic carboxylase. This enzyme is highly concentrated in the mitochondria of tissues capable of rapid gluconeogenesis (for example liver and kidney) but is of low activity in other tissues (muscle, brain). In animal tissue generally, the enzyme contains bound biotin

biotin ; N-carboxybiotin

and requires ATP. (Biotin is one of the B group of vitamins. It is characteristic of biotin-containing enzymes that they are inhibited by a protein from egg white called avidin, which binds biotin strongly.) The reaction sequence is

$$\text{enzyme–biotin} + \text{ATP} + \text{CO}_2 \xrightleftharpoons[]{\text{acetyl-CoA; Mg}^{2+}} \text{enzyme–biotin–CO}_2 + \text{ADP} + \text{Pi}$$

$$\text{enzyme–biotin–CO}_2 + \text{pyruvate} \rightleftharpoons \text{enzyme–biotin} + \text{oxaloacetate}$$

The enzyme has an almost complete dependence on acetyl coenzyme A as an allosteric activator (see p. 86). Thus, when the concentration of acetyl-CoA is high, the formation of oxaloacetate is stimulated. This, in turn, allows the excess acetyl-CoA to be oxidized in the citric acid cycle after combining with oxaloacetate to give citrate. Furthermore, high concentrations of acetyl-CoA will diminish the further supply of acetyl-CoA from glucose by facilitating the reconversion of pyruvate to glucose via oxaloacetate and phosphopyruvate. In this way acetyl-CoA formation from glucose will be slowed down until the excess has been removed by increased oxidation or by synthetic processes.

Oxaloacetate can be converted to phosphopyruvate in the mitochondria of liver and kidney (and in the liver cytoplasm of some animals) by phosphoenolpyruvate carboxykinase. GTP is the high-energy compound that is formed during

$$\begin{array}{c} \text{COOH} \\ | \\ \text{C=O} \\ | \\ \text{CH}_2 \\ | \\ \text{COOH} \end{array} + \text{GTP} \xrightarrow{\text{phosphoenolpyruvate carboxykinase}} \begin{array}{c} \text{COOH} \\ | \\ \text{C--OPO}_3\text{H}_2 \\ || \\ \text{CH}_2 \end{array} + \text{CO}_2 + \text{GDP}$$

the conservation of energy from succinyl-CoA in the citric acid cycle; it is also maintained in equilibrium with ATP by the enzyme nucleoside diphosphate kinase

$$\text{GDP} + \text{ATP} \rightleftharpoons \text{GTP} + \text{ADP}$$

Thus, if there is a plentiful supply of oxaloacetate and a high level of ATP arising from respiration, phosphopyruvate will be formed.

Pyruvic carboxylase and phosphoenolpyruvate carboxykinase have a capacity for phosphopyruvate synthesis that is ten to twenty times in excess of requirements and can operate at maximal activity using physiological concentrations of the substrates.

Besides permitting the formation of phosphopyruvate from oxaloacetate, the phosphoenolpyruvate carboxykinase reaction is important because it allows phosphoenolpyruvate produced during glycolysis to supply oxaloacetate when the latter is in short supply; for example, when citric acid cycle intermediates are being tapped off for synthetic purposes.

The second energy barrier to carbohydrate synthesis is encountered in the conversion of fructose-1,6-diphosphate to fructose-6-phosphate, and is bypassed by substituting an hydrolysis, catalysed by fructose-1,6-diphosphatase, for the reversal of the phosphofructokinase reaction. Fructose-1,6-diphosphatase is found mainly at the sites of gluconeogenesis, namely liver and kidney. The energy barrier between glucose-6-phosphate and glucose is also bypassed by a phosphatase, glucose-6-phosphatase, which hydrolyses glucose-6-phosphate to give glucose and phosphate.

The final point of divergence between the glycogenolytic and glyconeogenic pathways is at the stage of converting glucose-1-phosphate to glycogen. Despite the ease of reversal of the phosphorylase reaction ($K = 0.3$), the synthesis of glycogen from glucose-1-phosphate *in vivo* is by a group transfer reaction involving the high-energy compound uridine triphosphate or UTP. Besides driving the formation of glycogen, this route makes possible a more flexible control of glucogen synthesis and breakdown.

UTP combines with glucose-1-phosphate to give uridine diphosphate glucose (UDP-glucose) with the elimination of pyrophosphate, in a reaction catalysed by a specific pyrophosphorylase. The pyrophosphate is then split by pyrophosphatase, thereby pulling the freely reversible reaction in favour of UDP-glucose formation.

α-D-glucose-1-phosphate + uridine triphosphate ⇌ uridine diphosphate glucose (UDP-glucose) + HO–P(O)(OH)–O–P(O)(OH)–OH → 2 Pi + H_2O

The UDP-glucose transfers its glucose moiety to a terminal glucose residue in a glycogen molecule to lengthen the glucosyl chain. This transfer is effected by glycogen synthetase. Uridine diphosphate can be reconverted to UTP at the expense of ATP. Repetition of these reactions lengthens the chain of α-1,4 linked glucose residues (see p. 24). When this chain reaches a certain length during the synthesis of glycogen, a portion of the chain is transferred to an α-1,6 linkage in another part of the chain. In this way, 1,6 branch points are introduced into the molecule with a frequency of about one to every five or so glucose units. The enzyme catalysing this transfer is commonly called the branching enzyme. The reactions leading to glucose and glycogen synthesis from lactate are shown in Figure 4.1.

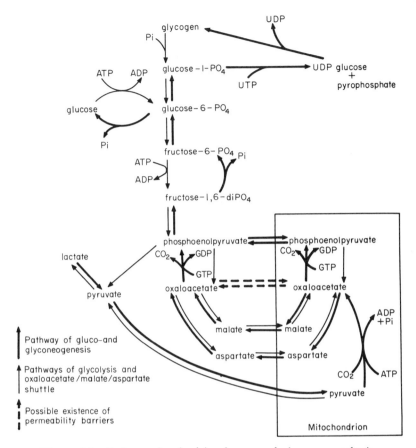

Figure 4.1 Pathways involved in glucose and glycogen synthesis

Oxaloacetate Formation from Propionate

Besides those already mentioned, there is another reaction involving CO_2 fixation that can provide a supply of C_4 dicarboxylic acids. Propionyl-CoA can react with CO_2 and ATP to give methyl malonyl-CoA which can be isomerized to form succinyl-CoA, an intermediate in the citric acid cycle. The carboxylase requires

$$ATP + CO_2 + CH_3 \cdot CH_2 \cdot COSCoA \xrightarrow[\text{biotin}]{Mg^{2+}} ADP + Pi + CH_3-\underset{\underset{COSCoA}{|}}{\overset{\overset{COOH}{|}}{CH}}$$

$$CH_3-\underset{\underset{COSCoA}{|}}{\overset{\overset{COOH}{|}}{CH}} \xrightleftharpoons[\text{vitamin B}_{12}]{} \underset{\underset{COSCoA}{|}}{\overset{\overset{COOH}{|}}{\underset{CH_2}{\overset{CH_2}{|}}}}$$

biotin, as a cofactor; this is converted into a carboxy derivative in a reaction involving ATP and CO_2. The enzyme-bound carboxybiotin can then transfer its CO_2 to propionyl-CoA. The enzyme responsible for converting methylmalonyl-CoA to succinyl-CoA is methylmalonyl-CoA isomerase which requires vitamin B_{12} as a cofactor.

Thus, if propionyl-CoA is available, liver and muscle can form succinyl-CoA by the combined action of propionyl-CoA carboxylase and methylmalonyl-CoA isomerase. The major source of propionyl-CoA in man is from fatty acids of odd chain length which account for some one to two per cent of the total fatty acids of the lipids stored.

Milk and Honey: the Metabolism of Galactose and Fructose

Besides glucose, fructose and galactose are two other monosaccharides that make an important contribution to the carbohydrate content of the diet. Fructose occurs in honey and is also a component of the disaccharide sucrose which is the

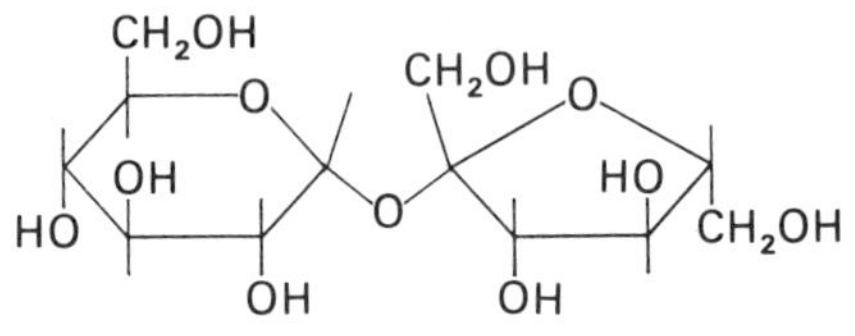

sucrose

widely used cane- or beet-sugar. Galactose is a constituent of lactose, the disaccharide present in milk. Sucrose and lactose are hydrolysed during digestion

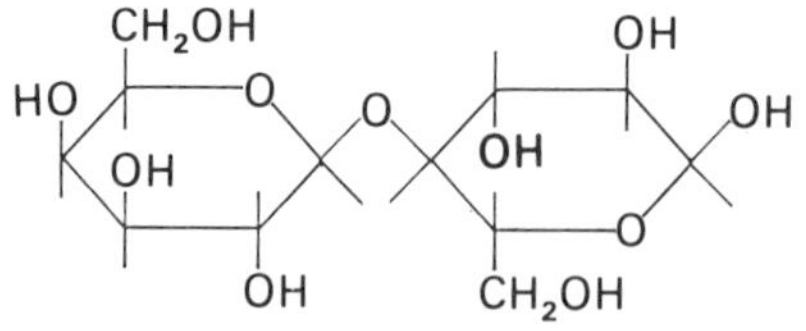

β-lactose

to give glucose and fructose, and glucose and galactose respectively. Fructose can be phosphorylated by hexokinase to fructose-6-phosphate which enters the glycolytic pathway. However, this reaction is slow and a special kinase, called fructokinase, is present in liver at very high activity which phosphorylates fructose to give fructose-1-phosphate. This enzyme is not found in the foetal liver, but develops soon after birth. Fructose-1-phosphate can be split by an aldolase to form dihydroxyacetone phosphate and glyceraldehyde. The former is an intermediate of the glycolytic pathway whilst the latter has to be phosphorylated before it can take part in glycolysis.

Fructose-1-phosphate can also be phosphorylated by 1-phosphofructokinase to give fructose-1,6-diphosphate, but again, this reaction is comparatively slow.

Galactose, like other hexoses, must be phosphorylated before it can be metabolized. Its conversion to galactose-1-phosphate is effected by galactokinase, an enzyme present in only small amounts in foetal liver but which increases in concentration after birth (see p. 375), when milk provides the diet.

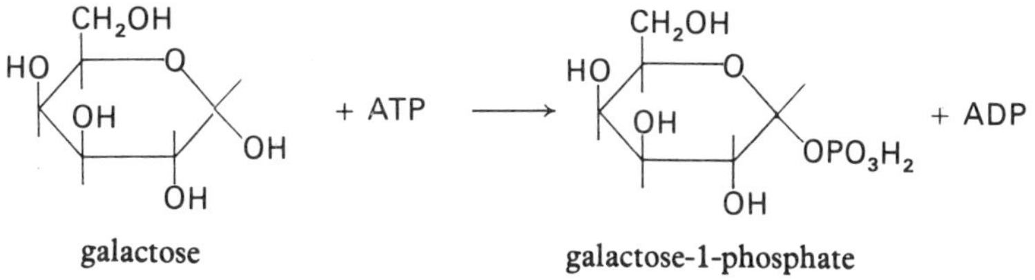

galactose + ATP → galactose-1-phosphate + ADP

Galactose-1-phosphate can react with UDP-glucose to give UDP-galactose and glucose-1-phosphate; the enzyme catalysing this reaction is galactose-1-phosphate uridyl transferase. UDP-galactose is epimerized to form UDP-glucose by an enzyme called UDP-galactose-4-epimerase which alters the configuration at carbon 4 in the galactose residue. The UDP-glucose formed in this way reacts with more galactose-1-phosphate and thereby initiates its conversion to glucose-1-phosphate so that it gains access to the glycolytic and other pathways.

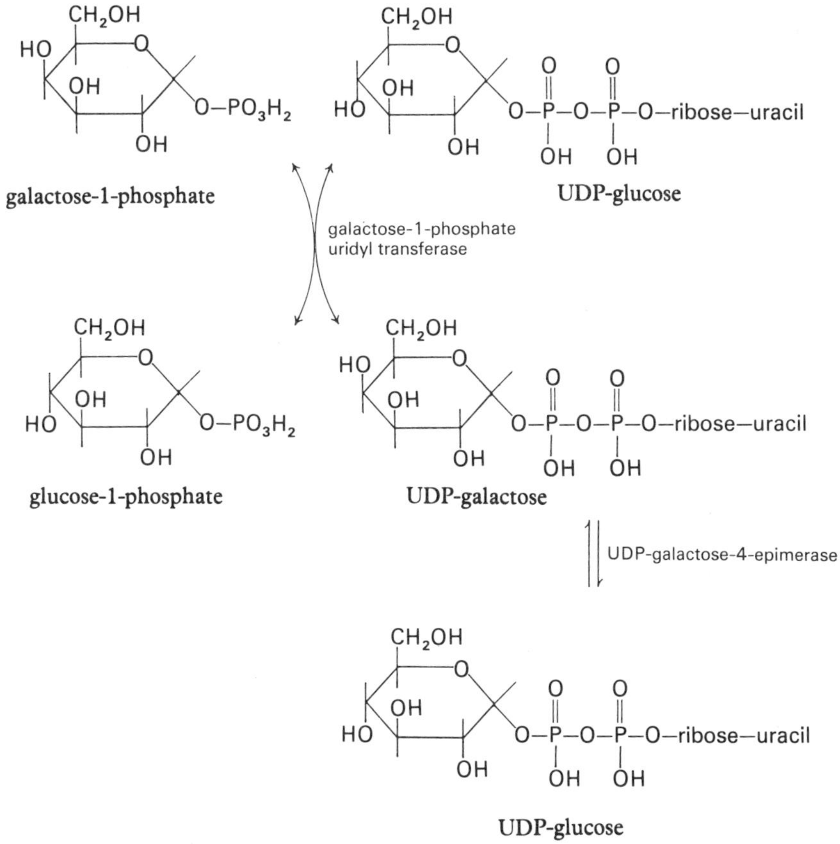

Diseases of Carbohydrate Metabolism

Fructosuria

This is a rare and harmless disorder of metabolism in which the ability to use fructose is impaired with the result that it is excreted in the urine. Feeding 50 g of fructose to a fructosuric patient produces a blood fructose level of 50 to 60 mg/100 ml, whilst in a normal individual the same dose gives rise to a blood fructose level of only 10 to 20 mg/100 ml. At present, the evidence suggests that the condition is caused by a deficiency of liver fructokinase.

Galactosaemia

This is an hereditary disease, transmitted by a recessive gene; the homozygote is unable to convert galactose to glucose. The symptoms of the patient are entirely due to the administration of galactose and if this is removed from the diet the symptoms disappear. As might be expected, the symptoms of galactosaemia appear soon after birth and are caused by the galactose component of lactose. The infants appear normal at birth, but rapidly become ill with vomiting and diarrhoea which lead to dehydration, and unless milk is removed from the diet the infant will die. However, if the diagnosis is made before the disease is too far advanced and galactose is removed from the diet, recovery may be complete. Any delay in treatment may be accompanied by irreversible mental deterioration of the child. Galactosaemia is caused by the congenital absence of galactose-1-phosphate uridyl transferase and may be diagnosed by estimating the transferase activity in the red cells of suspected galactosaemics. In the absence of this enzyme, galactose-1-phosphate accumulates in tissues. Why the accumulation of galactose-1-phosphate should cause such serious effects is not understood, although there are indications that it may inhibit glucose-6-phosphate dehydrogenase activity and thus impair the supply of NADPH + H^+ and pentose sugars (see p. 208).

Another enzyme metabolizing galactose-1-phosphate is UDP-galactose pyrophosphorylase which catalyses the reaction:

$$\text{UTP + galactose-1-phosphate} \rightleftharpoons \text{UDP-galactose + pyrophosphate}$$

This enzyme is absent in the newborn child but increases in activity in the adult; this increase in activity may explain the somewhat greater tolerance towards galactose shown by adult galactosaemics. Even in the adult the activity of the pyrophosphorylase is only about one-sixth of the normal activity of the transferase, so that adult galactosaemics must still avoid galactose in their diets.

Glycogen Storage Diseases

There are six, clearly recognized, hereditary diseases that result in the excessive deposition of glycogen in the liver or muscle which can be ascribed to the absence of a particular enzyme. In order to understand these diseases we must recall the way in which branches are introduced into glycogen molecules during synthesis (see p. 202) and removed during glycogenolysis (see p. 105). The branched structure of glycogen has been shown in Figure 2.11.

Type I: von Gierke's Disease

In this disease the abdomen is distended by the liver which is greatly enlarged owing to its excessively large content of glycogen. Patients suffering from the disease have abnormally low fasting levels of blood glucose (between 0 and 15 mg/100 ml against the normal value of 70 to 100 mg/100 ml). There is also a greatly elevated level of plasma lipids which may account for 10 per cent of the plasma volume. Biopsies show that the liver contains about 10 per cent of its wet weight as glycogen and that the structure of the polysaccharide is normal. The biochemical lesion responsible for the disease is the absence of glucose-6-phosphatase. Thus, whilst glycogen can be synthesized via the UDP-glucose pathway and broken down by phosphorylase, it cannot provide sufficient glucose to maintain the concentration of blood glucose at its normal value. The low level of blood glucose that is found is probably maintained by the hydrolytic action of the debranching enzyme and tissue amylases on the glycogen stores.

It is interesting to note that glycogen is not formed in the liver until shortly before birth when the branching enzyme makes its appearance. In the normal individual glucose-6-phosphatase activity develops at birth and the stored glycogen is used rapidly.

Type II: Pompe's Disease

This disease is characterized by extreme muscular weakness, a greatly enlarged heart, and a high concentration of glycogen in the heart, muscles and liver. It is believed to be caused by a congenital absence of lysosomal acid maltase which normally removes glucose residues from the external branches of glycogen. Thus it would appear that glycogen breakdown by the acid maltase of the lysosomes (see p. 49), as well as by phosphorylase, is of physiological importance.

Type III: Forbe's Disease

In contrast to von Gierke's disease, this condition is characterized by an excessive deposition of glycogen in the heart and muscles as well as in the liver. The disease is caused by the absence of the debranching enzyme. When glucose is available, the synthesis of glycogen proceeds normally and a branched molecule is formed. However, when the glycogen is broken down (by the combined action of phosphorylase and the transglucosylase) glucose residues can only be removed until the next branching point is reached because the debranching enzyme is absent. As glucose becomes available again chain extension continues, and, once another α-1,6 linkage is introduced, all the glucose residues between the two branching points are trapped and the glycogen molecule has irreversibly grown in size.

Type IV: Anderson's Disease

This is a rare disease believed to be caused by the absence of the branching enzyme.

Type V: McArdle's Disease

This disorder is characterized by the inability of skeletal muscles to carry out normal work, for example, under ischaemic conditions only 10 to 20 per cent of the normal amount of work can be performed by the forearm and hand muscles. Also the concentration of lactate and pyruvate falls in the venous blood coming from the muscle during exercise, whereas under normal conditions the concentrations of these two acids rise. The glycogen content of the muscles is greatly increased above the normal values and homogenates of muscle prepared from biopsy specimens are unable to form lactate from the glycogen stores, although they can do so from added glucose-1-phosphate. These observations are in complete agreement with the finding that phosphorylase is absent from the muscles. Since glycogen deposition is still proceeding, this disease provides evidence to support the view that phosphorylase is not involved in the biosynthesis of glycogen. Furthermore the pathway leading to glycogen synthesis via UDP-glucose is operative in patients suffering from McArdle's disease. Liver phosphorylase, in contrast to muscle phosphorylase, is unaffected and therefore the liberation of glucose into the blood from liver glycogen stores can still be elicited by the injection of adrenaline.

Type VI: Her's Disease

In this disease liver phosphorylase is missing whilst muscle phosphorylase activity is unaffected.

Congenital Absence of Glycogen Synthetase

This is a very rare condition and is characterized by an inability to maintain the level of blood glucose under fasting conditions owing to the lack of stored carbohydrates; the hypoglycaemia can be prevented by regular feeding.

The Pentose Phosphate Pathway

Besides the glycolytic pathway there is another important pathway that provides an alternative metabolic route for the degradation of glucose-6-phosphate. This is the pentose phosphate pathway, by whose overall activity three molecules of glucose-6-phosphate and six molecules of $NADP^+$ can be transformed into three molecules of CO_2, two of fructose-6-phosphate, one of glyceraldehyde-3-phosphate and six of NADPH + H^+. Thus, this method of dealing with glucose provides a supply of NADPH + H^+ which is necessary for, amongst other things, the synthesis of fatty acids. The successive transformations of glucose-6-phosphate involve several phosphorylated five-carbon sugars. Of these ribose-5-phosphate is of particular importance as it provides the ribose moiety of the ribonucleic acids.

The pentose phosphate pathway can most simply be considered in two parts, an oxidative part and a non-oxidative part. The oxidative part (Figure 4.2) consists of two dehydrogenations and a decarboxylation. Glucose-6-phosphate is dehydrogenated to give 6-phosphogluconic acid by glucose-6-phosphate dehydrogenase, with $NADP^+$ taking part as the hydrogen acceptor. The 6-phosphogluconate is then oxidatively decarboxylated in a reaction, again involving $NADP^+$ as hydrogen acceptor, to give D-ribulose-5-phosphate and CO_2. In the non-oxidative part of the pathway, some of the ribulose-5-phosphate is epimerized to form xylulose-5-phosphate and some is enolized to give ribose-5-phosphate. These last two sugars then react together in the presence of transketolase, which removes the ketol group

$$\begin{array}{l} CH_2OH \\ | \\ C=O \\ | \end{array}$$

from xylulose-5-phosphate and attaches it to the aldehyde group of ribose-5-phosphate, to form sedoheptulose-7-phosphate and glyceraldehyde-3-phosphate. Transketolase requires thiamine pyrophosphate, the pyrophosphate derivative of vitamin B_1, as the coenzyme that binds the ketol group. Transaldolase then catalyses the removal of the aldol group

$$\begin{array}{l} CH_2OH \\ | \\ C=O \\ | \\ CHOH \\ | \end{array}$$

from sedoheptulose-7-phosphate to glyceraldehyde-3-phosphate to make erythrose-4-phosphate and fructose-6-phosphate. The final reaction in the pathway is catalysed by transketolase and involves the formation of fructose-6-phosphate and glyceraldehyde-3-phosphate from erythrose-4-phosphate and xylulose-5-phosphate. The reactions of the pentose phosphate pathway are shown in Figure 4.2. The final products of the pathway, fructose-6-phosphate and glyceraldehyde-3-phosphate, can both enter the glycolytic pathway. NADPH + H^+ formed in the cytoplasm during the two dehydrogenations cannot be oxidized via the mitochondrial electron transport chain to any significant extent. $NADP^+$ is regenerated in the cytoplasm by reactions that use NADPH + H^+ as a reducing agent. Examples of such reactions are the reductions catalysed by fatty acid synthetase (see p. 255) and the reduction of oxidized glutathione (see pp. 280 and 354).

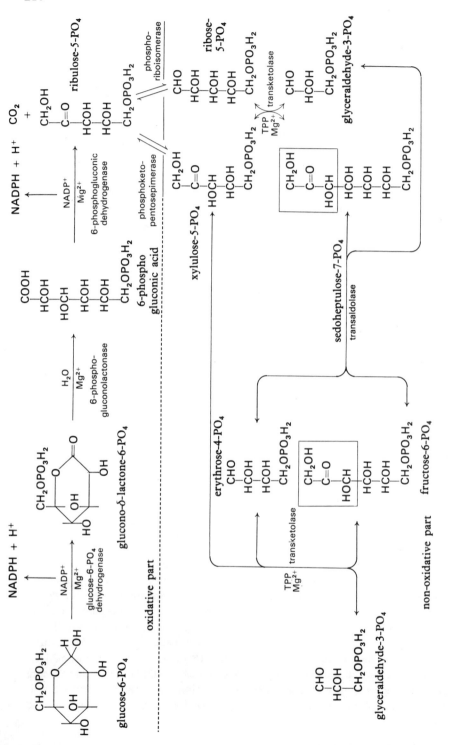

Figure 4.2 Pentose phosphate pathway

The carbon dioxide that arises from glucose during the oxidative decarboxylation step in the pentose phosphate pathway originates from carbon atom 1 of the molecule, whilst by the combined actions of glycolysis and the citric acid cycle both carbons 1 and 6 can be converted into CO_2 with equal probability. Thus, by allowing a tissue to metabolize glucose labelled in the carbon 1 position with ^{14}C and comparing the specific activity of the CO_2 evolved with that obtained when the carbon 6 position is labelled, the relative amounts of glucose being metabolized by the pentose phosphate and glycolytic pathways can be calculated. On the basis of such calculations it would appear that in liver about one molecule of glucose in sixteen passes through the pentose phosphate pathway. However, these calculations can only be regarded as approximations since progressive randomization of the carbon atoms, for example in the citric acid cycle, makes interpretation of the results very difficult.

Glucuronic Acid Formation

By contrast with the oxidation of the aldehyde group of glucose which yields gluconic acid (found in a phosphorylated form in the pentose phosphate pathway), oxidation of the primary alcoholic group at the C_6 position of D-glucose to a carboxyl group yields D-glucuronic acid. D-glucuronic acid is a com-

D-glucuronic acid

ponent of many mucopolysaccharides, for example, hyaluronic acid, chondroitin sulphate and heparin; it also forms conjugates with some steroid hormones and bilirubin prior to their excretion. The formation of glucuronic acid is essentially a divergence from the pathway of glycogen synthesis. UDP-glucose can be oxidized by a NAD^+-linked dehydrogenase to yield UDP-glucuronic acid. This compound, like the parent UDP-glucose, has a high free energy of hydrolysis and is able to take part in those reactions leading to polymer formation and detoxication processes. The overall pathway from glucose to glucuronic acid is:

glucose ⟶ glucose-6-phosphate ⟶ glucose-1-phosphate ⟶

UDP-glucose ⟶ UDP-glucuronic acid ⟶ glucuronic acid

The involvement of UDP-derivatives in the biosynthesis of polysaccharides illustrates an important aspect of biochemical energetics, namely that the formation of any polymer from its constituent monomers requires an input of energy. In biological systems this requirement is fulfilled by converting the monomers to derivatives that are able to condense together, in reactions having negative free-energy changes, to yield the polymer. In order to form such reactive monomers a nucleoside triphosphate is used to provide energy that is conserved in the activated monomers. The conserved energy is liberated when the activated monomers link together.

UDP-glucuronic acid can also be hydrolysed to give free glucuronic acid which, in many animals, can be converted into ascorbic acid (vitamin C). In man, the higher apes, guinea pigs and the fruit bat an enzyme involved in this conversion is missing so that these animals have what amounts to a genetic defect affecting the whole of their species and manifesting itself by a requirement for vitamin C.

In man, glucuronic acid can be converted, by a number of reactions including a decarboxylation, into L-xylulose. Most people are able to transform this sugar into D-xylulose. This transformation involves the reduction of L-xylulose by NADPH + H$^+$ to give the pentahydric alcohol xylitol which is then oxidized by NAD$^+$ to form D-xylulose:

$$
\begin{array}{ccccc}
\text{CH}_2\text{OH} & & \text{CH}_2\text{OH} & & \text{CH}_2\text{OH} \\
| & & | & & | \\
\text{C=O} & \text{NADP} & \text{HOCH} & \text{NAD} & \text{C=O} \\
| & \rightleftarrows & | & \rightleftarrows & | \\
\text{HCOH} & \text{NADPH}_2 & \text{HCOH} & \text{NADH}_2 & \text{HOCH} \\
| & & | & & | \\
\text{HOCH} & & \text{HOCH} & & \text{HCOH} \\
| & & | & & | \\
\text{CH}_2\text{OH} & & \text{CH}_2\text{OH} & & \text{CH}_2\text{OH} \\
\text{L-xylulose} & & \text{xylitol} & & \text{D-xylulose}
\end{array}
$$

Individuals suffering from the harmless hereditary condition known as essential pentosuria excrete large amounts of L-xylulose in their urine. These patients are unable to convert L-xylulose to xylitol with the result that the former passes into the urine. Pentoses can sometimes be detected in the urine of normal individuals, usually following overindulgence in plums or cherries; however, in such instances the chief sugar concerned is arabinose rather than L-xylulose.

CHAPTER 17

Nitrogen Metabolism

The Supply of Amino Acids

The chief supply of nitrogen in men comes in the form of protein and is supplied to the liver as the hydrolysed products of proteins, i.e. as amino acids (Figure 1.14). In the growing animal, these amino acids are used particularly for the synthesis of the proteins required for the increase in body size; the mature animal can afford to oxidize a large proportion of the amino compounds. In addition to those available from dietary sources, many amino acids are readily synthesized from the products of carbohydrate metabolism and ammonia. Thus aspartic acid, glutamic acid and alanine can be formed by amination of oxaloacetic, α-oxoglutaric and pyruvic acids respectively.

The Essential Amino Acids

Some amino acids cannot be synthesized at a rate fast enough to meet the needs of the body. Dietary requirements for these special amino acids were first defined for rats which grew normally only if ten particular amino acids were available in their food. These were, firstly, the branched chain amino acids, valine, leucine and isoleucine. The second group, more varied in their chemical structure, are methionine, phenylalanine, histidine, tryptophan, lysine, arginine and threonine. Arginine is included in this list because although rats (like humans) can live and grow without arginine, they develop faster if it is present in the diet.

Essential amino acids are necessary for the synthesis of most proteins and some important non-protein substances. Thus, phenylalanine (or tyrosine) is required for the production of the hormones adrenaline and thyroxine. The hormone insulin has 57 per cent of its constituent amino acids as essential amino acids and the pituitary hormone vasopressin also contains a substantial amount of essential amino acids. These (and other) hormones are such potent substances that they must be removed as soon as they have done their job; consequently, there is a continuing need for their synthesis, and if amino acids are not available from the diet they must be supplied by the breakdown of other body proteins. During starvation it becomes obvious that some tissues and proteins are less essential than others. For example, the liver can lose 50 per cent and skeletal muscle 30 per cent of its weight, while the heart loses only 3 per cent. Thus liver and muscle proteins can act as a store of amino acids.

When amino acids are finally degraded their amino groups are converted to urea (see p. 220), which is excreted in the urine. Therefore by measuring the amount of urea in the urine and the amount of nitrogen in the food, a nitrogen balance sheet can be drawn up. If the amount of nitrogen in the diet equals the amount of nitrogen excreted, the individual is said to be in nitrogen balance. The actual amount of protein required in the diet for nitrogen balance varies with the different types of protein ingested, as some proteins can provide larger amounts of essential amino acids than others and it is the intake of the essential amino acids that is critical. For man, about 30 g of first class protein are required per day to maintain nitrogen balance. Positive nitrogen balance, i.e. an excess of nitrogen intake over loss, is found during growth or when tissues are being repaired, for example during pregnancy or after burns. A negative nitrogen balance, when the nitrogen loss exceeds the intake, comes about during starvation and malnutrition. After surgical operation, there is always a period of net nitrogen loss when the liver and muscle protein is depleted.

Transamination and the Breakdown of the Carbon Skeleton of the Amino Acids

Amino acids may be oxidized to yield energy only after deamination (i.e. when the α-amino group has been removed). Experiments with rats fed with amino acids labelled with ^{15}N have shown that there is a very rapid exchange of the amino group amongst almost all the amino acids. This shuttling of the amino group back and forth between different carbon skeletons is called 'transamination' and is the most important way for initiating the degradation of amino acids. These transamination reactions enable the amino nitrogen of all the amino acids except lysine and threonine (which do not participate in transamination) to be transferred to oxaloacetate and α-oxoglutarate to form aspartate and glutamate; thereafter the amino group can be removed by either a deamination reaction (for example that catalysed by glutamic dehydrogenase) or by direct transfer from aspartic acid into the urea cycle (see p. 221).

The transaminases require a derivative of vitamin B_6 as a coenzyme. Vitamin B_6 as originally defined, is pyridoxine, but the coenzymes are the related compounds pyridoxal phosphate and pyridoxamine phosphate:

pyridoxine

pyridoxal

pyridoxamine

pyridoxal phosphate

pyridoxamine phosphate

In the transamination reactions, the α-amino groups of the amino acids are attached to the aldehyde group of pyridoxal phosphate to form a Schiff's base which breaks down to give a keto acid and pyridoxamine phosphate. The amine of pyridoxamine phosphate can then be transferred to another keto acid (Figure 4.3; see also Figure 1.19).

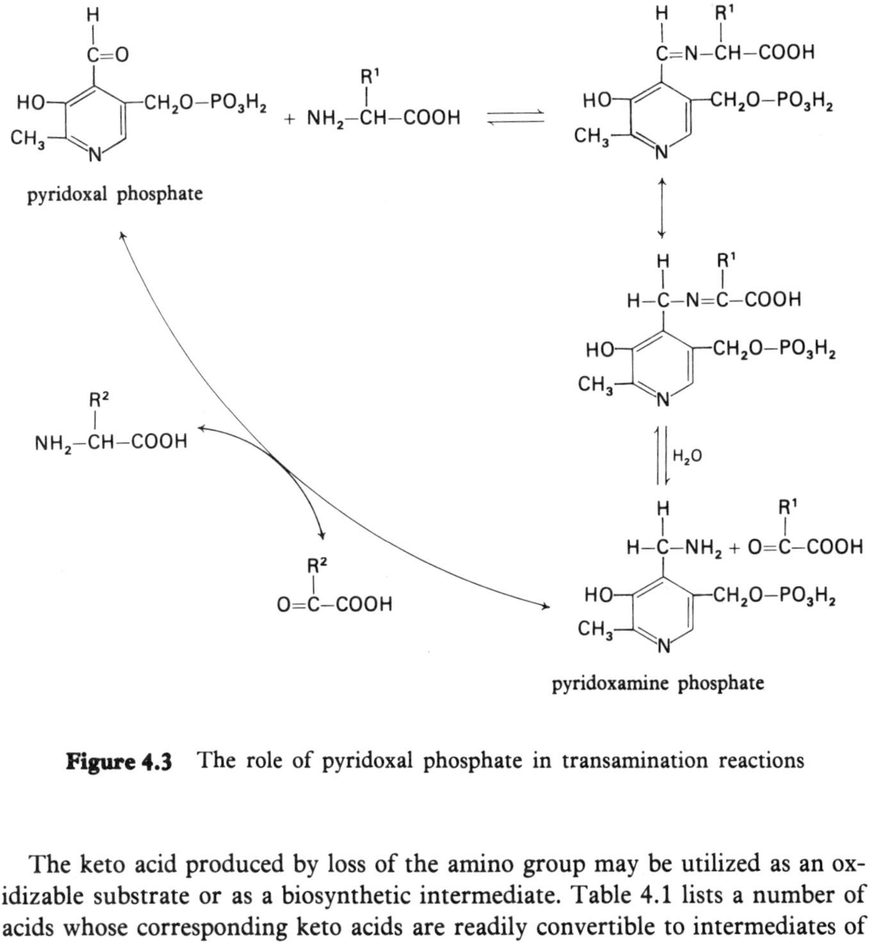

Figure 4.3 The role of pyridoxal phosphate in transamination reactions

The keto acid produced by loss of the amino group may be utilized as an oxidizable substrate or as a biosynthetic intermediate. Table 4.1 lists a number of acids whose corresponding keto acids are readily convertible to intermediates of oxidative pathways already considered.

The amino acids are traditionally divided into the glucogenic and the ketogenic (see Table 4.1). The glucogenic acids yield carbon skeletons which can be used to synthesize glucose. The most obvious examples are alanine, aspartic acid and glutamic acid; the other amino acids are not so clearly linked with precursors of glucose and their metabolism is often complex.

Transaminases present in mitochondria can also be important in determining the rate of glutamate oxidation. In many tissues glutamic dehydrogenase is not

Table 4.1 Products of breakdown of amino acids

Glucogenic amino acids		Ketogenic amino acids	
acid	product	acid	product
Glycine	Pyruvate	Leucine	Acetoacetate
	Succinate		
		Isoleucine	Acetyl-CoA
Serine	Pyruvate		
			Propionyl-CoA
Threonine	Propionate		
		Phenylalanine	Fumarate
Valine	Propionate		
		Tyrosine	Acetoacetate
Histidine	α-Oxoglutarate		
Arginine	α-Oxoglutarate		
Lysine	α-Oxoglutarate		
Isoleucine	Acetyl-CoA		
	Propionyl-CoA		
Cystine	Pyruvate		
Proline	α-Oxoglutarate		
Hydroxyproline	Pyruvate		

very active but glutamate can still be oxidized at a high rate. This is accounted for by the following series of reactions. A little of the glutamate is oxidized to α-oxoglutarate by glutamic dehydrogenase:

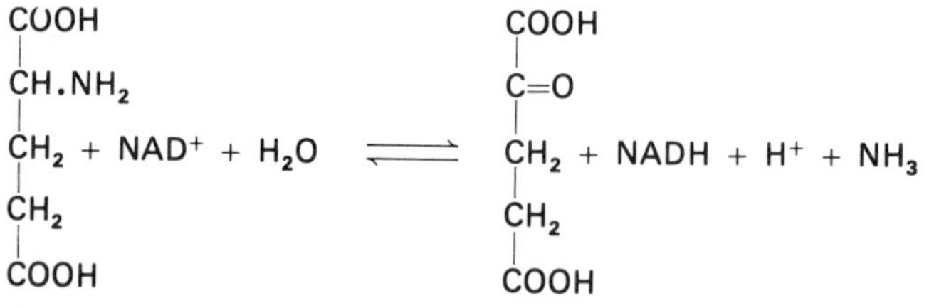

$$\begin{array}{c} COOH \\ | \\ CH.NH_2 \\ | \\ CH_2 \\ | \\ CH_2 \\ | \\ COOH \end{array} + NAD^+ + H_2O \rightleftharpoons \begin{array}{c} COOH \\ | \\ C=O \\ | \\ CH_2 \\ | \\ CH_2 \\ | \\ COOH \end{array} + NADH + H^+ + NH_3$$

glutamic acid → α-oxoglutaric acid

The α-oxoglutarate is then oxidized rapidly in the citric acid cycle to oxaloacetate which transaminates with more glutamate to yield aspartate and α-oxoglutarate:

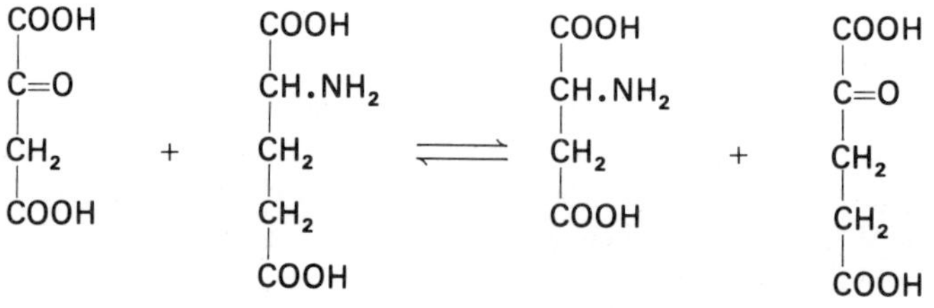

$$\begin{array}{c} COOH \\ | \\ C=O \\ | \\ CH_2 \\ | \\ COOH \end{array} + \begin{array}{c} COOH \\ | \\ CH.NH_2 \\ | \\ CH_2 \\ | \\ CH_2 \\ | \\ COOH \end{array} \rightleftharpoons \begin{array}{c} COOH \\ | \\ CH.NH_2 \\ | \\ CH_2 \\ | \\ COOH \end{array} + \begin{array}{c} COOH \\ | \\ C=O \\ | \\ CH_2 \\ | \\ CH_2 \\ | \\ COOH \end{array}$$

oxaloacetic acid glutamic acid aspartic acid α-oxoglutaric acid

The latter, after oxidation to oxaloacetate, can again accept amino groups from glutamate. This mechanism achieves simultaneously a rapid oxidation of glutamate and the preservation of amino groups: it is of special importance in tissues like mammary gland and liver where amino groups are required for the synthesis of milk and plasma proteins.

It has been shown that excess dietary amino acids (i.e. those not needed for growth and replacement) are degraded in preference to carbohydrates and fat; the two latter fuels can be stored in the body (as glycogen or triglyceride) but the amino acids leaving the intestine accumulate in the blood and tissues of the body. Nitrogen excretion rises, and the carbon skeletons of the amino acids are either used as a fuel or converted to carbohydrate and fat.

This preferential increase in the degradation of amino acids is largely a consequence of the fact that the affinities of the transaminases and other degrading enzymes are relatively low (K_m values are in the range 1–50 mM), so that there is an almost proportional increase in catabolic reactions as the concentration of the amino acids rises. Moreover, the amounts of the enzymes which initiate attack on most of the common amino acids increase in response to the nutritional state (see p. 446). For the enzymes degrading the essential amino acids, these changes are particularly large, so that when the supply of these amino acids is low the key degradative enzymes virtually disappear—an obvious protective device.

CHAPTER 18

Nitrogen Excretion

Ammonia Metabolism

Ammonia, carbon dioxide and oxygen are the three gaseous substrates of the body. Virtually the sole function of oxygen is to combine with the hydrogens removed from substrate during catabolic reactions to yield water that may leave the body by the lungs, skin or kidneys. The other two gases are concerned in both catabolic and anabolic reactions according to the needs of the body. Carbon dioxide leaves the body by the lungs and the kidneys, whereas ammonia leaves only by the kidneys. Both carbon dioxide and ammonia are closely monitored since excess of either may have toxic effects on the whole body. The role of carbon dioxide in maintaining the acid–base balance of the body is referred to on p. 316. Whereas carbon dioxide (excreted by the lungs) and bicarbonate (excreted by the kidney) are freely interconvertible, most of the ammonia excreted by the kidney has already been transformed irreversibly into a different chemical compound (urea) by the liver.

The concentration of ammonia in the tissues ranges from $0 \cdot 2$ mM in the heart muscle to $0 \cdot 9$ mM in abdominal muscle and the kidney, with brain and thigh muscle ($0 \cdot 3$ mM) and liver ($0 \cdot 7$ mM) occupying intermediate positions. The concentration of ammonia in the blood, apart from the portal blood (see later) is only about $0 \cdot 05$ mM and ammonia as such is not the main form in which the excess ammonia of the tissues is transported to the liver. Within the cells of tissues other than the liver the ammonia can be removed by two reactions:

1. α-oxoglutarate + NADH + H$^+$ + NH$_3$ $\xrightarrow{\text{glutamate dehydrogenase}}$ glutamate + NAD$^+$ + H$_2$O

2. glutamate + NH$_3$ + ATP $\xrightarrow{\text{glutamine synthetase}}$ glutamine + ADP + Pi

The glutamate in reaction 1 is able through transamination reactions to aminate any suitable ketoacid to yield a different amino acid

glutamate + pyruvate $\xrightleftharpoons{\text{transaminase}}$ alanine + α-oxoglutarate

Substances like pyruvate which are present in relatively plentiful supply can, therefore, be used as a depository for excess amino groups. Ammonia is transported therefore as amino acids having a total concentration in the blood plasma of about 3–4 mM. Glutamine (0·4 mM) glutamic acid (0·23 mM) and alanine (0·4 mM) are the most plentiful and of these glutamine and alanine penetrate into the liver most easily.

Ammonia is produced in muscle and in the brain because of the relatively high activity of enzyme adenylate deaminase

$$AMP + H_2O \longrightarrow IMP + NH_3$$

In the case of muscle, much of the ammonia may enter the blood stream as such and find its way to the liver. In the case of the brain the high activity of glutamine synthetase keeps the standing concentration of the free ammonia low and glutamine enters the blood to transport the ammonia to the liver. The IMP left within the tissues may be reaminated by the following reactions:

$$aspartate + GTP + IMP \longrightarrow adenylsuccinate + GDP + P$$

$$adenylsuccinate \longrightarrow AMP + fumarate$$

The liver receives ammonia by two main routes, from the portal blood as free ammonia and amino acids and from the systemic blood as amino acids (mainly glutamine and alanine). The amount of free ammonia (formed by bacterial action in the gut) and of amino acids in the portal blood depends on the diet. In the normal man the portal blood contains about 0·18 mM ammonia which at the pH of the blood will be 98 per cent in the ammonium form. The venous outflow from the liver contains 0·06 mM ammonia and it can be calculated from the arteriovenous difference that about one-fifth of the daily output of the urea comes from ammonia produced in the gut. Within the liver ammonia can arise from the action of glutaminase

$$glutamine + H_2O \longrightarrow glutamate + ammonia$$

or from AMP by the action of adenylate deaminase. These two reactions are irreversible. Ammonia can also arise from glutamate catalysed by glutamic dehydrogenase, but this reaction is reversible and is in equilibrium with most other amino acids because of transamination reactions. Thus in times of ammonia overloading it will be removed from the system first forming glutamate and thence by transamination forming alanine from the pyruvate and lactate that are always present. The combination of glutamate dehydrogenase and the transaminase system thus results in an overall buffering of the ammonia concentration within the liver. The standing concentration of ammonia in the liver is about 0·7 mM which is equal to the K_m for ammonia of carbamyl phosphate synthetase. Thus this enzyme which starts the production of urea will also contribute towards buffering the ammonia concentration.

In the formation of urea the two nitrogens arise from two different sources, ammonia and aspartate. The amounts of these two nitrogen donors must be kept in balance. With excess of free ammonia glutamate will be formed which will transaminate with available oxaloacetate to form the necessary aspartate. The α-

oxoglutarate liberated from the transamination may either be reaminated or oxidized to provide oxaloacetate for the production of more aspartate. Thus the production or removal of ammonia is automatically adjusted to the metabolic needs of the cell.

The Urea Cycle

We have already noted that most of the ammonia excreted by the kidney is in the form of urea, synthesized primarily in the liver. The daily elimination of urea in the urine is about 25 g, nearly 90 per cent of the nitrogenous excretion. Since

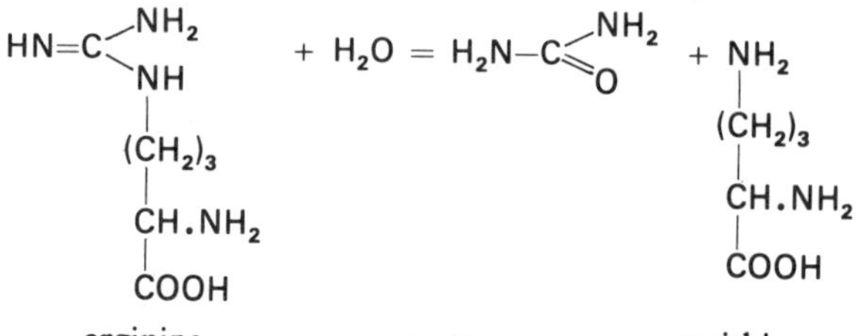

$$HN=C\begin{smallmatrix}NH_2\\NH\end{smallmatrix} \;+\; H_2O \;=\; H_2N-C\begin{smallmatrix}NH_2\\O\end{smallmatrix} \;+\; NH_2$$
$$\underset{\text{arginine}}{\underset{|}{(CH_2)_3}-\underset{|}{CH.NH_2}-COOH} \qquad \underset{\text{urea}}{} \qquad \underset{\text{ornithine}}{\underset{|}{(CH_2)_3}-\underset{|}{CH.NH_2}-COOH}$$

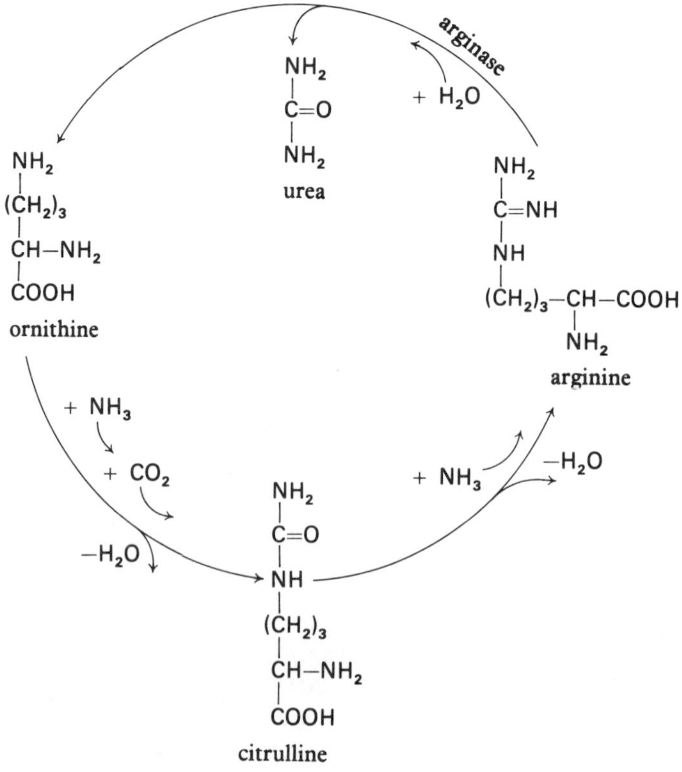

urea is formed in the liver, hepatectomy prevents its production and causes a constantly mounting level of blood ammonia. It was known from the beginning of the present century that the enzyme arginase, which splits arginine to ornithine and urea is concentrated in the liver. However, when an individual is in nitrogen balance almost all the nitrogen ingested as protein is excreted as urea, so that all the other amino acids must contribute to urea synthesis as well as arginine. Using thin slices of liver incubated with an oxidizable substrate to provide energy, it was found that urea could be synthesized from ammonia and CO_2. In the presence of small amounts of ornithine, citrulline or arginine the rate of urea synthesis from ammonia and CO_2 was stimulated and much more urea was formed than anticipated from the amount of amino acid added. Other common amino acids did not act catalytically although their amino groups could be converted into urea stoichiometrically. In the absence of an oxidizable substrate to provide energy only arginine gave rise to urea. These results led to the formulation of the ornithine cycle as the pathway of urea synthesis. The reactions involved in the ornithine cycle have been examined more closely and its individual steps can be described in greater detail. The first step is the formation of carbamylphosphate from one molecule of ammonia and one of CO_2 in a reaction catalysed by carbamylphosphate synthetase.

$$NH_3 + CO_2 + 2\,ATP \rightleftharpoons H_2N.CO.O.PO_3H_2 + 2\,ADP + Pi$$

This reaction uses two molecules of ATP to drive it in the direction of synthesis and to keep the concentration of ammonia below the toxic level. The activity of the enzyme is increased by N-acetyl glutamic acid which takes no direct part in the reaction but which may act as an allosteric activator (see p. 86). Ornithine transcarbamylase then catalyses the transfer of the carbamyl residue to ornithine to form citrulline and to liberate inorganic phosphate. The second amino group

$$\begin{array}{c}NH_2\\|\\(CH_2)_3\\|\\CH.NH_2\\|\\COOH\end{array} + \begin{array}{c}CO.NH_2\\|\\O.PO_3H_2\end{array} \rightleftharpoons \begin{array}{c}NH.CO.NH_2\\|\\(CH_2)_3\\|\\CH.NH_2\\|\\COOH\end{array} + H_3PO_4$$

ornithine　　　　　　　　　　　　citrulline

of urea is derived from aspartic acid which reacts with citrulline, forming argininosuccinic acid. This reaction, catalysed by argininosuccinate synthetase,

$$\begin{array}{c}NH\\\|\\C.OH\\|\\NH\\|\\(CH_2)_3\\|\\CH.NH_2\\|\\COOH\end{array} + \begin{array}{c}COOH\\|\\H_2N.CH\\|\\CH_2\\|\\COOH\end{array} + ATP \rightleftharpoons \begin{array}{cc}NH & COOH\\\| & |\\C-NH-CH\\|\quad\quad\quad|\\NH\quad\quad CH_2\\|\quad\quad\quad|\\(CH_2)_3\quad COOH\\|\\CH.NH_2\\|\\COOH\end{array} + AMP + PP$$

citrulline
(enol form)　　　aspartic acid　　　　　　　argininosuccinic acid

requires ATP and liberates pyrophosphate which is degraded to two molecules of inorganic phosphate by pyrophosphatase. Thus this reaction involves the cleavage of two high-energy bonds and is essentially irreversible. Argininosuccinate is split to form arginine and fumaric acid by argininosuccinase. Arginine is

$$\begin{array}{c} \text{NH} \\ \| \\ \text{C-NH-CH} \\ | \quad\quad | \\ \text{NH} \quad \text{CH}_2 \\ | \quad\quad | \\ (\text{CH}_2)_3 \;\; \text{COOH} \\ | \\ \text{CH.NH}_2 \\ | \\ \text{COOH} \end{array} \quad \rightleftharpoons \quad \begin{array}{c} \text{NH} \\ \| \\ \text{C-NH}_2 \\ | \\ \text{NH} \\ | \\ (\text{CH}_2)_3 \\ | \\ \text{CH.NH}_2 \\ | \\ \text{COOH} \end{array} \;+\; \begin{array}{c} \text{HC.COOH} \\ \| \\ \text{HOOC.CH} \end{array}$$

argininosuccinic acid   arginine   fumaric acid

finally hydrolysed by arginase to give urea and ornithine which can re-enter the cycle by reacting with carbamylphosphate. Fumarate formed during the cycle can be reconverted to aspartate via oxaloacetate and transamination. Each turn of the

$$\begin{array}{c} \text{HC.COOH} \\ \| \\ \text{HOOC.CH} \end{array} \quad \underset{}{\overset{\pm H_2O}{\rightleftharpoons}} \quad \begin{array}{c} \text{CH}_2\text{COOH} \\ | \\ \text{CHOH.COOH} \end{array} \quad \underset{\text{reduction}}{\overset{\text{oxidation}}{\rightleftharpoons}}$$

fumaric acid   malic acid

$$\begin{array}{c} \text{CH}_2\text{COOH} \\ | \\ \text{CO.COOH} \end{array} \quad \overset{\text{transamination}}{\rightleftharpoons} \quad \begin{array}{c} \text{CH}_2\text{COOH} \\ | \\ \text{CH.NH}_2 \\ | \\ \text{COOH} \end{array}$$

oxaloacetic acid   aspartic acid

cycle consumes the equivalent of four molecules of ATP, three of which can be produced by oxidative phosphorylation during the conversion of fumarate to oxaloacetate. Thus the formation of one molecule of urea leads to the net consumption of one molecule of ATP.

The process of urea formation is a good example of cooperation between different cell compartments. Carbamyl phosphate synthetase and citrulline synthetase are located in the mitochondria whilst argininosuccinate synthetase, argininosuccinase and arginase are extramitochondrial. Therefore ornithine must pass into the mitochondria to be converted to citrulline which must then enter the cytoplasm to be converted to argininosuccinate. The oxidation of fumarate to oxaloacetate occurs in the mitochondria, but the transaminases converting aspartate to oxaloacetate are found in both the mitochondrial and the soluble fraction of the cell. Thus, besides the chemical cycle, there is also a cyclic movement of components within the cell.

The capacity of the urea cycle is always sufficient to ensure the rapid removal of any surplus amino groups and therefore the overall rate depends on the availability of the precursors. The operation of the cycle may also be regulated by:

(i) The availability of stoichiometric amounts of aspartate and carbamyl phosphate. In each molecule of urea, one nitrogen is derived from each of these two compounds. They may be derived from ammonia thus

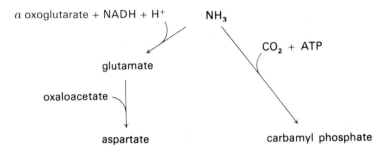

and from glutamate (as the collecting point for the nitrogen of amino acids after transamination) thus

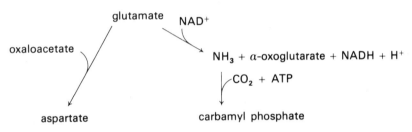

While the mechanism ensuring the coordination of the supplies of aspartate and carbamyl phosphate is not understood, it seems likely that a key regulatory element is the activation of carbamyl phosphate synthetase by acetyl N-glutamate as, under physiological conditions, the activity of the synthetase is almost directly proportional to the concentration of the activator. Moreover, it is synthesized and located in the mitochondria, along with the carbamyl phosphate synthetase and ornithine transcarbamylase.

(ii) The availability of ornithine. The K_m for ornithine of the transcarbamylase is 1·4 mM, and the normal tissue concentration of ornithine is about 0·3 mM suggesting that *in vivo* the concentration of ornithine may also be an important rate controlling factor.

Urea Cycle Enzymes in the Brain

It should be noted that although the brain possesses most of the enzymes of the urea cycle it lacks both the mitochondrial, N-acetyl glutamate-activated, carbamyl phosphate synthetase and ornithine transcarbamylase; hence it is unable to convert ammonia to urea. The brain is, however, able to synthesize pyrimidines (see p. 185) and to this end possesses the cytoplasmic carbamyl phosphate synthetase that uses glutamine rather than ammonia as the $-NH_2$ source.

Other Nitrogenous Constituents of Urine

Creatinine

Apart from urea, the main sources of urinary nitrogen are creatinine (about 1–4 g daily) and uric acid (0·7 g daily). Creatinine is the cyclic anhydride of creatine (see p. 102). Under normal physiological conditions there is a slow spontaneous conversion of creatine to creatinine in the body and the amount of

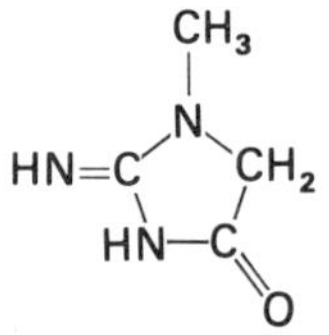

$$HN=C\begin{matrix}\\ \diagdown\\ HN\end{matrix}\begin{matrix}CH_3\\ |\\ N\\ \diagup\diagdown\\ CH_2\\ \diagup\\ -C\diagdown\\ O\end{matrix}$$

creatinine

creatinine excreted by an individual is almost constant (bearing some relationship to the total muscle mass of the body). Normally, creatine itself does not appear in the urine, though it does so if large doses are fed and also during childhood when the muscle mass is changing. There are two important abnormal conditions which also lead to a large excretion of creatine in the urine, amputation of a limb and muscular dystrophy. After amputation, there is a lag before the liver decreases its synthesis of creatine to a level that can be accommodated within the reduced muscle mass available. With the progressive muscular dystrophies the excretion of creatine continues as the muscle mass diminishes.

Uric Acid

The formation of uric acid is described on pp. 192–193.

CHAPTER 19

Detoxication

In the course of its life an organism may ingest substances which are useless and may in fact be poisonous. A whole series of enzymes has evolved which transforms these substances into something less harmful; for example, ammonia, which is very toxic, is converted to urea which is non-toxic—this transformation occurs continually because of the necessity for metabolizing protein. Originally it is probable that most of the detoxifying enzymes had to deal only with toxic substances formed by bacterial action in the gut and with hormones which had fulfilled their function. For the human being, the chemical situation has become much more complicated because of the battery of new drugs to which the body is frequently subject and the presence in the environment of toxic insecticides and other newly synthesized substances which are being encountered for the first time. Clearly the organism cannot have enzymes in store to deal specifically with all the new chemicals produced by the organic chemist; the enzymes already available must serve as effectively as possible. Sometimes their action makes matters worse—as with fluoroacetate (p. 144).

One of the main difficulties in studying detoxication is to know what is meant by a toxic substance. Many compounds develop toxicity if the dose is sufficiently great and some species are quite resistant to substances which are extremely toxic to others. Also the time of exposure to the toxic substance is important, as is the route by which it enters the body. An adequate understanding of the toxicity of a compound can be obtained only after a most exhaustive investigation of its interaction with the enzyme complexes of various organs and tissues.

The reactions most commonly encountered in detoxication are hydroxylations, oxidations, reductions and conjugations.

Hydroxylations

The hepatic endoplasmic reticulum contains an enzyme system capable of hydroxylating a variety of compounds. This system uses $NADPH + H^+$ as a reductant, and passes reducing equivalents via a reductase enzyme to a specific cytochrome (cytochrome P450) and thence to the substrate; molecular oxygen is also required and the substrate is hydroxylated in the process (see also p. 404).

The system appears to interact with a very wide variety of substrates—aliphatic, aromatic and unsaturated compounds, as well as those containing sulphur and nitrogen. Common drugs hydroxylated include (a) phenobarbitone (a sedative); (b) antipyrine (an analgesic and antipyretic); (c) amphetamine (a stimulant); (d) heroin (a narcotic); (e) meprobomate (a tranquillizer); (f) acetanilide

$$\text{NADPH} + \text{H}^+ \longrightarrow \underset{\text{reductase}}{\text{flavoprotein}} \longrightarrow \text{Fe-protein} \downarrow$$

$$\text{cytochrome P450} \underset{\text{H}_2\text{O}}{\overset{\text{O}_2}{\rightleftarrows}} \begin{array}{c} \text{substrate} \\ \text{hydroxylated substrate} \end{array}$$

(which has no action). The products of the hydroxylation may be inactive (a, e, b), of altered activity (c, d) or more active than the original (f).

[Structure: acetanilide → p-acetyl aminophenol]

acetanilide   p-acetyl aminophenol

A striking feature of the hydroxylating system is the extent to which increased activity can be induced by treatments with drugs. The physiological effect of such induction can be readily seen, for example, by the decreased sleeping time produced by an anaesthetic administered to rats which have been previously exposed to the same drug. Induction of this kind can occur with a very large range of compounds, including anaesthetic gases, sedatives, tranquillizers, analgesics, interneural blocking agents, antihistamine compounds, central nervous stimulants, psychomotor stimulators, insecticides and carcinogenic hydrocarbons. Ethanol too increases the enzyme activity (as well as the amount of cytochrome P450 and endoplasmic reticulum). These changes produce a more rapid hydroxylation of the drug, as a consequence of increased amounts of the cytochrome P450 and of some, but not all, of the other microsomal enzymes. There is, in fact, a general hypertrophy of the smooth endoplasmic reticulum in the cell.

The consequence of hydroxylation is an easier elimination of unwanted foreign compounds because of their conversion from lipid soluble molecules into more polar substances. Hydroxylations also play a part in the metabolism and elimination of endogenous steroids (including the steroid hormones). *In vitro* there is competitive inhibition between such compounds as cortisol, testosterone and estradiol and the drug hexobarbital for hydroxylating enzymes. Such competitive interactions may explain the observation that pregnant women have a lowered capacity for metabolizing and deactivating drugs, as during pregnancy the concentration of circulating steroid hormones increases considerably. While the same competitive inhibition can be demonstrated *in vitro* with the oral contraceptive steroids, there is no evidence that even long-term use of these agents interferes seriously with the capacity to metabolize drugs.

Oxidation

A common way of degrading unwanted material is by oxidizing it. Often the products are organic acids. For example, benzylamine is converted to benzoic acid by oxidation of the amino group (presumably by an amine oxidase and aldehyde dehydrogenase).

benzylamine → benzoic acid

Reduction

Aromatic nitro groups can be reduced to the corresponding amines. Nitrobenzene (a very toxic substance rapidly absorbed through the skin) and picric acid are both treated in this way, giving *para*-aminophenol and picramic acid respectively.

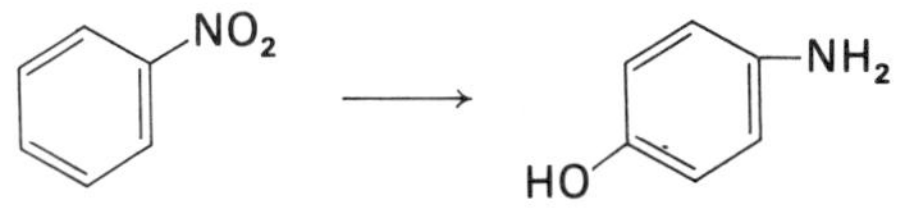

nitrobenzene → *p*-aminophenol

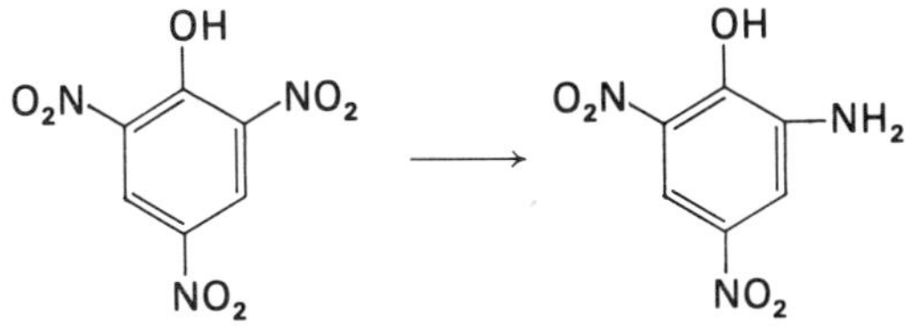

picric acid → picramic acid

Conjugation

Probably the most versatile method available to the body for removing unwanted substances is to combine them with another compound, in the process of conjugation. Glycine is frequently used as the conjugating substance, reacting with carboxyl groups of acids through its amino group to form a substituted amide. The commonest of these is hippuric acid, the conjugate with benzoic acid. The peptide bond formed in this conjugate requires a preliminary conversion of the benzoic acid to benzoyl-CoA. This activation requires ATP. The ability to synthesize hippuric acid is the basis of a test for liver function. Sodium benzoate is administered orally or intravenously and the amount of hippuric acid appearing in the urine over a standard period is measured; in the oral test normally

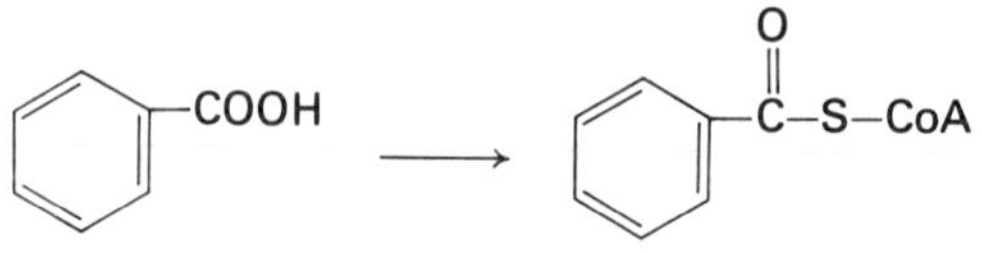

benzoic acid → benzoyl coenzyme A

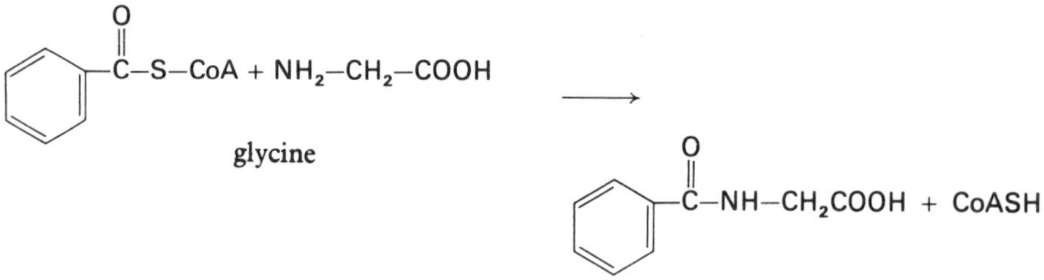

benzoyl-S-CoA + NH$_2$—CH$_2$—COOH (glycine) → C$_6$H$_5$—C(=O)—NH—CH$_2$COOH + CoASH (hippuric acid)

about 50 per cent of the 6 g administered is excreted in 4 hours. Salicylic acid (*ortho*-hydroxybenzoic acid)

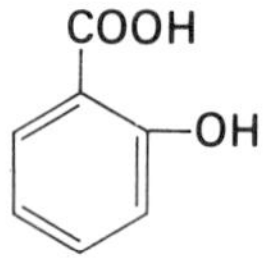

[structure: salicylic acid, COOH and OH on benzene ring]

is also excreted as the glycine conjugate; this acid may be formed in the body by hydrolytic deacylation of aspirin (acetylsalicylic acid).

Many substances, including unwanted steroid hormones, may be conjugated as sulphate esters; for example, phenol is excreted in this form. The synthesis of the ester can occur only after activation of the sulphate by ATP. The activated sulphate, 3'-phosphoadenosine,5'-phosphosulphate (PAPS—see Figure 4.4) is

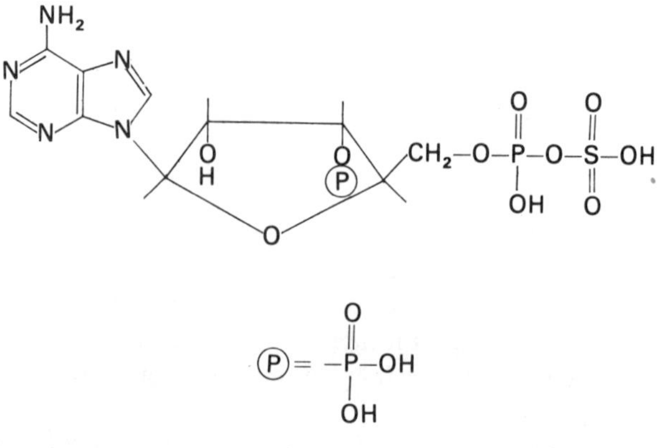

[structure of PAPS showing adenine, ribose with 3'-phosphate (P), and 5'-CH$_2$—O—P(=O)(OH)—O—S(=O)$_2$—OH]

$$\text{(P)} = -\overset{O}{\underset{OH}{\overset{\|}{P}}}-OH$$

Figure 4.4 3'-Phosphoadenosine,5'-phosphosulphate (PAPS)

formed at the expense of two molecules of ATP, one to provide the AMP residue for attachment to the sulphate, the second to phosphorylate the 3-position of the ribose of the adenosine. PAPS is used not only in detoxication, but also for inserting sulphate into mucopolysaccharides and in forming the sulpholipids of the brain.

Glucuronic acid may also be used as a conjugating substance, reacting with phenols through its aldehyde group. The reactive form of the conjugating agent is UDP-glucuronic acid, which can be derived from UDP-glucose by the oxidation of the alcohol group at C_6 of the sugar. Most of the corticosteroids are excreted in the urine as inactive glucuronides, as are substances like bilirubin (p. 279).

Some important detoxications involve acetylation, as for example with sulphonamides. Unfortunately, the acetyl derivatives of the earliest sulphonamide drugs (for example sulphanilamide) were very much less soluble than the drugs themselves so with heavy doses and an insufficient intake of water it was possible to suffer kidney damage by deposition of the insoluble acetyl derivative.

Finally, conjugation may involve methylation of the toxic substances. This occurs, for example, with pyridine. Methionine acts as the methyl donor after activation with ATP (see Figures 4.8 and 4.9).

All the enzymes involved in detoxication are part of the normal complement of the body and may function to remove a metabolically active substance formed during the normal metabolism of the tissues. Detoxication is thus a part of the natural regulatory mechanism of the body. When a foreign substance is presented to the cell it is largely a matter of chance as to whether the enzymic manipulations that occur produce a substance which is more or less harmful. For example, the oxidation of the side chains of barbiturates and the deamination of amphetamine both produce less toxic substances but the oxidation of the insecticide parathion, itself non-toxic, produces an intensely poisonous anticholinesterase. Similarly, substances like tetraethyl tin or tetraethyl lead, both non-toxic, have one of the ethyl groups removed to give the poisonous triethyl tin and triethyl lead. Thus while it may be accurate to think of the basic reactions of detoxication as 'purposeful' in that they have evolved as part of the complex chemical mechanism of the body, it is not correct to suppose that an organism can deal with new foreign substances in a purposeful way. In this sense the term 'detoxication' is a misnomer.

CHAPTER 20

Fat Metabolism

Liver synthesizes triglyceride from glucose arriving in the portal blood by the same pathway as that described for adipose tissue (see Section 5). Once formed the triglycerides are not stored within the liver but are exported in combination with protein (β-globulin), phospholipids and cholesterol as very low density lipoproteins (VLDL). Triglycerides exported from the liver in this way can, and do, contribute to the triglyceride stores of the adipose tissue following hydrolysis by membrane bound lipoprotein lipase (see p. 253) and re-esterification within the fat cell. They also act as oxidizable substrates for many tissues, including muscle but excluding brain. The utilization of triglycerides by tissues is dependent upon the activity of their clearing factor lipase which varies with nutritional and hormonal status (see p. 249).

Although triglycerides circulating as VLDL certainly can act as a source of energy for many tissues, it seems unlikely that they provide more than 20 per cent of the body's energy needs. Their most important role is probably to convey triglyceride synthesized in the liver to the adipose tissue for storage.

The liver also receives long chain free fatty acids from the adipose tissue in amounts that can vary widely, depending upon circumstances. These can be metabolized in the liver in a number of ways. The simplest of these is esterification to triglyceride followed by export as VLDL. They can also be oxidized to acetyl-CoA to provide energy. Acetyl-CoA so formed can either be oxidized further to CO_2 and H_2O or converted to the ketone bodies, acetoacetate and β-hydroxybutyrate, which are exported to other tissues for oxidation.

Acetyl-CoA formed from both glucose and fatty acids also acts as the precursor for cholesterol synthesis (see p. 239) in the liver.

Oxidation of Fatty Acids

Prior to their oxidation, long chain fatty acids have to be converted into esters of coenzyme A by fatty acyl-CoA synthetase, or thiokinase, in a reaction involving ATP (see Figure 4.6). Subsequent passage of the activated fatty acids through the inner-mitochondrial membrane to the sites of their oxidation in the matrix is effected by the carnitine acyl transferase system whose activity appears to be rate limiting in fatty acid oxidation. At the outer face of the inner mitochondrial membrane, the fatty acyl residue is transferred from coenzyme A to carnitine and traverses the membrane as shown in Figure 4.5. On arrival at the inner face of the membrane the fatty acid residue reacylates coenzyme A.

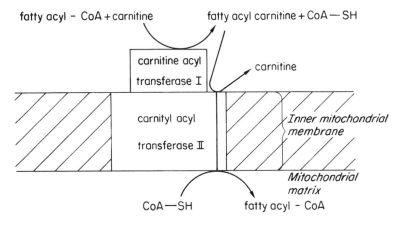

Figure 4.5 Carnitine and fatty acid penetration of the inner mitochondrial membrane

$$(CH_3)_3N^+-CH_2-\underset{\underset{OH}{|}}{CH}-CH_2-\overset{\overset{O}{\|}}{C}-OH$$
carnitine

Within the mitochondrial matrix fatty acids are oxidized to units of acetyl-CoA which can subsequently be oxidized via the TCA cycle or converted to ketone bodies (Figure 4.6).

The activated fatty acid is dehydrogenated by acyl-CoA dehydrogenase with FAD as hydrogen acceptor to form an $\alpha:\beta$ unsaturated acyl-CoA derivative. As with the succinic dehydrogenase reaction in the citric acid cycle, methylene hydrogen atoms do not have a sufficiently negative redox potential to reduce NAD^+ so that a hydrogen acceptor of more positive potential [FAD] has to be used.

Water is then added across the double bond to give an L-(+)-β-hydroxyacyl-CoA compound which in turn is oxidized by β-hydroxyacyl-CoA dehydrogenase with NAD^+ as hydrogen acceptor to form the corresponding β-ketoacyl-CoA. This reaction is formally similar to the malate dehydrogenase reaction. In the final reaction of the sequence, catalysed by β-ketothiolase, coenzyme A reacts with the β-ketoacyl-CoA to yield acetyl-CoA and an acyl-CoA that is two carbon atoms shorter than the original fatty acid. Acetyl-CoA can enter the citric acid cycle and be completely oxidized to H_2O and CO_2, whilst the shortened acyl-CoA can reenter the pathway of fatty acid oxidation. Eight repetitions of this 'β-oxidation' pathway allow stearic acid (C_{18}) to be converted into nine molecules of acetyl-CoA.

Only a limited number of fatty acid molecules are activated at any given time and no more are attacked until the original molecules are completely degraded. This is because there is only a limited amount of coenzyme A available for the activation and β-ketothiolase reactions. Thus, there is an orderly consumption of

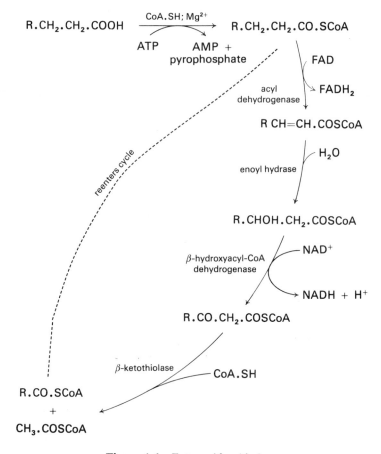

Figure 4.6 Fatty acid oxidation

fatty acids and not a random attack in which all the acids become somewhat shorter.

Comparison of the Pathways of Synthesis and Oxidation of Fatty Acids

The degradation pathway, although theoretically reversible, is of minor importance in fatty acid synthesis because the concentrations of reduced coenzymes (NADH + H$^+$ and FADH$_2$) needed to drive it are not available in mitochondria. Although the equilibrium of the β-ketothiolase reaction is also unfavourable, the formation of acetoacetyl-CoA from two molecules of acetyl-CoA by reversal of this reaction may be readily accomplished because the concentration of acetyl-CoA available is sufficiently high. However, during the next turn of the cycle, when butyryl-CoA has to react with more acetyl-CoA to give β-ketohexanyl-CoA, the equilibrium is against synthesis because butyryl-CoA is present in only low concentrations. Thus, further chain extension by reversal of the oxidative pathway becomes progressively hampered owing to the fact that successive long chain acyl-CoAs are increasingly rare. In contrast, in fatty acid synthesis the condensation step is accompanied by a decarboxylation which pulls the reaction in the direction of synthesis and allows the synthetic pathway to engage in chain extension beyond four carbon units. Another difference is that the reducing power for synthesis is provided by NADPH + H$^+$ rather than by NADH + H$^+$. Finally, in fatty acid biosynthesis the intermediates acylate the SH groups of the synthetase rather than that of coenzyme A.

As with glycogen synthesis and breakdown, the existence of separate pathways for the oxidation and synthesis of fatty acids allows flexibility of control.

Energy Yield of Triglyceride Oxidation

During each turn of the fatty acid oxidation spiral a molecule of FAD and a molecule of NAD$^+$ are reduced. The reoxidation of these coenzymes by the electron transport chain can be coupled to the formation of 5 molecules of ATP. Thus, during the eight turns of the spiral needed to convert stearic acid to acetyl-CoA, 40 molecules of ATP can be produced, giving a net yield of 39 since one molecule of ATP is required to activate the fatty acid initially. Moreover, as each acetyl-CoA is oxidized a further 12 molecules of ATP can be formed. Thus the net yield is 39 + (12 × 9) = 147 molecules of ATP per molecule of stearic acid oxidized. This may be compared with 117 molecules of ATP formed during the complete oxidation of three glucosyl residues from glycogen, which also contain 18 carbon atoms.

Glycerol liberated by adipose tissue can be oxidized by the liver with a net yield of 22 molecules of ATP. However, during starvation the glycerol of triglycerides assumes more importance as a source of glucose to maintain the level of plasma glucose.

Ketone Bodies

Acetyl-CoA residues derived from fatty acids can condense to form acetoacetyl-CoA. This can be deacylated in the liver to yield acetoacetate both directly and via β-hydroxy-β-methylglutaryl-CoA (though the latter is quantitatively the more important pathway; Figure 4.7). Acetoacetate so formed leaves the liver (which is unable to reactivate it because the appropriate enzyme is not present) and passes to other tissues able to reconvert it to acetoacetyl-CoA and use it as a metabolic fuel. In heart, the enzyme thiophorase (3-oxoacid-CoA transferase) reactivates acetoacetate by transferring the CoA group from succinyl-CoA (formed in the citric acid cycle) to acetoacetate. Many tissues, for example, heart muscle, can obtain as much as 70 per cent of their total energy by the oxidation of acetyl-CoA residues derived from acetoacetate. Also acetoacetate decarboxylates spontaneously to give acetone and can be reduced by β-hydroxybutyrate dehydrogenase to D-(−)-β-hydroxybutyrate with NADH + H$^+$ as hydrogen donor. Acetoacetate, β-hydroxybutyrate and acetone are the 'ketone bodies'. It

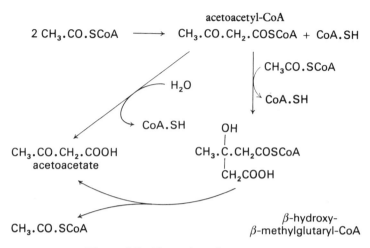

Figure 4.7 Formation of acetoacetate

should be noted that the β-hydroxybutyryl-CoA formed during fatty acid oxidation is the L (+) isomer whilst the ketone body and the β-hydroxybutyryl derivative of fatty acid synthetase are the D (−) isomers. These observations indicate that the ketone body D-(−)-β-hydroxybutyrate cannot be derived directly from the L-(+)-β-hydroxybutyryl-CoA formed during fatty acid oxidation.

Ketone body production is related to the rate at which free fatty acids arrive at the liver from the adipose tissue. Under conditions when glucose is readily available, the flow of free fatty acids is matched by the liver's ability to oxidize them to CO_2 and H_2O and to convert them to triglycerides. However, as the rate of arrival of free fatty acids rises, as happens during heavy work, starvation, high fat feeding, and at times when the sympathetic system is activated, they are diverted more and more towards ketone body formation. The physiological significance of this is that an insoluble fuel unable to enter the cells of the brain is transformed into a water soluble material able to leave the capillaries and enter cells with ease. Ketone bodies are able to supply much of the brain's energy requirements during starvation (see p. 326) and are oxidized in preference to glucose and fatty acids by many other tissues. As the level of ketone bodies rise they stimulate insulin release and thereby decrease the rate of fatty acid release from adipose tissue. That is to say, when ketone bodies are being produced at a rate in excess of their consumption, they are able to reduce the supply of their precursors to the liver. This conserves triglyceride stores and minimizes ketone body and hence fuel loss in the urine. Furthermore it prevents the ketoacidosis typical of uncontrolled ketone body production. In diabetic animals, fatty acid release from adipose tissue is greatly elevated and in consequence the liver overproduces ketone bodies which are unable to regulate fatty acid release from the adipose tissue owing to deficiencies in insulin release or activity.

Despite the long-standing association of ketone bodies with diabetes, they should not be regarded as pathological entities: they are important metabolic fuels under certain conditions; it is their uncontrolled overproduction that is pathological.

Fatty Livers

Under normal conditions the rate of triglyceride synthesis by the liver is balanced by the rate at which triglycerides are released into the blood in the form of VLDL. Thus fatty livers may arise when there is an imbalance between triglyceride synthesis and the synthesis of β-globulin.

During periods of starvation free fatty acids (FFA) arrive at the liver in larger amounts than usual since the rate of FFA release from adipose tissue increases as the level of blood glucose falls. To some extent the liver is able to compensate for this increased arrival of FFA by increasing the rates of the various pathways able to metabolize fatty acids, including triglyceride synthesis and export. However, during even moderate starvation there is a tendency for triglyceride to accumulate in the liver.

Similarly in uncontrolled diabetes the liver is again faced with the problem of processing large amounts of FFA and, despite increased metabolic activity, triglycerides accumulate causing a fatty liver to develop.

Reduced synthesis of β-globulins, even with normal plasma fatty acid concentrations, will also result in fatty livers. An example of this is the fatty liver occurring in carbon tetrachloride poisoning. This damages the endoplasmic reticulum and inhibits protein synthesis. Similarly, the absence of choline from the diet can lead to the development of a fatty liver. This is probably because choline can act as a source of methyl groups to reconvert homocysteine to methionine, an essential amino acid required for β-globulin synthesis.

CHAPTER 21

Synthesis of Phospholipids and Cholesterol

Phospholipids

The triglycerides discussed in the preceding chapters are important as energy stores. The phospholipids are essential components of the lipoprotein membranes of subcellular structures. These lipids differ from triglycerides in possessing both polar and non-polar regions which enable the molecules to be orientated at interfaces with the polar, water-soluble groups in the aqueous phase and the non-polar groups in the non-aqueous phase. The formulae of some phospholipids containing glycerol, with their water-soluble groups indicated, are shown below. Each is an ester of a phosphatidic acid and a water-soluble alcohol. Studies with radioactive phosphate have shown that in many tissues phospholipid molecules are continually turning over. The rate of turnover varies markedly with changes in the physiological activity of the cell. For example, the rate of incorporation of phosphate into phosphatidyl inositol is stimulated about ten-fold during periods of amylase secretion by the acinar cells of the pancreas.

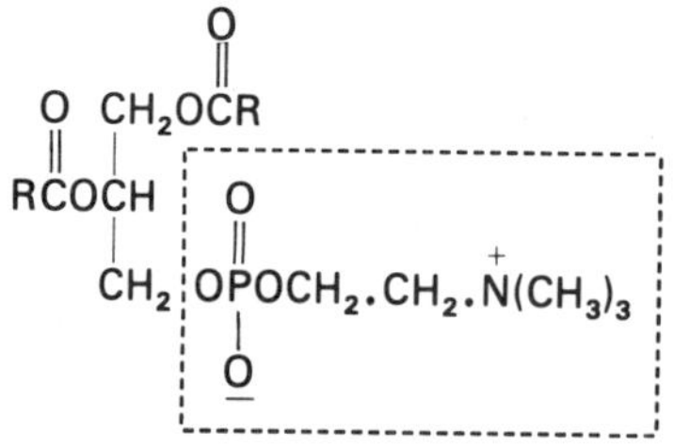

phosphatidyl choline (α-lecithin)

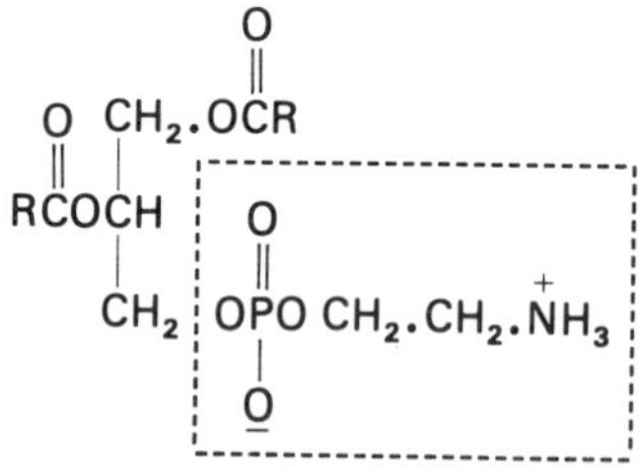

phosphatidyl ethanolamine (cephalin)

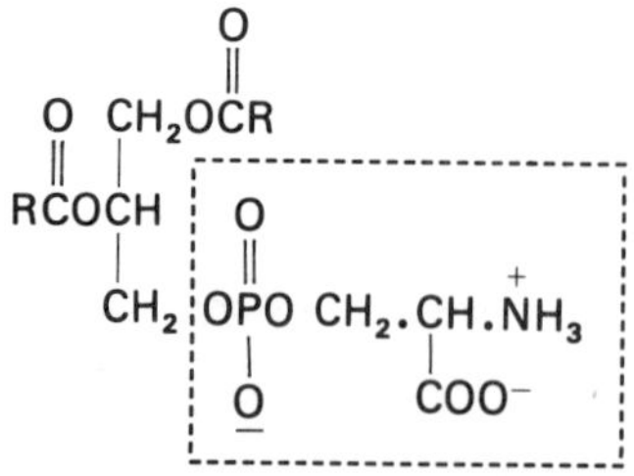

phosphatidyl serine (cephalin)

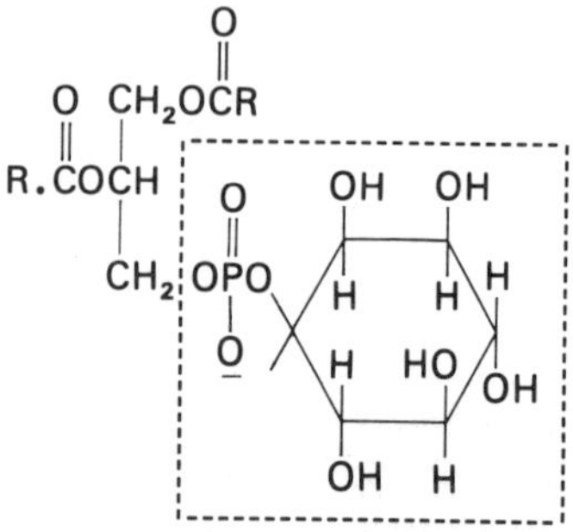

phosphatidyl inositol

Dashed lines enclose the water-soluble groups.

The pathway of biosynthesis of glycerol-containing phospholipids is the same as that of triglycerides as far as the formation of a 1,2-diglyceride. This then reacts, not with an activated fatty acid, but with activated phosphorylcholine or phosphorylethanolamine to yield the corresponding phosphatide. Choline and ethanolamine are activated by the sequence of reactions shown below for choline:

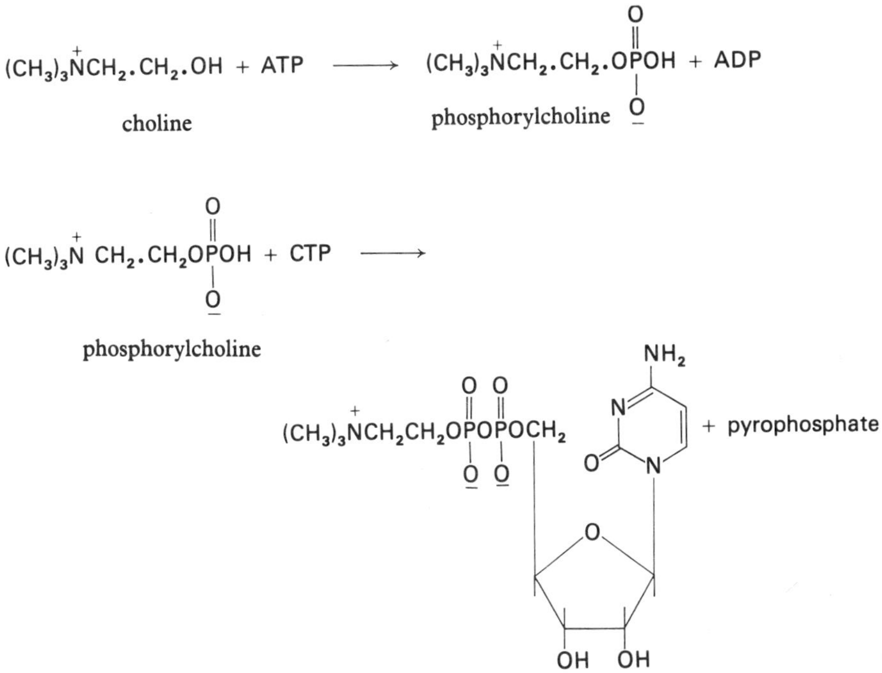

$$(CH_3)_3\overset{+}{N}CH_2.CH_2.OH + ATP \longrightarrow (CH_3)_3\overset{+}{N}CH_2.CH_2.O\underset{\underset{\underline{O}}{|}}{\overset{\overset{O}{\|}}{P}}OH + ADP$$

choline phosphorylcholine

$$(CH_3)_3\overset{+}{N}\,CH_2.CH_2O\underset{\underset{O^-}{|}}{\overset{\overset{O}{\|}}{P}}OH + CTP \longrightarrow$$

phosphorylcholine

$$(CH_3)_3\overset{+}{N}CH_2CH_2O\underset{\underset{\underline{O}}{|}}{\overset{\overset{O}{\|}}{P}}O\underset{\underset{\underline{O}}{|}}{\overset{\overset{O}{\|}}{P}}OCH_2\text{—[cytosine-ribose]} + \text{pyrophosphate}$$

cytidine diphosphate choline (CDP-choline)

Choline reacts with ATP to give phosphorylcholine which is then converted into cytidine diphosphate choline by further reaction with cytidine triphosphate (CTP). CTP is a high-energy phosphate compound similar to ATP but having the adenine group replaced by the pyrimidine cytosine. The reactions involved in choline activation are expensive from the energetic standpoint because three pyrophosphate bonds are broken (pyrophosphate formed during the synthesis of CDP-choline is cleaved by pyrophosphatase). This high cost of activation is identical with that incurred during the conversion of glucose to glycogen. The activated intermediates for phospholipid synthesis are handled as the high-energy cytidine diphosphate derivatives whilst in carbohydrate synthesis uridine diphosphate derivatives are used. Presumably this specificity for cofactors allows control to be exercised over one pathway more or less independently of the other.

The interconversion of ethanolamine and choline and of phosphatidylethanolamine and phosphatidylcholine requires a methyl donor. This is provided by methionine that has been activated by reaction with ATP. In this reaction all the phosphate groups of ATP are lost and the adenosine is attached to the sulphur of

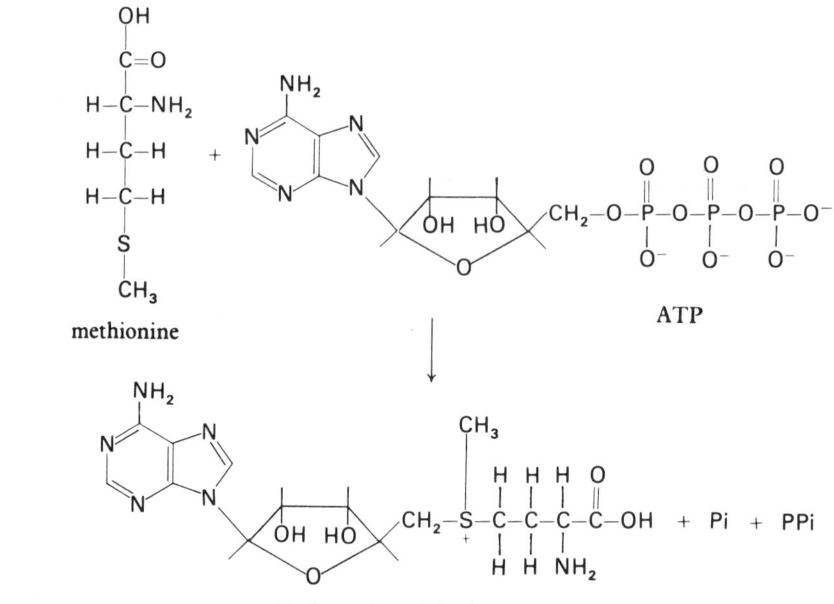

S-adenosyl methionine

Figure 4.8 Formation of *S*-adenosyl methionine

the amino acid, forming *S*-adenosyl methionine (Figure 4.8). The transfer of a methyl group from *S*-adenosyl methionine gives rise to *S*-adenosyl homocysteine (Figure 4.9). Homocysteine may be remethylated by a methyl group attached to tetrahydrofolic acid (see p. 189). The methyl group can arise by the transfer, and subsequent reduction, of the CH_2OH group of serine to tetrahydrofolic acid with the concomitant formation of glycine.

The formation of phosphatidylserine differs from the pathway outlined above in that serine displaces ethanolamine from phosphatidyl ethanolamine.

Certain toxins, for example, some snake venoms and the toxin of the gas gangrene bacillus, contain the enzyme lecithinase. This degrades lecithin to lysolecithin by removing the acyl residue in the 2 position and so produces a powerful detergent able to dissolve cell membranes. Much of the danger of snake bites comes from the haemolysis that occurs when lysolecithin is formed both in the plasma and in the red-cell membranes from preexisting lecithin.

Besides the phospholipids derived from phosphatidic acids there are others derived from the long chain bases sphingosine and dihydrosphingosine. These

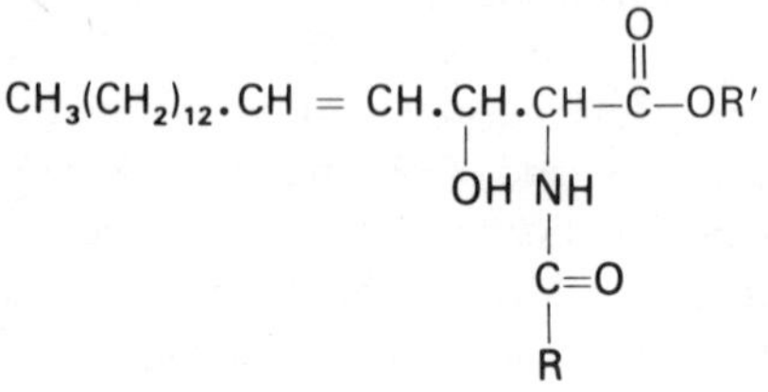

sphingomyelin R = long chain fatty acid
R′ = phosphorylcholine

cerebroside R = long chain fatty acid
R′ = galactose

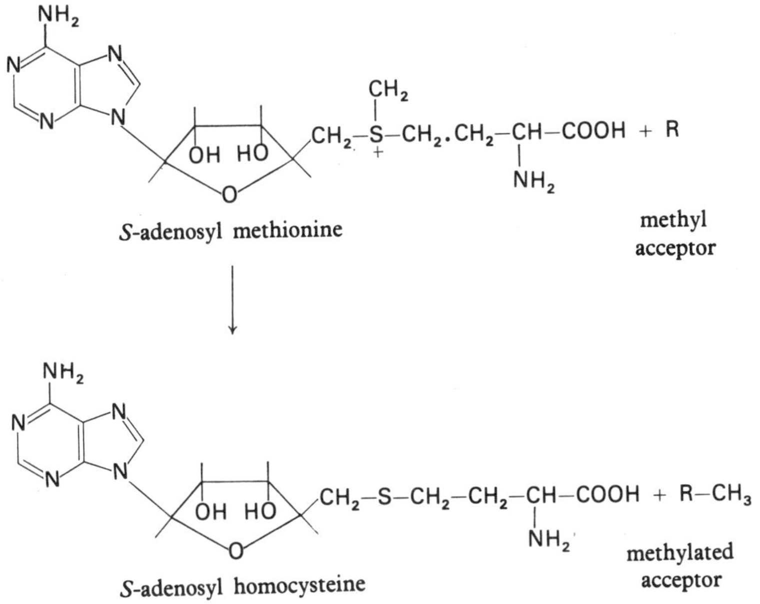

Figure 4.9 Transfer of a methyl group from S-adenosyl methionine

sphingomyelins, as their name suggests, are especially abundant in nerve and brain. They contain a single fatty acid in an amide linkage with the amino group of sphingosine, and are linked through a phosphodiester bridge to choline. Besides phospholipids there are other lipids based on glycerol and sphingosine which do not contain phosphate. These include cerebrosides, gangliosides and sulpholipids.

Some of the pathways involved in phospholipid and triglyceride synthesis are described in Figure 4.10.

Cholesterol

Cholesterol synthesized in the liver is secreted into the circulation as a component of very-low-density lipoproteins and into the bile together with the bile acids which are derived from it.

All the carbon atoms of cholesterol are derived from acetate (Figure 4.11) by the following series of reactions (Figure 4.12). Acetoacetyl-CoA reacts with acetyl-CoA to form β-hydroxy-β-methylglutaryl-CoA (HMGCoA). This compound is reduced to mevalonic acid and then converted, via three phosphorylations and a decarboxylation, into Δ^3-*iso* pentenylpyrophosphate. Three of these pyrophosphorylated isoprenoid units condense, head to tail, to form farnesyl pyrophosphate. Two molecules of this then condense, tail to tail to give squalene. Squalene is converted to lanosterol, the first steroid precursor of cholesterol, via a complicated ring closure and is subsequently transformed into cholesterol.

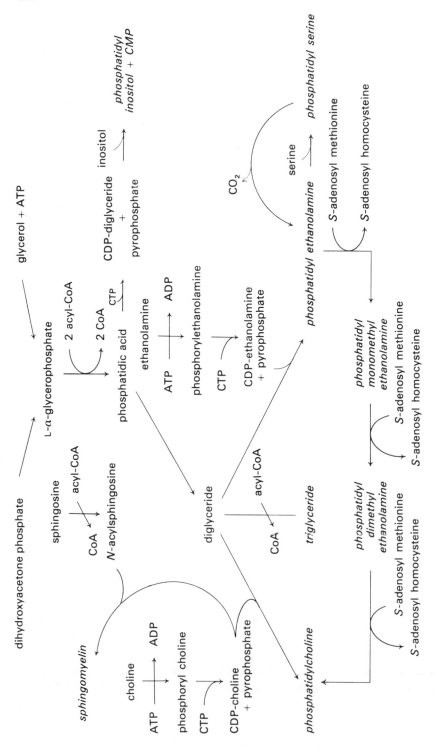

Figure 4.10 Pathways involved in the biosynthesis of triglycerides and phospholipids

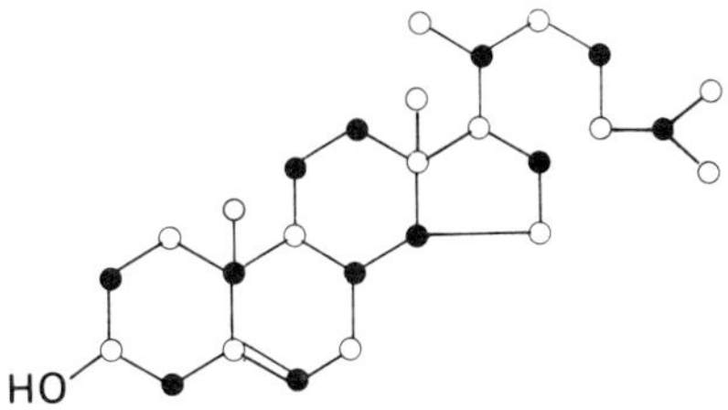

Figure 4.11 Source of carbon atoms in cholesterol. ○ derived from CH₃ group of acetate; ● derived from COOH group of acetate

Cholesterol and Atherosclerosis

Besides being synthesized in the liver and appropriate endocrine organs, cholesterol also forms part of the diet, both as the free alcohol and as the esters of fatty acids. The disease atherosclerosis is characterized by the deposition of cholesterol in the arterial walls and it has been widely held that a high dietary intake of cholesterol can increase this deposition. Furthermore, there is evidence to suggest that such deposition can be reduced if the diet contains an adequate supply of polyunsaturated fatty acids. Whilst cholesterol deposition is clearly of great importance in the development of atheromatous lesions, the factors leading to its occurrence are poorly understood and are almost certainly more complex than the existence of a dietary excess of cholesterol coupled with a deficiency of essential fatty acids. There seems to be no absolute requirement for dietary cholesterol and the level of plasma cholesterol in many people is unaltered by their dietary intake. This is because the ingestion of large quantities of cholesterol leads to the formation of a cholesterol lipoprotein complex that depresses the rate of endogenous synthesis by inhibiting the reduction of HMGCoA to mevalonic acid by a reductase located in the microsomal membranes. This feedback mechanism is absent in some cases of human hepatoma. It appears that in most instances a high level of plasma cholesterol is a consequence of an innate high rate of synthesis rather than of an excessively large intake.

Bile Acids

The cholesterol content of bile is reasonably constant whilst its content of bile acids, such as cholic and deoxycholic acids formed by the oxidation of cholesterol, is more variable and reflects the dietary intake of cholesterol. The bile acids are conjugated with taurine and glycine to form the anions of bile salts such as taurocholate and glycocholate. Cholesterol is soluble in hepatic bile because it is able to form water-soluble complexes with the bile acids. In the gall bladder, water and bile salts are reabsorbed so that the concentration of cholesterol in the stored bile increases. If this concentration proceeds too far, cholesterol crystallizes out of solution and can give rise to gall stones.

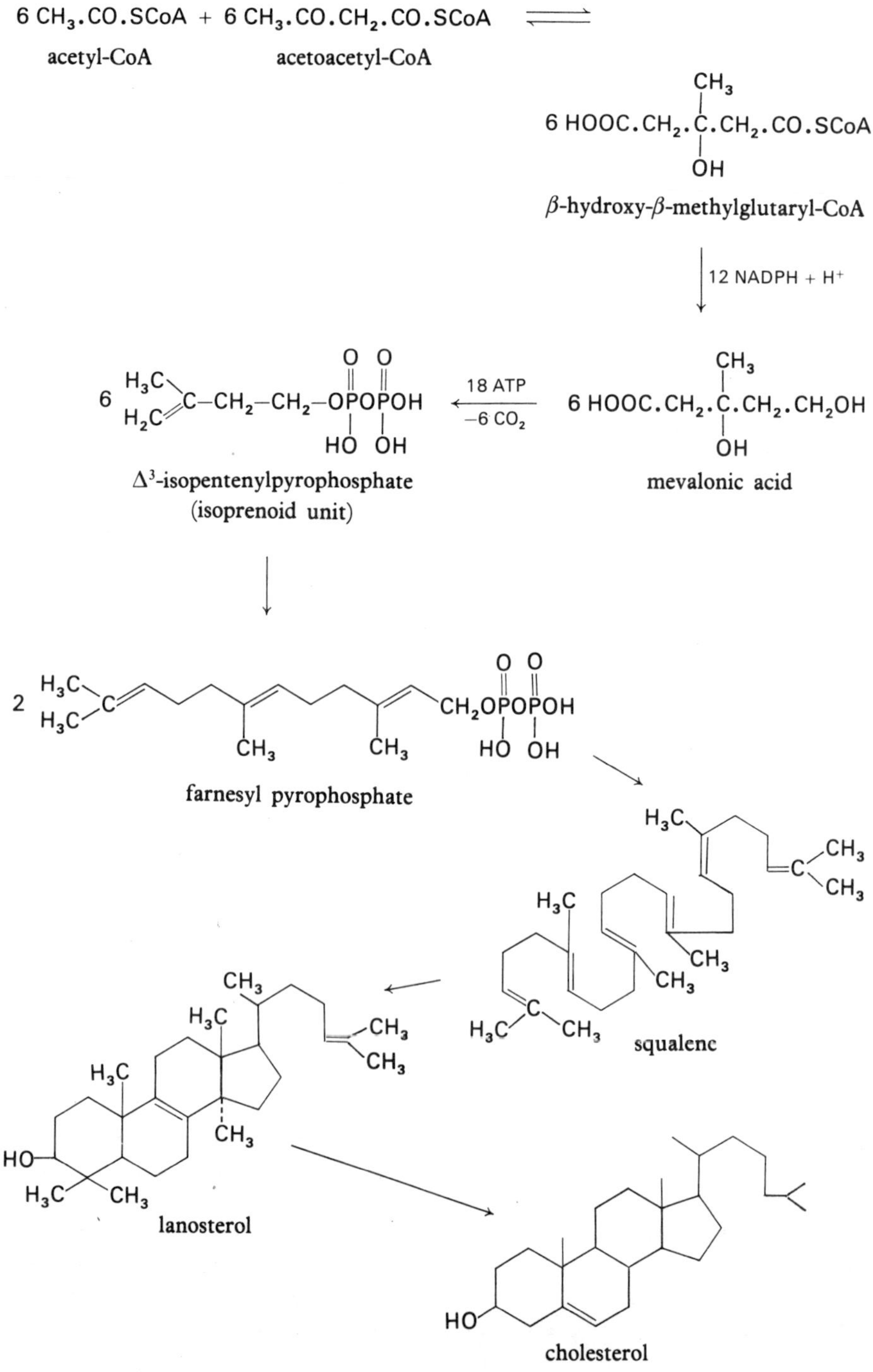

6 CH$_3$.CO.SCoA + 6 CH$_3$.CO.CH$_2$.CO.SCoA ⇌
acetyl-CoA   acetoacetyl-CoA

6 HOOC.CH$_2$.C(CH$_3$)(OH).CH$_2$.CO.SCoA

β-hydroxy-β-methylglutaryl-CoA

↓ 12 NADPH + H$^+$

6 HOOC.CH$_2$.C(CH$_3$)(OH).CH$_2$.CH$_2$OH

mevalonic acid

6 H$_2$C=C(CH$_3$)—CH$_2$—CH$_2$—OP(O)(OH)OP(O)(OH)OH ⟵ 18 ATP, −6 CO$_2$

Δ^3-isopentenylpyrophosphate
(isoprenoid unit)

↓

2 (CH$_3$)$_2$C=CH—CH$_2$—CH$_2$—C(CH$_3$)=CH—CH$_2$—CH$_2$—C(CH$_3$)=CH—CH$_2$OP(O)(OH)OP(O)(OH)OH

farnesyl pyrophosphate

squalene

lanosterol

cholesterol

Figure 4.12 Biosynthesis of cholesterol

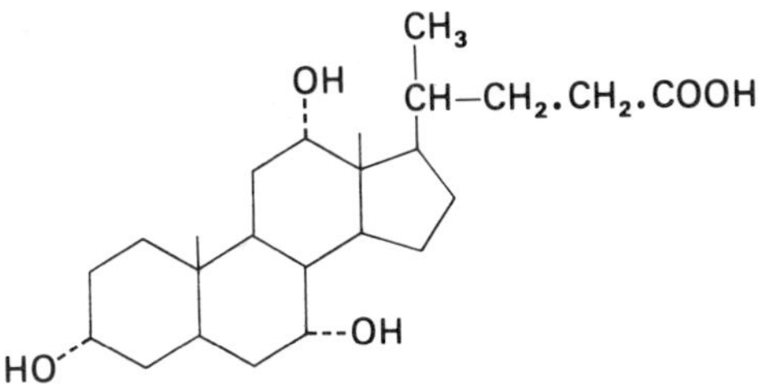

cholic acid

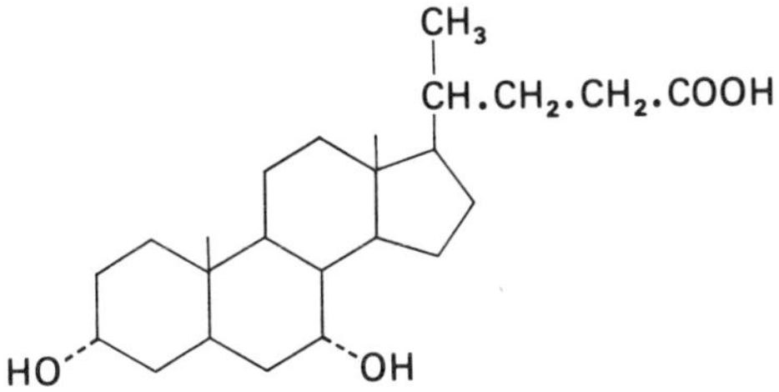

chenodeoxycholic acid

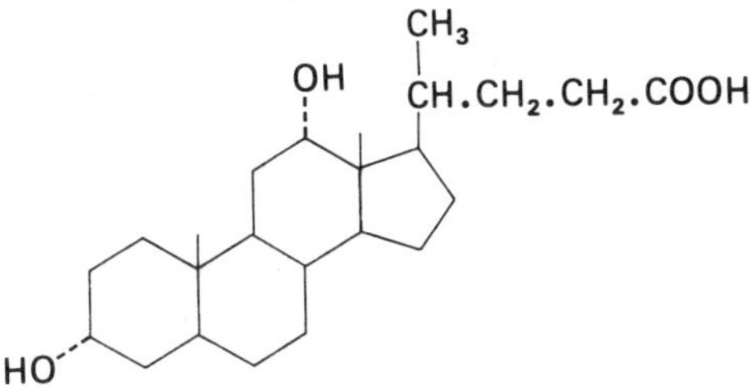

deoxycholic acid

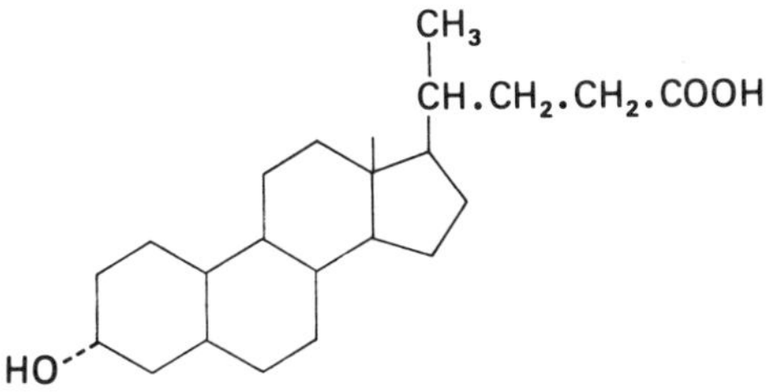

lithocholic acid

Ultrastructure of Liver Parenchyma Cells

Figure 4.13 shows a liver parenchyma cell which is responsible for the metabolic processes discussed in this section. Mitochondria produce ketone bodies from fatty acids, ribosomes bound to the endoplasmic reticulum synthesize proteins such as β-globulins and albumins for export. The endoplasmic reticulum also carries enzymes involved in detoxication and lipid biosynthesis.

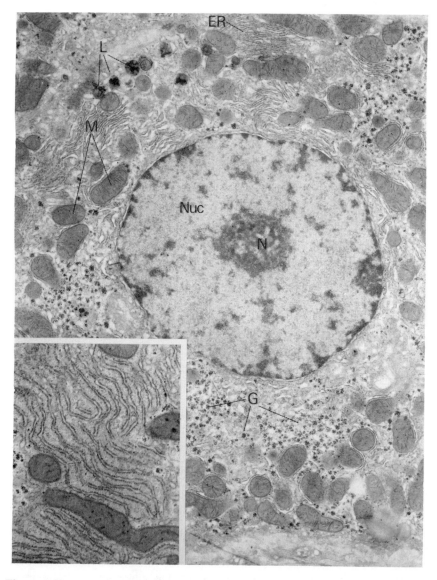

Figure 4.13 An electron micrograph of a liver parenchyma cell showing mitochondria (M), lysosomes (L), glycogen granules (G), endoplasmic reticulum (ER), nucleus (Nuc) and nucleolus (N). The insert, at a higher magnification, shows profiles of the endoplasmic reticulum bearing ribosomes on their outer surface. By kind permission of Mr J. H. Kugler

References

Regulation in Metabolism. Newsholme, E. A. and Start, C. John Wiley and Sons Ltd., London, 1973.

The Metabolism of Animal Phospholipids and their Turnover in Cell Membranes. Dawson, R. M. C. *Essays in Biochemistry,* **2**, 1966.

SECTION 5

Adipose Tissue

CHAPTER 22

The Biochemistry of Adipose Tissue

The weight of adipose, or fatty, tissue in the 70 kg man is variable but usually lies in the range 9 to 13 kg. Of that, 80 per cent or about 7·5–12 kg is accounted for by triglyceride. An individual with a 10 kg store of triglyceride carries a potential source of energy equivalent to 93,000 kcal or 37 days supply assuming an energy expenditure of 2500 kcal per day. In contrast, the total glycogen reserves of the liver and muscles amount to only 190 g, equivalent to 513 kcal or about 5 hours energy expenditure. An obese 140 kg man may well have triglyceride stores of 60 to 70 kg and a daily calorie requirement of only about 2000 kcal. His fat stores would thus be sufficient to supply all his energy for about 11 months. The best results that could be expected if his food intake were reduced to 1000 kcal/day would be a weight loss of 0·75 kg/week. In many instances the subject compensates for a reduced calorie intake by reducing his energy expenditure so that the observed weight loss is less than expected. It can be readily appreciated that such small losses in weight for such a major reduction in the pleasure of eating can discourage obese patients and persuade them to return to their usual diet.

Adipose tissue is located in the abdominal cavity around the kidneys and between the mesentery (omental fat), beneath the skin and between skeletal muscle fibres. In addition to its role as an energy source, it also provides thermal insulation and mechanical protection. Within the cells of white adipose tissue, triglyceride is stored as a single large droplet surrounded by a thin band of cytoplasm which contains a peripherally placed nucleus (Figure 5.1).

Contrary to older ideas, the fat in much of the adipose tissue is in a state of continuous metabolic flux involving the uptake of triglyceride and glucose, the synthesis and degradation of triglyceride and the release of glycerol and free fatty acids.

Carbon Inputs to the Adipose Tissue

The adipose tissue receives triglyceride in the form of chylomicra from the gut and very-low-density lipoproteins (VLDL) from the liver. These triglyceride molecules do not enter the adipose tissue as such; they are first hydrolysed to fatty acids and glycerol by a lipoprotein lipase (the clearing factor lipase) which is located on the surface of either the capillary endothelium cells or the fat cells. The fatty acids are then taken into the cells to be resynthesized to triglycerides

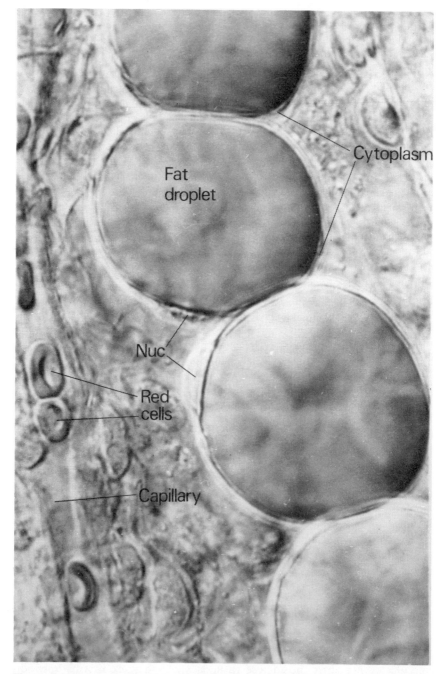

Figure 5.1 Micrograph of living fat cells taken by the differential interference contrast method. The spherical lipid droplets account for most of the volume of the cells, the cytoplasm forms a thin band around the margin of the cells. Nuc = nucleus. By kind permission of Mr J. H. Kugler

with L-α-glycerophosphate derived from glucose. Glycerol is not assimilated by adipose tissue to any notable extent because of the low activity of glycerol kinase in the fat cells.

Chylomicra and VLDL consist of a water insoluble core of triglyceride surrounded by a hydrophilic layer composed of protein, phospholipid and cholesterol. Loss of triglyceride as a result of clearing factor lipase activity decreases the size and increases the density of the particles. The size balance between the hydrophobic interior and the hydrophilic shell is maintained by lecithin–cholesterol acyl transferase which converts the hydrophilic hydroxyl group of cholesterol into a hydrophobic acyl ester able to enter the core of the particle. In this way, chylomicra and VLDL are converted into low-density lipoproteins (LDL) in the extra cellular spaces of the adipose tissue. Eventually the LDL are taken up by cells of the reticuloendothelial system which release their cholesterol into the blood where it associates with the high-density lipoproteins (HDL) formed in the liver and is conveyed to the liver for incorporation into VLDL.

During starvation lipoprotein lipase disappears from the adipose tissue; this ensures that any VLDLs exported from the liver are not returned to the fat depots but are made available to the heart and skeletal muscles which still possess the enzyme. Lipoprotein lipase activity in the adipose tissue of the rat falls during pregnancy whilst that of the mammary gland increases at parturition so that triglycerides in chylomicra and VLDL are directed towards milk production rather than storage in adipose tissue.

Glucose enters adipose tissue by an insulin-sensitive mechanism and is used as a source of carbon and reducing power for triglyceride synthesis.

After a meal, the level of glucose in the portal circulation may exceed the renal threshold. Thus much of the absorbed glucose would be lost via the kidneys if it were not taken rapidly into cells and converted into some non-diffusible storage material. The primary responsibility for doing this rests with the liver, which converts glucose into glycogen and triglyceride before it can escape into the peripheral circulation. Glucose is also removed by muscles which store it as glycogen and it is sequestered, unchanged, in the skin. In the post-prandial condition the adipose tissue is under the influence of insulin and therefore able to remove glucose from the circulation for conversion to triglyceride. In addition, the adipose tissue takes up triglyceride from VLDL formed from glucose in the liver. In humans the relative importance of these two processes is uncertain.

Triglyceride Synthesis

Triglycerides are synthesized from long chain fatty acids and L-α-glycerophosphate by the following series of reactions (see p. 252).

Before they can react with L-α-glycerophosphate, fatty acids derived from chylomicra and VLDL are converted into their reactive acyl-CoA forms at the expense of ATP. This reaction probably proceeds via the intermediate formation of acyladenylate derivatives which remain bound to the enzyme, thiokinase, in much the same way as do the amino acyladenylates involved in protein synthesis

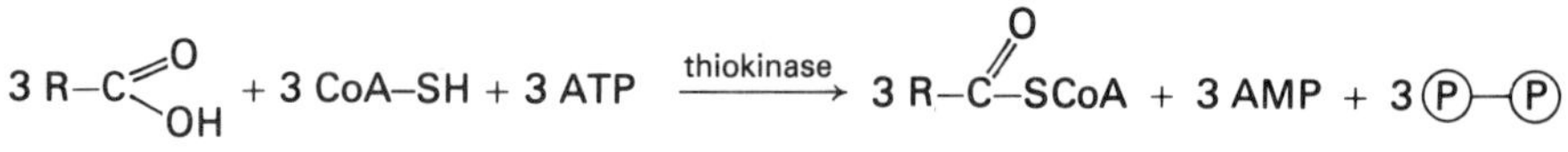

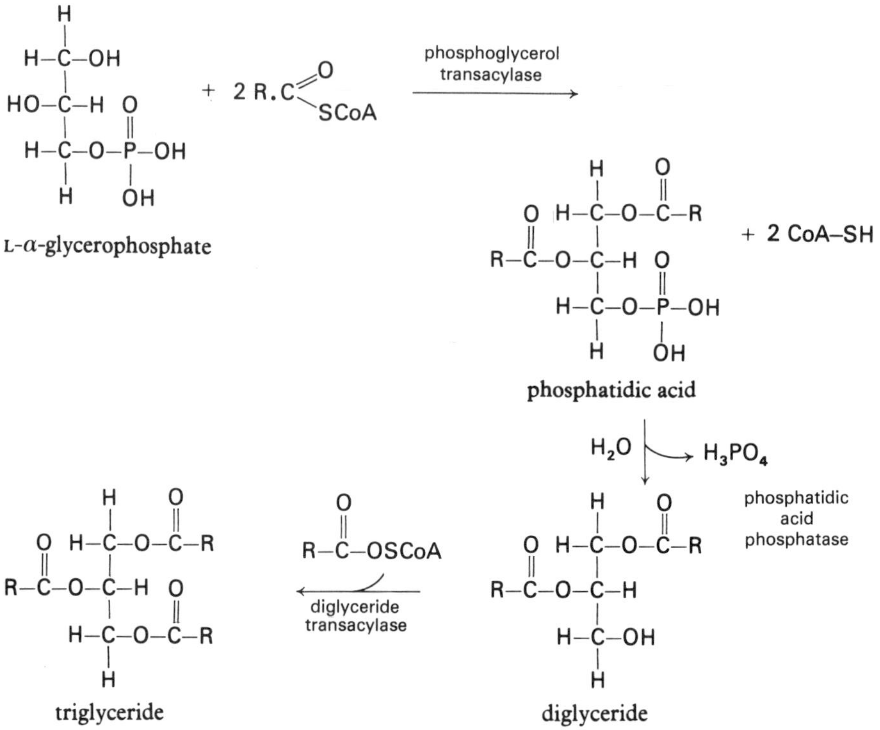

$$3\ R-C\overset{O}{\underset{OH}{}} + 3\ CoA-SH + 3\ ATP \xrightarrow{\text{thiokinase}} 3\ R-\overset{O}{\underset{}{C}}-SCoA + 3\ AMP + 3\ \textcircled{P}-\textcircled{P}$$

L-α-glycerophosphate + 2 R−C(=O)SCoA → (phosphoglycerol transacylase) → phosphatidic acid + 2 CoA–SH

phosphatidic acid → (phosphatidic acid phosphatase, +H₂O, −H₃PO₄) → diglyceride

diglyceride + R−C(=O)−OSCoA → (diglyceride transacylase) → triglyceride

(see p. 178). Two molecules of activated fatty acids then react with a molecule of L-α-glycerophosphate to form a 1,2-glycerophosphatidic acid (phosphatidic acid) which in turn loses its phosphate group to yield a 1,2-diglyceride. By reacting with a further molecule of activated fatty acid, the diglyceride is converted into a triglyceride. The three acyl groups in a triglyceride frequently differ from one another in both chain length and degree of saturation.

In adipose tissue, glycerol is not converted to L-α-glycerophosphate at a significant rate because of the low activity of glycerol kinase. Thus the main source of L-α-glycerophosphate is by reduction from dihydroxyacetone phosphate formed during glycolysis.

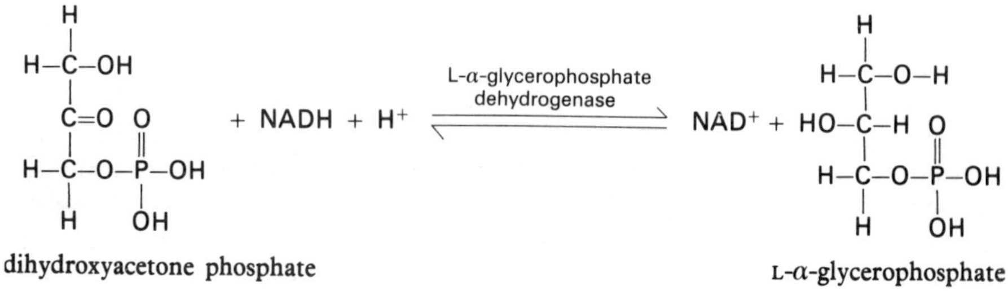

dihydroxyacetone phosphate + NADH + H⁺ ⇌ (L-α-glycerophosphate dehydrogenase) ⇌ NAD⁺ + L-α-glycerophosphate

Fatty Acid Synthesis

The basic unit from which long chain fatty acids are formed in the cytoplasm is acetyl-CoA formed from glucose by glycolysis and the action of pyruvic dehydrogenase.

Acetyl-CoA is first converted into malonyl-CoA by an ATP-dependent carboxylation that requires biotin (one of the B vitamins) as a coenzyme for acetyl-CoA

$$\text{enzyme–biotin} + CO_2 + ATP \longrightarrow \text{enzyme–biotin–}CO_2 + ADP + P_i$$

$$\text{enzyme–biotin–}CO_2 + CH_3.COSCoA \longrightarrow \begin{array}{c} COSCoA \\ | \\ CH_2 \\ | \\ COOH \end{array} + \text{enzyme–biotin}$$

carboxylase. This enzyme may be important in controlling the overall rate of fatty acid synthesis as its activity is less than that of fatty acid synthetase, the next enzyme in the pathway. The activity of acetyl-CoA carboxylase is enhanced by citrate, the precursor of cytoplasmic acetyl-CoA. The entire process of conversion of malonyl-CoA to long chain saturated fatty acids is catalysed by a multienzyme system called fatty acid synthetase. This enzyme complex bears two sulphydryl groups concerned with the reaction sequence; the system is primed when acetyl-CoA reacts with one of them and malonyl-CoA with the other (Figure 5.2). The acetyl group is transferred to the malonyl group which condenses with it and simultaneously loses carbon dioxide. In this way an acetoacetyl derivative of the enzyme is formed in a reaction that is rendered irreversible by the accompanying decarboxylation. The acetoacetyl enzyme is reduced by $NADPH + H^+$ to give the D-(−)-β-hydroxybutyryl derivative. This is dehydrated and then reduced by $NADPH + H^+$, to yield the butyryl enzyme. The butyryl group is next transferred back to the —SH group with which the acetyl-CoA originally reacted. The cycle is repeated once the enzyme has combined with more malonyl-CoA. This sequence of reactions is continued until a fatty acyl chain containing about sixteen carbon atoms has been formed. At that stage reaction with coenzyme A leads to the release of long chain fatty acyl-CoA.

Fatty Acid Elongation and Desaturation

Molecules of saturated fatty acyl-CoA formed by the soluble synthetase system can be elongated both within mitochondria and by enzymes bound to the endoplasmic reticulum. In the mitochondrial pathway acetyl-CoA units are added and reduced by what is essentially a reversal of the β-oxidation pathway (see p. 230) except that the reduction of the $\alpha:\beta$ double bond is effected by $NADPH + H^+$ rather than by $FAD.2H$. The microsomal pathway uses malonyl-CoA as a source of acetyl units.

Palmitic and stearic acids can be dehydrogenated between carbons 9 and 10 to introduce a *cis* double bond with the formation of palmitoleic and oleic acids

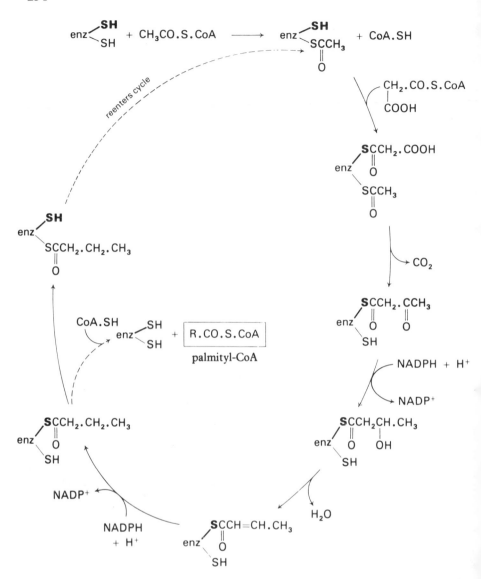

Figure 5.2 Fatty acid synthesis

respectively. These dehydrogenations, which require oxygen and NADPH +H$^+$, are carried out by mixed function oxygenases (see pp. 225 and 404).

Stearyl–CoA + NADPH + H$^+$ + O$_2$ ⟶ Oleyl–CoA + NADP$^+$ + 2 H$_2$O

Supply of Acetyl-CoA to the Cytoplasm for Fatty Acid Synthesis

Acetyl-CoA is formed from pyruvate in the mitochondria but the site of fatty acid synthesis lies in the cytoplasm; yet acetyl-CoA is unable to leave the mitochondria sufficiently rapidly via the carnitine acetyl transferase system to sustain the observed rate of fatty acid synthesis. This problem is solved by acetyl-CoA condensing with oxaloacetate to form citrate which can pass through mitochondrial membranes relatively freely. Once in the cytoplasm citrate can be converted to acetyl-CoA and oxaloacetate by ATP-citrate lyase (citrate cleavage enzyme). Thus mitochondrial acetyl-CoA is made available for fatty acid syn-

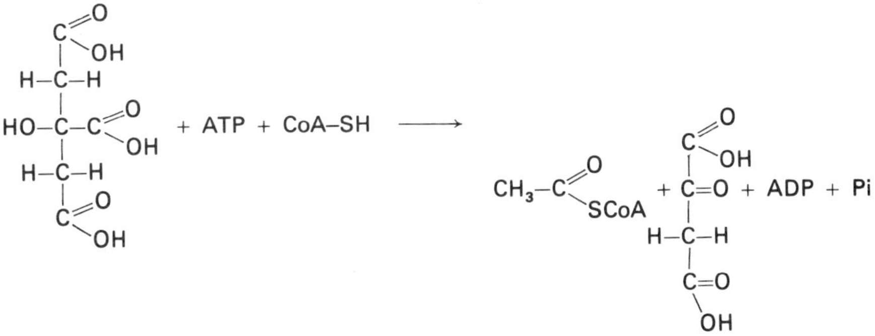

$$\text{citric acid} + ATP + CoA\text{-}SH \longrightarrow \text{acetyl-CoA} + \text{oxaloacetic acid} + ADP + Pi$$

thesis in the cytoplasm by an ATP-dependent process involving citrate cleavage enzyme (ATP-citrate lyase).

Provision of Reducing Power for Fatty Acid Synthesis

In addition to a supply of acetyl-CoA, the synthesis of fatty acids requires reducing power. Although NADH + H$^+$ is produced during the formation of acetyl-CoA from glucose via glycolysis and pyruvic dehydrogenase, it cannot be used for fatty acid synthesis as fatty acid synthetase has a specific requirement for NADPH + H$^+$. Administration of insulin, which stimulates fatty acid synthesis greatly, can increase the carbon flow through the pentose phosphate shunt until it accounts for about a quarter of the glucose metabolized by adipose tissue. Even so, the resulting NADPH + H$^+$ is sufficient to supply only about 60 per cent of the total reducing power required for the observed fatty acid formation. The deficit is made up by converting NADH + H$^+$ (generated in the cytoplasm during the transformation of glucose into pyruvate) into NADPH + H$^+$. This is effected by means of a cyclic series of reactions which is also concerned with moving citrate from the mitochondria into the cytoplasm (Figure 5.3). Oxaloacetate formed from citrate in the cytoplasm is first reduced to malate by a cytoplasmic malic dehydrogenase with NADH + H$^+$ and then oxidatively decarboxylated by malic enzyme, which is very active in adipose tissue, to give pyruvate and NADPH + H$^+$. Pyruvate, on entering the mitochondria, can be carboxylated to oxaloacetate by pyruvic carboxylase which is also present in adipose tissue. The reaction catalysed by malic dehydrogenase, malic enzyme and pyruvic carbox-

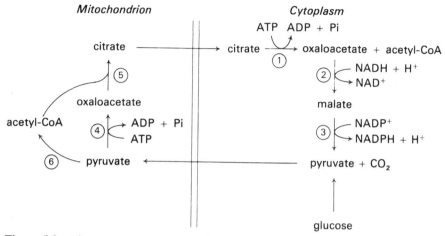

Figure 5.3 The pyruvate–malate cycle. 1. Citrate cleavage enzyme; 2, malic dehydrogenase; 3, malic enzyme; 4, pyruvic carboxylase; 5, citrate synthetase; 6, pyruvic dehydrogenase

ylase constitutes an ATP-requiring transhydrogenase converting $NADH + H^+$ to $NADPH + H^+$.

$$oxaloacetate + NADH + H^+ \longrightarrow malate + NAD^+$$
$$malate + NADP^+ \longrightarrow pyruvate + NADPH + H^+ + CO_2$$
$$pyruvate + CO_2 + ATP \longrightarrow oxaloacetate + ADP + Pi$$

Sum:

$$NADH + H^+ + NADP^+ + ATP \longrightarrow NAD^+ + NADPH + H^+ + ADP + Pi$$

The necessity for balancing the supply of reducing power with a particular carbon source is illustrated by the fact that when acetate alone is supplied to adipose tissue *in vitro* it is unable to be used for fatty acid synthesis because reducing power can be provided neither by the pentose phosphate pathway nor by the pyruvate–malate cycle. If glucose is added in addition to acetate, it all flows through the pentose phosphate pathway to produce the $NADPH + H^+$ required for fatty acid synthesis.

In addition to stimulating glucose uptake, fatty acid synthesis and carbon flow through the pentose phosphate shunt, insulin increases oxygen consumption and decreases the flux through the citric acid cycle. The rise in oxygen consumption is caused by the increased oxidation of pyruvate to acetyl-CoA and provides ATP for the transhydrogenation and citrate cleavage described above. This ATP is required because the two molecules of ATP formed during the conversion of one molecule of glucose to two of pyruvate are only sufficient to account for the carboxylation of two molecules of acetyl-CoA to malonyl-CoA in the cytoplasm. The reduced carbon flow through the citric acid cycle ensures that citrate acts as a precursor of cytoplasmic acetyl-CoA rather than as an oxidizable substrate.

Lipolysis

Within the cells of adipose tissue triglycerides are continuously being hydrolysed to fatty acids and glycerol by a set of three enzymes, triglyceride, diglyceride and monoglyceride lipases. Of these, triglyceride lipase appears to be the least active and therefore rate limiting. As its activity is controlled by a number of hormones it is frequently described as the hormone-sensitive lipase. If the rate of triglyceride breakdown exceeds the rate at which the liberated fatty acids can be resynthesized to triglyceride, there is a net release of fatty acids for export to other tissues (provided that albumins, with which fatty acids form water-soluble complexes, are present in the extracellular fluid). Persons suffering from a congenital absence of plasma albumins, have an impaired ability to mobilize fatty acids. Even when lipolysis is balanced by re-esterification there is a continuous loss of glycerol from adipose tissue since it cannot be rephosphorylated.

There are strains of obese mice that have high levels of glycerol kinase in their adipose tissue; so far as we know no similar inborn error has been described in humans. It would appear that the absence of glycerol kinase from normal adipose tissue ensures that fatty acids are not irreversibly trapped in the tissue.

At any particular time the relative activities of the lipolytic and synthetic pathways determine whether there is net lipolysis and fatty acid mobilization or triglyceride storage.

Mobilization of fatty acids can be brought about by hormones such as adrenaline, noradrenaline, glucagon and ACTH and by noradrenaline released from the endings of sympathetic neurones that innervate adipose tissue. These hormones activate adenyl cyclase in the membranes of fat cells and thereby increase the cytoplasmic concentration of c-AMP (see p. 402) which activates a protein kinase responsible for converting triglyceride lipase into a more active, phosphorylated form. Growth hormone given together with corticosteroids increases lipolysis after a lag of a few hours and also enhances the effectiveness of adrenaline. These effects are probably consequences of an increased synthesis of adenyl cyclase during the lag period.

Insulin increases the rates of fatty acid synthesis and esterification and decreases the rate of lipolysis. On account of the low activity of glycerol kinase in adipose tissue, it was once thought that the rate of esterification was controlled by the supply of L-α-glycerophosphate from glucose and that the antilipolytic action of insulin could be explained by its ability to increase glucose entry into the cells and thereby to increase the concentration of L-α-glycerophosphate. However, it is now by no means certain that the rate of esterification in the absence of insulin is normally limited by the availability of L-α-glycerophosphate. More recently it has been shown that insulin reduces the content of c-AMP in adipose tissue (possibly by activating phosphodiesterase or inhibiting adenyl cyclase) thereby reducing the activity of the hormone sensitive lipase and lowering the rate of lipolysis.

Insulin also increases glucose flow through the pentose phosphate pathway and the rate of fatty acid synthesis. Control over fatty acid synthesis may, in part, be a consequence of insulin's ability to transform pyruvic dehydrogenase into a more active, dephosphorylated, form and thereby to increase the supply of acetyl-CoA from glucose. Insulin also increases the activity of acetyl-CoA carboxylase, but the mechanism is unknown.

During starvation the fall in insulin levels mobilizes fatty acids for oxidation by other tissues by decreasing lipogenesis and increasing lipolysis. Similarly in diabetes there is a net release of free fatty acids from adipose tissue.

Obesity

Carbohydrate in excess of what is needed for growth and energy, is laid down as fat as there is no other means for disposing of it (see also p. 465). Indeed in a state of nature food is scarce and must be conserved; to excrete it would be improvident. Hence a small, regular, excess of calorie intake over expenditure leads slowly but inexorably to obesity. Feeding young rats hypercaloric diets induces obesity which can be abolished by dietary restriction. However, on returning to a normal diet, such animals have a greater tendency to deposit fat than littermates who were never allowed to become obese and who received the normal diet throughout. Such experiments suggest that the efficiency with which fat cells can remove glucose from the circulation can be enhanced by overfeeding young animals. If this information can be transposed to humans, it suggests that to allow young children to become overweight may predispose them to obesity in later life even if their juvenile obesity is corrected.

The ability of adipose tissue to store fat is also governed by the frequency with which meals are taken. Persons eating one or two meals a day have a greater tendency to lay down fat than those consuming the same amount of food in four or five meals. If all the day's food is consumed in a single meal, the carbohydrate intake will exceed the glycogen storage capacity of the liver and muscles so the remainder will be converted to fat until it is required to supply energy.

Brown Adipose Tissue

Whereas white adipose tissue acts as a store of triglyceride which can be mobilized and exported for oxidation elsewhere in the body, brown adipose tissue can oxidize its stored fat during periods of exposure to cold to produce heat. The tissue is particularly well-developed in hibernating mammals and in new-born mammals which, on exposure to cold, produce heat without shivering; new-born humans come into this second category. Brown fat tissue is located between the muscles of the neck and back, in the axillae and groins, and around abdominal and thoracic viscera. The red-brown colour which gives the tissue its name is only evident when it is totally depleted of fat and is a consequence of the high concentration of mitochondrial cytochromes; white adipose tissue is much poorer in mitochondria than brown. The tissue also has a rich blood supply and is heavily innervated by noradrenergic fibres of the sympathetic nervous system. Exposure of non-shivering new-born animals to cold results in the release of noradrenaline from nerve endings on the brown fat cells and from the adrenal medulla. Noradrenaline activates adenyl cyclase in the membrane and thereby increases the intracellular concentration of c-AMP. In turn, this leads to the activation of a triglyceride lipase and the breakdown of stored fat. Glycerol is lost from the cells as the activity of glycerokinase is very low, in the new born at least, and fatty acids are oxidized with the evolution of heat and consumption of oxygen. Heat produced by the brown adipose tissue warms neighbouring tissues and is carried further afield by the circulation. Calculations suggest that a cycle of triglyceride hydrolysis and resynthesis powered by ATP formed during fat oxidation is not responsible for heat generation although it would constitute an exothermic ATP-ase. It is more widely thought that fatty acid oxidation is only loosely coupled to ADP phosphorylation in brown adipose tissue so that much of the free energy of oxidation, which in most tissues would be conserved as ATP, is released as heat. Possibly the loosening of electron transport from phosphorylation is related to the uncoupling action of long chain fatty acids.

Although brown adipose tissue undoubtedly produces heat during cold exposure, it is not responsible for all the extra heat generated by non-shivering animals or released in response to infusion of noradrenaline. Nevertheless, it seems that the extra heat produced by other tissues under these conditions depends upon the possession of brown adipose tissue. Observations of this kind have led to the suggestion that the tissue may produce an endocrine secretion in addition to acting as a source of heat. Attempts to investigate this interesting possibility by injecting extracts of brown adipose tissue into non-cold-adapted animals have, however, not yielded conclusive results.

References

The Human Adipose Cell. Galton, D. J. Butterworths, London, 1971.
Regulation of Fatty Acid Synthesis in Higher Animals. Numa, S. *Ergebnisse der Physiologie/Reviews of Physiology,* **69**, 53–96, 1974.

SECTION 6

Skin and Structural Tissues

CHAPTER 23

The Biochemistry of Skin and the Structural Tissues

Although it is convenient to consider the body as a set of interconnected organs, this approach leaves out the important question of the shape of the body and the positional arrangement of the organs within the body. The human is draped over its bones and wrapped up in its skin and it is on the type of draping and wrapping that we base most of our ideas of the beauty of the human body. Since the skeleton is the only rigid tissue, it provides the basic shape and points of attachment for other tissues. Much of the material making the necessary attachment between tissues is fibrous, as is much of the wrapping material. The fibrous proteins, collagen and elastin, make most of the connections where organs and bones are held together, and the skin is characterized by another fibrous protein called keratin. On the other hand the ends of bones are kept apart by other polymers of a carbohydrate nature, capable of resisting pressure. These are the mucopolysaccharides as found, for example, in cartilage. Even individual cells are often kept apart by a thin layer of a ground substance, composed mainly of mucopolysaccharide. Since the connective tissue fulfils a passive role it has a rather low rate of metabolism; indeed much of the tissue is extracellular. Fibroblasts, or cells resembling fibroblasts, are the main cellular components and it would appear that these cells could not be highly aerobic, since they are often embedded in an avascular fibrous material. Starting from the outside we shall first consider the biochemistry of the skin.

The Skin

The skin, which comprises some 7 to 8 per cent of the body weight, insulates the body against the hostile external environment. The maintenance of the skin's defensive role against various forms of trauma (abrasion, actinic rays, bacterial, fungal and insect colonization) requires a constant regeneration process. Thus the skin is an active biochemical entity synthesizing new material (often very specialized) and using energy and basic substrates for this constant synthesis. In many ways the skin is very comparable to the intestine (which as pointed out in Chapter 13 is the outer defence of the body in the chemical sense). Both tissues have an outer layer of epithelium which is constantly and completely regenerating against a hostile environment and both have subepithelial tissues of considerable complexity which have major physiological roles; for example, both tissues are endowed with blood vessels whose capacity varies enormously according to the body's needs.

Working from the outside inwards, the skin is divided into three layers (1) the outer thin layer of epidermis (about 0·1 mm thick on most of the body, greater than 0·5 mm on palms and soles) which has the very special property of synthesizing the protein keratin and the skin lipids; (2) the dermis, between 2 and 4 mm thick, which synthesizes the mucopolysaccharides and the proteins collagen and elastin; and (3) the subcutaneous fat layer which is very variable in quantity but which will not be dealt with further in this chapter (see Adipose Tissue, Section 5). Besides these comparatively clear-cut layers there are the skin appendages, hair and nails (dead keratinized cells), the sebaceous glands and the sweat glands. This then is the versatile tissue in which we are wrapped.

The Epidermis and Keratin Synthesis

This fibrous protein (which is characteristic of the epidermis and structures derived from it, such as hair and nails) contains a very high proportion of glycine, serine and cystine (each about 15–20 per cent of the total amino acids). It is the cystine which, by cross-linking the protein chains, ensures high tensile strength and low solubility. Treatment of keratin with suitable reducing agents, such as thioglycolic acid, will cleave the disulphide linkages, greatly increasing the solubility of the protein and reducing its tensile strength. Such softening of hair by reducing agents is used in the artificial waving or straightening of hair, the disulphide links being reestablished after the hair has been arranged in the required form. Unlike collagen and elastin, keratin is synthesized and retained within the cell. During the process of keratinization, cell division takes place parallel to the surface of the skin and the cells move further from the source of nourishment. In the epidermal cells the first sign of keratin formation is the appearance of delicate fibrils. These thicken to about 60 Å, aggregate and coalesce to form the fibrils visible in the light microscope. As the fibrils grow, amorphous granules make their appearance and, in the last stage of formation of the keratin, associate with the fibrils. As with collagen, a soluble monomer, called prekeratin (molecular weight of about 640,000) may be extracted from skin; it contains SH groups but no disulphide bridges. Polymerization (the formation of the fibrils) involves the establishment of these bridges. It appears that a disulphide bridge occurs roughly at every 18 amino acid residues.

The Dermis and Synthesis of Collagen, Elastin and Mucopolysaccharides

The main cells of the dermis are the fibroblasts that secrete the monomers of the extracellular fibrous proteins collagen and elastin and also synthesize the mucopolysaccharides that make up so much of the so-called 'ground substance'—the amorphous gel between the cells and the fibres of protein.

Collagen

Threads of fibrous protein appear throughout the ground substance of almost all organs. Such threads are particularly plentiful in the dermis and in other connective tissues, including bone, and are usually embedded in an amorphous

Table 6.1 Approximate amounts of collagen in some human tissues

	Collagen content (mg/g wet wt.)	Total quantity of collagen in the total tissue (g)
Skeletal muscle	12·1	363
Liver	1·3	2·2
Kidneys[a]	31·4	9·1
Skin	197	965
Bone[b]	69	830
Lungs	39	39

[a] Analysis for rat kidney. Total quantity calculated for human kidney assuming the same content as in the rat.

[b] Analysis for rat bone. Total quantity calculated for human bone assuming the same content as in the rat.

matrix of mucopolysaccharides. Most of the fibres are rather inelastic and have a high tensile strength. These inelastic fibres, which comprise some 80 per cent of the dry weight of tendons, consist of the protein collagen. Quantitatively collagen is the most abundant protein in the body. Values of from 20 per cent to 40 per cent of the total body protein have been obtained by different workers. In contrast to keratin, virtually all the collagen in the body is extracellular and, as with keratin, devoid of enzyme activity. Because it is outside the cell, collagen is not easily accessible to enzymes and is broken down by the body only very slowly except in the involuting uterus (see p. 371). Table 6.1 shows the approximate quantities of collagen found in some tissues.

As might be expected, skin and bone contain most of the body's collagen, but a substantial amount is also found in skeletal muscle. As a percentage of dry weight, collagen contributes most to tendon, up to 86 per cent, cordae tendinae 84 per cent, skin 79 per cent and cornea 78 per cent. These are all tissues where great toughness is required.

Formation of Collagen

It seems clear that the cells responsible for the production of collagen are the fibroblasts. Figure 6.1 shows collagen fibres from rat tail. These large and insoluble fibres are assembled *in situ* and are not exported as finished products from the cell; rather, a soluble monomer is released which polymerizes outside the cell to give the fibres as illustrated in Figure 6.2. The monomeric molecules are tropocollagen (see p. 67); they consist of three coiled helixes of polypeptide chains. When released from the cell they line themselves up in overlapping ranks forming the fibres which are first stabilized by hydrogen bonding and then by covalent links between the ranks of monomers. The staggered alignment of the tropocollagen molecules (each 2900 Å long) results in a periodicity (at 700 Å) in the collagen fibres, responsible for the striated appearance. The way in which the extracellular tropocollagen molecules are instructed in the direction in which they should point, or in the shape of the final fibre, is not at all clear.

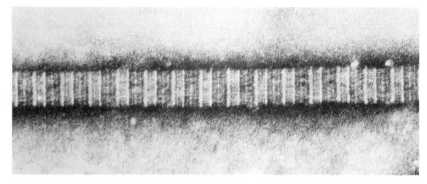

Figure 6.1 An electron micrograph of a single collagen fibre from rat tail; the periodicity results from the staggered alignment of chains of tropocollagen subunits. By kind permission of Dr J. Moore

The amino acid composition of collagen is most unusual. Not only are some amino acids present in much higher amounts than in any other protein, but some unusual amino acids not found in other proteins contribute substantially to the polypeptide chain. Thus one third of the amino acids of the collagen of human skin is glycine. Alanine, proline and hydroxyproline each amount to 10 per cent of the total; thus these three, together with glycine, comprise 60 per cent of the amino acid content of collagen. Aspartic acid and glutamic acid together comprise another 10 per cent and arginine and serine about another 5 per cent each. The remaining 20 per cent is shared amongst the other amino acids, including the unusual amino acid hydroxylysine. Tryptophan and cysteine are absent. It is generally believed that the hydroxy amino acids contribute very largely to the hydrogen bonding mechanism that maintains the collagen as a continuous fibre. Other covalent bonds, possibly ester bonds, provide cross-links which further stabilize the molecule and which increase in number as do the hydrogen bonds, as the collagen ages.

Biosynthesis of Collagen

Exchange between the amino acids of the metabolic pool and those of the collagen molecule is usually very slow. A net increase in the amount of collagen can be induced experimentally—for example when a wound in the skin heals or when a polysaccharide such as carageenan (extracted from seaweed called Irish

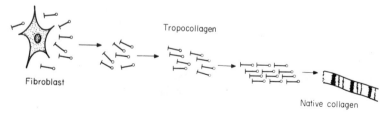

Figure 6.2 Polymerization of tropocollagen

Moss) is injected subcutaneously. In the latter case granulomas develop at the site of injection and the fibroblasts within rapidly synthesize collagen to about 10 per cent of the weight of the granuloma. There seems no doubt that the protein is synthesized by the usual mechanism but ascorbic acid plays an essential but obscure role in the process. Many observations have shown that in the scorbutic subject wound healing is slow or does not occur, and that in scorbutic guinea pigs collagen is lacking in the skin, particularly in regenerating skin. Also in the granulomas induced experimentally by carageenan there is only a small synthesis of collagen if the animal is scorbutic. There seems no doubt that proline and lysine respectively are the precursors of the hydroxylated compounds (hydroxyproline and hydroxylysine) within the collagen molecule. There is, however, some doubt about the stage in the synthesis of the protein at which the hydroxylation occurs and the reasons why some proline or lysine molecules are selected for hydroxylation and not others remains obscure. What is clear is that both oxygen and ascorbic acid are necessary for the hydroxylation process. Under anaerobic conditions, or in the absence of vitamin C, proline is incorporated into collagen but is not converted to hydroxyproline.

Ageing of Collagen

It has been established experimentally that the collagen of older people is less soluble than that in younger people. Also, the newly formed collagen laid down in skin during wound healing is more soluble than the older collagen in the surrounding skin. Part of the ageing process appears to be the establishment of more and more cross-links between chains with a consequent stiffening of the molecules. Such changes are thought to be responsible for the degeneration of the elastic tissues in old age, which is irreversible because the collagen of these tissues does not turn over. However, in other tissues collagen does turn over. In rats, for example, the turnover of the collagen in the liver is about one-half that of the other protein of the tissue and there seems to be some relationship between the overall metabolic activity of a tissue and the rate of its collagen turnover.

Elastin

The elastic fibres which are characteristic of ligament are composed largely of the elastic protein, elastin. Elastin is an extracellular, fibrous protein apparently produced by the same cells as collagen. However, it differs in almost every particular from collagen. The fibres are elastic, branched and yellow in colour. The elasticity depends on the presence of water; when dried elastin is brittle. The distribution is also different from that of collagen; elastin is sparse in skin and tendon, of roughly equal quantity to the collagen in lung and artery, and provides about 80 per cent of the protein in ligament. The amino acid composition is similar to that of collagen, one third of the amino acids being glycine. The four amino acids proline, glycine, alanine and valine comprise 80 per cent of the amino acids but hydroxyproline is present in very small amounts and cysteine is absent. Almost 95 per cent of the side chains of the protein are non-polar.

Degradation studies have shown that the yellow colour of elastin is due to the presence of two unusual amino acids desmosine and isodesmosine. These substances form the bridges between the polypeptide chains of elastin. From isotope studies with lysine it has been shown that the desmosines are formed by reaction between lysine moieties present in the elastin precursor. In copper deficiency

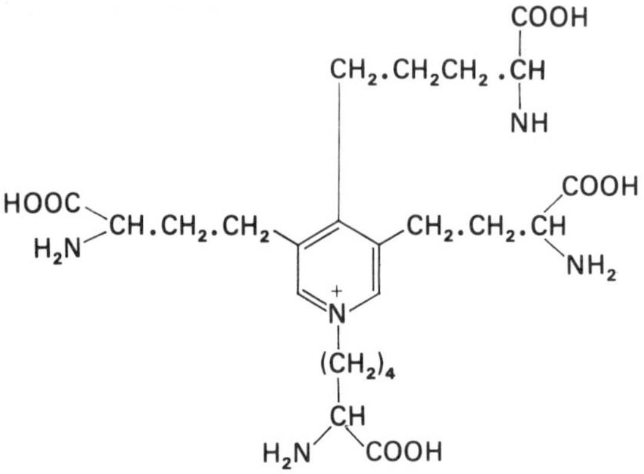

desmosine

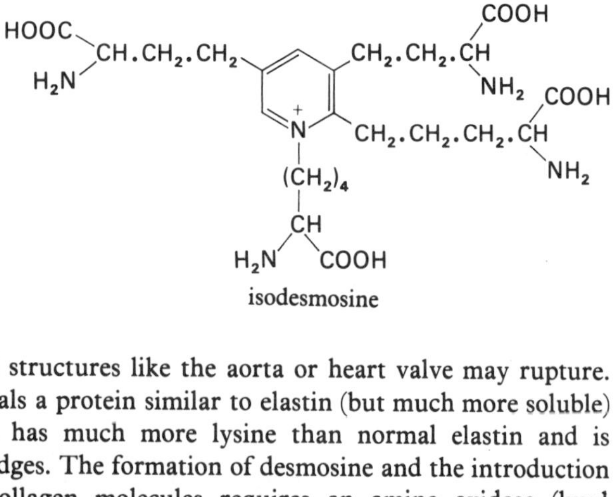

isodesmosine

elastin is not formed and structures like the aorta or heart valve may rupture. From such deficient animals a protein similar to elastin (but much more soluble) may be extracted which has much more lysine than normal elastin and is deficient in desmosine bridges. The formation of desmosine and the introduction of cross-links between collagen molecules requires an amine oxidase (lysyl oxidase) having copper and pyridoxal phosphate as cofactors.

Lathyrism

The sweet pea (*Lathyrus odoratus*) contains β-aminopropionitrile. This substance administered to an animal causes loss of tensile strength in the tissues with increased solubility of both collagen and elastin. In both cases the cross-linking between the protein chains fails to occur. It may be that β-aminopropionitrile reacts with the carbonyl group of pyridoxal phosphate which is involved in forming the cross-links in both elastin and collagen.

The Ground Substance—Mucopolysaccharide Synthesis

The ground substance is the amorphous gel that exists in the dermis between fibroblasts, blood vessels and fibres. The material (about one-tenth of the dry weight of the tissue) may be extracted by exhaustive treatment with 0·15 M NaCl which leaves cells and fibres apparently intact. Most of the extract consists of proteins which in composition match the plasma proteins. About 10–12 mg of protein (equivalent to 0·2 ml plasma) are present in each gram of skin. Of the mucopolysaccharides which make up the rest of the saline extract of skin more than 70 per cent is hyaluronate (0·2 mg/g dermis or 0·5 μmole hexosamine/g dermis). Water is bound to the mucopolysaccharide and the gel-like structure in combination with the collagen and elastic fibres produces the resilient properties of the dermis. In the female monkey there is a massive increase in the hyaluronic acid content of the sex skin during oestrous with a large increase in the water content and consequent swelling. Such oedematous changes may occur in the human during menstruation (but generally, not locally).

The mucopolysaccharides are synthesized by the fibroblasts, or their analogues the chondrocytes. One special mucopolysaccharide, heparin, is synthesized by a special cell (the mast cell) and retained within the cell in granules together with histamine for secretion when required (histamine is synthesized by decarboxylation of histidine, cf. biosynthesis of catecholamines). Other mucopolysaccharides are secreted continuously as they are formed.

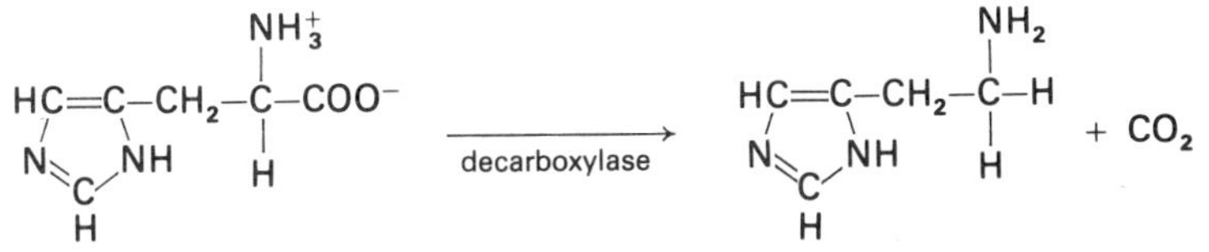

$$\text{L-histidine} \xrightarrow{\text{decarboxylase}} \text{histamine} + CO_2$$

There is no clear definition of a mucopolysaccharide, since historically this group of substances (often mucus-like in nature) has comprised polymers containing hexosamine residues, or lacking hexosamine and associated with protein, or not associated with protein. More commonly the term mucopolysaccharides is applied only to those substances which contain a uronic acid residue and hexosamine. Of the many mucopolysaccharides, only the acidic mucopoly-

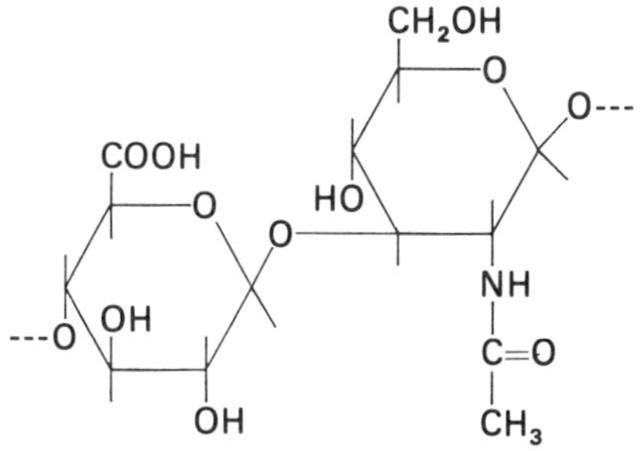

hyaluronic acid

saccharides will be considered, especially the hyaluronic acids and the chondroitin sulphates.

Hyaluronic acid is composed of alternating units of glucuronic acid and *N*-acetylglucosamine. Chondroitin sulphate contains glucuronic acid alternating

chondroitin sulphate A

with *N*-acetylgalactosamine in which one of the hydroxyl groups is sulphated as shown. It is likely that mucopolysaccharides are linked more or less firmly with protein *in vivo*; frequently serine appears to be the linking amino acid. The mechanism of the synthesis of the protein–mucopolysaccharide complex is not clear, but it appears that the fibroblast synthesizes the complete molecule. The hyaluronic acid of the synovial fluid has been extensively studied both in the normal and in the arthritic subject. Whilst the normal knee joint may contain only 0·4 ml of synovial fluid, an arthritic joint may contain 20–30 ml of fluid containing a relatively unpolymerized hyaluronic acid and a disproportionately high content of protein.

Synthesis of the Protein–Mucopolysaccharide Complex

The following mechanism has been suggested. Firstly the protein is synthesized at the ribosome, this is followed by the addition of the sugars and sulphate at the endoplasmic reticulum with an accumulation of the final product in vesicles whose contents are finally exported by reverse pinocytosis.

General Biochemistry of the Skin

The oxygen consumption of the skin is roughly equally divided between dermis and epidermis and since the epidermis is only about one-twentieth of the skin, it follows that it is much more active than the dermis. Epidermis *in vitro* has an oxygen consumption of about 0·5 μmole/min/g, virtually unaltered by the addition of substrates, and the rate of glycolysis is very similar. Probably *in vivo* the rate is somewhat less, being limited by diffusion of oxygen into the avascular tissue.

Apart from the liver, most tissues contain very little free glucose. However, the skin is an exception, containing up to approximately 3 g of glucose (the blood contains approximately 4–5 g). The concentration of skin glucose reflects the blood concentration (values about 70 per cent of the blood) but changes after the blood glucose so that skin glucose remains high for a longer period than blood glucose. This results in the skin buffering the blood glucose, absorbing glucose during hyperglycaemia and contributing to the blood during hypoglycaemia. Assuming the glucose concentration in the extracellular space is the same as the plasma concentration, it follows from the total quantity of glucose in the skin that much must be intracellular at a concentration roughly one-third of that in the plasma. Although the cells of the skin contain the enzymes necessary for glycogen synthesis, there is little or none found unless the skin is damaged. The utilization of glucose by the skin is largely anaerobic yielding lactate, with most probably being formed via the pentose phosphate pathway. The carbon of the glucose can also be synthesized into lipids and steroids. Skin can also carry out the same syntheses when supplied with acetate. Most of the lipid synthesis is carried out by the epidermis, the rate of the synthesis being related to the skin's concentration of glucose. Something of the order of 1 μmole of glucose is transformed into lipids per gram of epidermis per hour. All the lipid classes are synthesized by the epidermis and fatty acids up to C_{30} have been found. The unusual features of the skin fatty acids include a large percentage of $C_{16:\Delta 6}$ monoenoic acid in place of the normal $C_{18:\Delta 9}$ (oleic acid), about 3 per cent of branched chain fatty acids, and about 10 per cent of odd chain fatty acids. The branched chain acids may originate from branched chain amino acids but may also be formed by the utilization of methyl malonyl-CoA in the elongation of the fatty acid chain (see p. 253). As with other tissues of the body, the skin needs to be supplied with preformed essential fatty acids. The lipids synthesized by the skin are spread in a thin film on the surface. The film increases the smoothness of the skin and it is generally believed, without much experimental evidence, to act as a barrier to microorganisms and to water. Certainly the growth of *Trichophyton* (the fungus responsible for athletes foot) is inhibited by the free fatty acids in the skin's lipid film. The synthesis of sterols and their esterification by essential fatty acids is apparently a necessity for normal keratinization and imbalance in this process is associated with various scaly conditions of the skin. The surface lipids of the skin contain some 10 per cent of squalene. This intermediate of cholesterol synthesis (see p. 239) does not normally accumulate in cells. Finally, there is the well-known conversion of 7-dehydrocholesterol into vitamin D by the effect of sunlight.

Steroid Metabolism of the Skin

The nature of the skin is modified by the sex hormones. Androgens stimulate the activity of the sebaceous glands and influence the sexual distribution of hair. Eunuchs do not suffer from acne or baldness, whereas an imbalance of androgens in women can be associated with the growth of facial hair. Many skin complaints are now treated with corticosteroids. The skin also metabolizes steroids presented to it, sometimes transforming them to more active substances and sometimes to less active or inactive materials and this independently modifies the response to a particular hormone.

Role of Vitamin A in the Metabolism of Cartilage and Bone

It has been long known that overdosage of vitamin A in growing animals produces softening of the bones and degeneration of cartilage. Vitamin A induces these changes in two ways. The first is stimulation of the breakdown of the lysosomes in the cartilage with release of the hydrolytic enzymes which degrade the proteins and mucopolysaccharides of the young cartilage. The second is inhibition of the synthesis of components of cartilages such as hexosamine and hydroxyproline.

Calcium Phosphate Deposition in Bone

The main supporting material of the body is bone, which owes its strength to the presence of calcium phosphate in the form of hydroxyapatite. The calcium is constantly being laid down by the osteoblasts and eroded by the osteoclasts. The formation of the calcium phosphate is brought about by the enzyme alkaline phosphatase, which liberates phosphate from organic phosphates masking the surface of bone crystals. Phosphate in the extracellular fluid then combines with calcium to form insoluble calcium phosphate which extends the bone crystals. In the inherited disease of hypophosphatasia the bone does not calcify properly and the alkaline phosphatase of the tissues is low. The network of collagen within bone appears to act as a former for the crystallization of calcium salts. Abnormalities of collagen arrangement can lead to abnormalities of the bone structure.

Physical Properties of Structural Tissues

By comparing values in Tables 6.1 and 6.2, it can be shown that the amount of mucopolysaccharide in the tissue is only one-tenth to one-hundredth of the amount of the fibrous proteins that they envelop. However, because the mucopolysaccharides trap water within their gel-like structure they probably occupy as large a volume as the structural proteins. Such a mucopolysaccharide gel containing embedded protein fibres may be compared to one of the modern plastics like polythene incorporating within it glass fibres and rubber strands. This material

Table 6.2 Approximate amounts of acid mucopolysaccharides in some tissues

Tissue	Mucopolysaccharide content (mg/g of fresh tissue)	Total quantity of mucopolysaccharides in the total tissue (g)
Skeletal muscle	0·1	3
Liver	0·14	0·23
Kidneys	0·36	0·1
Plasma	0·15	0·4
Skin	2·02	9·9
Bone	0·88	10·56
Lungs	0·15	1·54

would have considerable tensile strength but would be able to alter its shape and transmit pressures throughout its bulk with ease. Modification of the basic tissue 'plastic' is made by varying the proportions of chondroitin sulphate (rigid molecules, see p. 24) to hyaluronic acid (flexible molecules, see p. 24), by altering the amounts of inelastic fibres (collagen) to elastic fibres (elastin) and finally by adding inorganic filler (hydroxyapatite) to give bulk and rigidity. With these five components structures can be made as different as cartilage (low friction, good elasticity and rigidity due to a high proportion of chondroitin sulphate), synovial fluid (high lubrication and high shock absorption due to a high proportion of hyaluronic acid), tendon (high tensile strength and low elasticity due to a high proportion of collagen) and bone (high load bearing due to high concentration of hydroxyapatite filler).

References

Potentials in Exploring the Biochemistry of Human Skin. Hsia, S. L. *Essays in Biochemistry*, **7**, 1–39, 1971.

Elastic Fibres in the Body. Ross R. and Bornstein, P. *Scientific American*, 1971, June, p. 44.

Keratins. Fraser, R. D. B. *Scientific American*, 1969, August, p. 86.

Structure and Metabolism of Connective Tissue Proteins. Gallop, P. M., Blumenfeld, O. O. and Seifter, S. *Annual Review of Biochemistry*, **41**, 617–673, 1972.

The Amino Sugars Vol. IIA. Balags, E. A. and Jeanloz, R. W. (eds.). Academic Press, New York, 1965.

SECTION 7

The Blood

CHAPTER 24

Red Blood Cells

Haemopoiesis

Red cells arise from mesenchymal cells present in bone marrow. This process of differentiation involves the conversion of cells that are relatively unspecialized, with a wide range of metabolic activities, into highly specialized cells laden with haemoglobin, lacking a nucleus and having a meagre anaerobic metabolism. The absence of a nucleus, and hence of DNA (which can act as a template for the *m*-RNA needed for protein synthesis), means that mature red cells are unable to repair the damage which they suffer during passage through the capillaries. Consequently red cells have a limited life span of about 120 days. Since a 70 kg man contains some $2 \cdot 5 \times 10^{13}$ erythrocytes, it follows that roughly $2 \cdot 5 \times 10^6$ must be formed every second to replace those that are removed from the circulation. The rate of production can greatly exceed this value when the oxygen tension of the blood is reduced following haemorrhage or exposure to low atmospheric tensions of oxygen. Low oxygen tensions in the blood cause a protein activator of haemopoiesis, haemopoietin, to be released into the circulation from the kidneys. Recent work suggests that haemopoietin stimulates the differentiation of mesenchymal cells by acting as a 'genetic derepressor'. That is to say, in the presence of haemopoietin, DNA which was previously unable to function, is able to act as a template for *m*-RNA synthesis and so provides the information necessary for the synthesis of certain proteins including globin, the protein component of haemoglobin. The stimulatory effect of haemopoietin on haemoglobin and RNA formation by bone marrow is inhibited by the antibiotic actinomycin D (see p. 181) which is a specific inhibitor of DNA-dependent RNA synthesis.

Synthesis of Haem

The protoporphyrin ring of haem, the pigment of haemoglobin, is synthesized from glycine and succinyl-CoA. These two compounds react together to give δ-amino levulinic acid, two molecules of which combine to form the pyrrole derivative porphobilinogen. Subsequently four molecules of porphobilinogen polymerize to form a cyclic tetrapyrrole which undergoes modification of its side groups to yield protoporphyrin IX; the parent compound of haem.

Finally the four nitrogen atoms of the tetrapyrrole are able to chelate ferrous iron to form haem (see Figure 2.24). Apparently the body is unable to convert all the porphobilinogen into protoporphyrin and some of it always ends up as

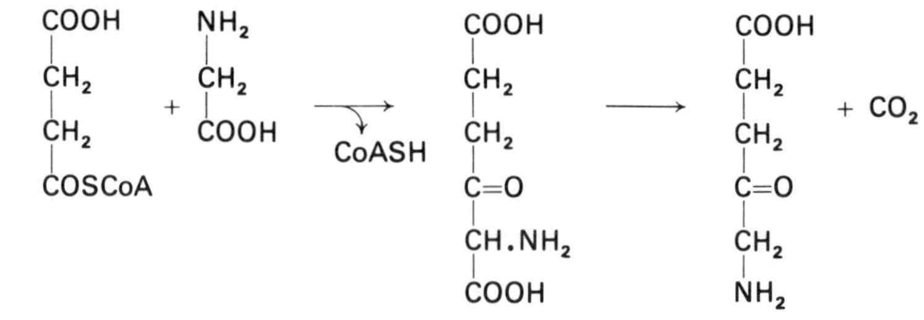

$$\underset{\text{succinyl-CoA}}{\begin{array}{c}\text{COOH}\\|\\\text{CH}_2\\|\\\text{CH}_2\\|\\\text{COSCoA}\end{array}} + \underset{\text{glycine}}{\begin{array}{c}\text{NH}_2\\|\\\text{CH}_2\\|\\\text{COOH}\end{array}} \xrightarrow{\text{CoASH}} \underset{\substack{\alpha\text{-amino-}\beta\text{-oxo}\\ \text{adipic acid}}}{\begin{array}{c}\text{COOH}\\|\\\text{CH}_2\\|\\\text{CH}_2\\|\\\text{C}=\text{O}\\|\\\text{CH.NH}_2\\|\\\text{COOH}\end{array}} \longrightarrow \underset{\substack{\delta\text{-amino-}\\\text{levulinic acid}}}{\begin{array}{c}\text{COOH}\\|\\\text{CH}_2\\|\\\text{CH}_2\\|\\\text{C}=\text{O}\\|\\\text{CH}_2\\|\\\text{NH}_2\end{array}} + CO_2$$

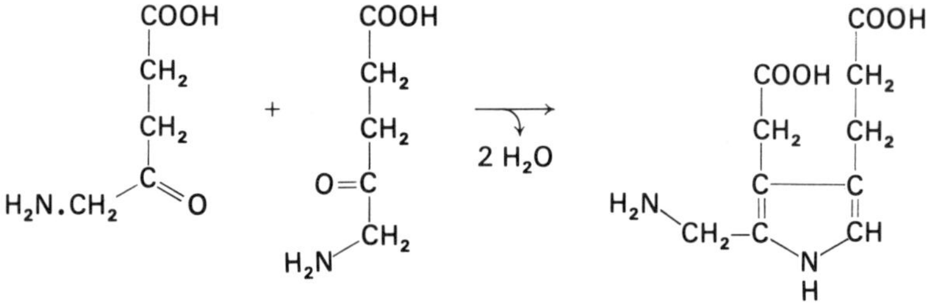

[Reaction of two δ-aminolevulinic acid molecules losing 2 H₂O to form porphobilinogen]

coproporphyrin which is useless for the synthesis of haem and is excreted in the urine. The amount of coproporphyrin so excreted is a reasonably constant fraction of the amount of haem synthesized. Therefore when the rate of haemoglobin synthesis is increased, following haemorrhage for example, the amount of coproporphyrin in the urine also increases. There are conditions (some hereditary, some acquired) termed porphyrias, in which porphyrins are synthesized at an excessively high rate so that their level in the blood rises and the

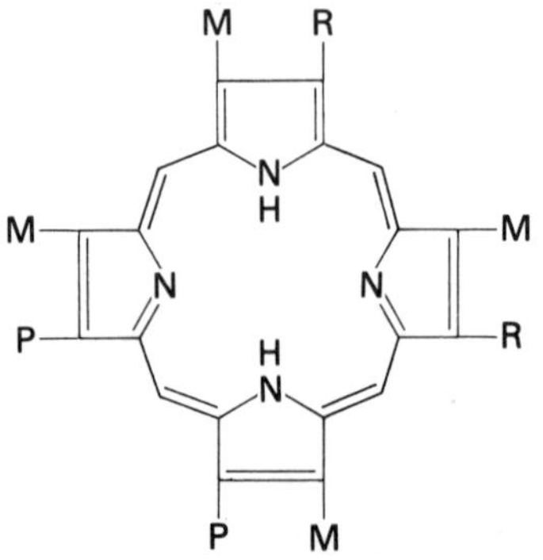

[Porphyrin ring structure with substituents M, R, P labeled around the four pyrrole rings]

$M = -CH_3$; $R = -CH=CH_2$; $P = -CH_2 . CH_2 . COOH$ in protoporphyrin IX
$R = -CH_2 . CH_2 . COOH$ in coproporphyrin II

urine becomes dark red in colour. Often in these conditions porphyrins are deposited under the skin which then becomes sensitive to light and ulcerates. There can also be mental disturbances and recently it has been suggested that the apparent madness of George III was a result of porphyria.

Bile Pigments

When haemoglobin is broken down, the porphyrin ring is opened and converted into the bile pigments biliverdin and bilirubin which are excreted in the

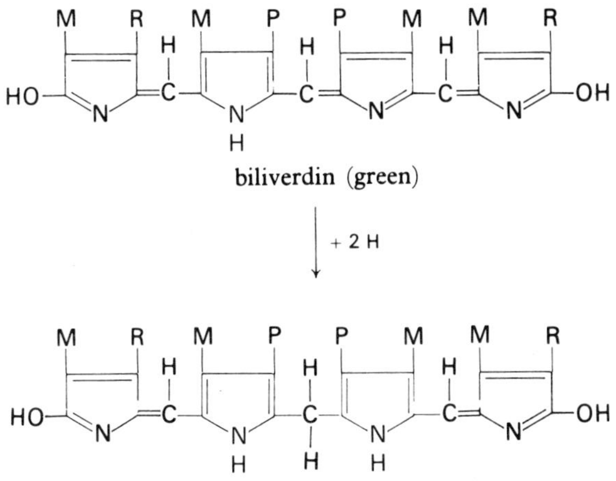

biliverdin (green)

$\downarrow$ + 2 H

bilirubin (yellow)

faeces. A 70 kg man destroys about 6·25 g of haemoglobin daily giving a total excretion of bile pigments around 200 mg per day.

Carbohydrate Metabolism and Drug Sensitivity of Red Cells

Intermediate between the nucleated normoblasts and enucleate red cells are the reticulocytes, which, although not possessing an organized nucleus, retain some basophilic material. These cells contain mitochondria allowing them to obtain succinyl-CoA and energy by oxidative phosphorylation, both of which are required for the synthesis of haemoglobin. During periods of very rapid red cell production, large numbers of these immature red cells enter the circulation. In contrast to the reticulocytes, mature red cells lack mitochondria and meet their small energy requirement by glycolysing glucose at a rate of 1·5 mmoles/kg/hr. Some of the ATP supplied in this way is used to drive the sodium pump which maintains the differential distribution of sodium and potassium ions across the erythrocyte membrane.

Besides the glycolytic pathway, red cells contain the enzymes of the pentose phosphate pathway and therefore are able to form NADPH + H^+ in the reactions catalysed by glucose-6-phosphate dehydrogenase and 6-phosphogluconate dehydrogenase. Lowering the activity of glucose-6-phosphate dehydrogenase and hence the ratio (NADPH + H$^+$/NADP$^+$), decreases the stability of red cell membranes and increases their tendency to lyse. For many years it had been known that some individuals in the Mediterranean and Middle East developed a haemolytic anaemia, termed favism, after eating fava ('broad') beans. The administration of the antimalarial drug primaquine was sometimes accompanied by haemolysis of the red cells and subsequently it was found that sensitive patients also suffered from a haemolytic attack when given acetanilide or sulphanilamide. It was discovered eventually that the concentration of glutathione (both total and reduced) in the red cells of drug-sensitive individuals was lower than normal. Reduced glutathione, a tripeptide, is the principal compound con-

$$\underset{O=\overset{|}{C}-OH}{H_2N-\overset{|}{C}H-CH_2-CH_2-\overset{O}{\overset{\|}{C}}-N-\underset{H}{\overset{|}{C}H}-\overset{O}{\overset{\|}{C}}-N-CH_2-\overset{O}{\overset{\|}{C}}-OH}$$

glutathione (γ-glutamylcysteinylglycine)

taining SH groups in red cells and a certain minimum concentration of it is required to maintain the integrity of the cell membrane and to prevent it from lysing (see also p. 353). Oxyhaemoglobin converts reduced glutathione into the oxidized form which is broken down by red cells and the addition of one of the sensitizing drugs mentioned above accelerates the rate of this oxidation. Normally, red cells have the capacity to synthesize sufficient NADPH + H$^+$ to reduce oxidized glutathione as soon as it is formed and so prevent its destruction. However, whilst the activity of glutathione reductase in drug-sensitive red cells is normal, that of glucose-6-phosphate dehydrogenase is only about one-tenth of the normal value. Hence, the biochemical lesion causing the drug sensitivity of red cells is an inability to form sufficient NADPH + H$^+$ to reduce glutathione and so prevent its degradation. Drug-sensitivity can now be predicted by measuring the glucose-6-phosphate dehydrogenase activity of red cells. NADPH + H$^+$ is also used to reduce methaemoglobin (that is haemoglobin having its iron in the ferric state) back to haemoglobin. Primaquine-sensitive red cells have a diminished ability to carry out this reduction as they lack an adequate supply of NADPH + H$^+$ and recently this defect has also been used to identify persons likely to suffer a haemolytic attack if treated with the drug.

One curious observation was that the haemolysis of the red cells in susceptible patients ceased if administration of primaquine was continued. This is because only old cells are haemolysed; young cells can produce NADPH + H$^+$ at a rate sufficient to keep their glutathione reduced. Thus when all the old cells are lysed, lysis stops. It is known that the glucose-6-phosphate dehydrogenase of red cells declines with age, presumably because the cells, lacking a nucleus, are unable to replace lost enzyme molecules. Since the deficiency of glucose-6-phosphate dehydrogenase is a genetic defect, it might be expected that other types of cells would also be deficient in the enzyme. This is so, but the relative deficiency is not as great as in red cells, presumably because nucleated cells can constantly synthesize new enzyme as the old is broken down.

The role of glucose-6-phosphate dehydrogenase in red-cell metabolism is shown in Figure 7.1

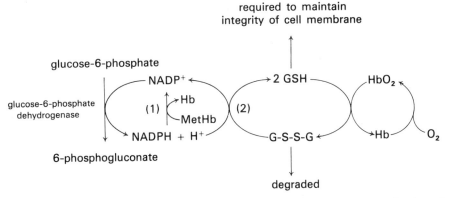

Figure 7.1 Role of glucose-6-phosphate dehydrogenase in red cells. (1) Reduction of methaemoglobin (Fe^{3+}) to haemoglobin (Fe^{2+}) by methaemoglobin reductase. (2) Reduction of oxidized glutathione to reduced glutathione by glutathione reductase

Oxygen and Carbon Dioxide Carriage by Haemoglobin

Haemoglobin is one of the few proteins whose amino acid sequence and three-dimensional structure are known (Figure 1.34). It has a molecular weight of about 67,000 and consists of four coiled polypeptide chains. Each subunit contains a cleft formed by the folding of the polypeptide chain into which a haem group is set edgeways and held by hydrophobic bonds and a dative covalent link between a histidine residue and the iron atom.

There are four different species of normal haemoglobin polypeptides in human beings, designated as α, β, γ and δ. Most haemoglobin molecules contain two α-chains and two β-, γ or δ-chains; thus haemoglobins A, A_2 and foetal haemoglobin have the composition of $\alpha_2\beta_2$, $\alpha_2\delta_2$ and $\alpha_2\gamma_2$ respectively. Each of the four polypeptides is inherited through a separate gene although they are very similar in structure. Many abnormal haemoglobins have been characterized in which a single amino acid in a particular polypeptide chain is replaced by another owing to a mutation in the gene responsible for determining the amino acid sequence of the polypeptide. The best known of these abnormal haemoglobins is haemoglobin S which occurs in people who suffer from sickle-cell anaemia and those bearing the sickle cell trait (see p. 385).

Oxygen is carried by red cells in combination with haemoglobin; in the capillaries of the lungs, oxygen attaches to the iron of the haem group to form oxyhaemoglobin. In the capillaries of other tissues, where the partial pressure of oxygen is low, oxyhaemoglobin dissociates to give free oxygen. It should be emphasized that in the gain and loss of oxygen by haemoglobin the iron always remains in the ferrous state. The extent to which haemoglobin is converted to oxyhaemoglobin depends, in a sigmoid fashion, on the partial pressure of oxygen (Figure 7.2). Such an S-shaped dissociation curve indicates that haemoglobin can be almost completely saturated by oxygen over a wide range of oxygen partial pressures. Such behaviour affords some protection against exposure to low oxygen tensions.

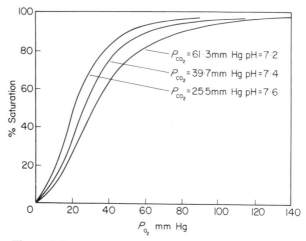

Figure 7.2 Effect of CO_2 tension on the dissociation of oxyhaemoglobin in blood. From White, Handler and Smith, *Principles of Biochemistry*, 3rd ed., McGraw–Hill, 1964

The sigmoid shape of the dissociation curve results from a cooperative interaction between the four haem groups which are carried on separate polypeptide chains. The four haem groups are not in contact with one another so that the haem–haem interaction upon which the shape of the oxygen dissociation curve depends must involve very subtle interactions between the four polypeptide chains (see p. 72). With increasing partial pressures of CO_2 the oxygen dissociation curve is displaced towards the right (Figure 7.2) This phenomenon, known as the Bohr effect, means that oxyhaemoglobin dissociates as the pCO_2 is increased; such behaviour is, of course, well suited to the unloading of oxygen in the tissues where the pCO_2 is high.

Besides carrying 97 per cent of the oxygen present in blood, haemoglobin carries some 20 per cent of the blood CO_2 bound to amino groups (the carbamino-bound CO_2)

$$Hb-NH_2 + CO_2 \rightleftharpoons Hb-NH.COO^- + H^+$$

Haemoglobin is also responsible for buffering a further 60 per cent of the total CO_2. Oxyhaemoglobin is a stronger acid than haemoglobin, thus, when blood is oxygenated in the lungs, protons are released from haemoglobin, i.e.

$$HHb + O_2 \rightleftharpoons HbO_2 + H^+$$

These protons react with bicarbonate ions to form carbonic acid which is converted rapidly to CO_2 and H_2O by the carbonic anhydrase (see p. 316) present in red cells

$$H^+ + HCO_3^- \rightleftharpoons H_2CO_3 \underset{\text{carbonic anhydrase}}{\rightleftharpoons} H_2O + CO_2$$

Owing to the low pCO_2 in the lungs, the blood is able to unload much of its bicarbonate as CO_2 in this way. In the tissues oxyhaemoglobin dissociates and in doing so takes up protons to form reduced haemoglobin which is a weaker acid than oxyhaemoglobin. These protons are derived from the carbonic acid formed by the action of red cell carbonic anhydrase on the H_2O and CO_2 produced by the metabolic activities of the tissue. Thus the pH of the blood is not disturbed by the high pCO_2 in the tissues because the protons originating from carbonic acid are tightly bound by haemoglobin. Not only does the dissociation of oxyhaemoglobin into oxygen and reduced haemoglobin permit the buffering of H^+ ions but the dissociation is itself facilitated by an increase in hydrogen ion concentration. Thus CO_2 buffering by haemoglobin facilitates oxygen unloading by oxyhaemoglobin and vice versa. The bicarbonate ions formed from CO_2 in the red cells are neutralized by K^+ ions that formerly neutralized oxyhaemoglobin.

As a result of the transformations described above, most of the CO_2 entering erythrocytes in the capillaries is converted to HCO_3^- ions with the result that the ratio $[HCO_3^-][Cl^-]$ in red cells is larger than the corresponding ratio in the plasma. This imbalance is corrected by the diffusion of HCO_3^- ions from the red cells into the plasma and their replacement by Cl^- ions moving in from the plasma. The chloride shift is reversed when the red cells enter the pulmonary capillaries and begin to take up oxygen and discharge CO_2.

The carriage of oxygen and carbon dioxide by erythrocytes is summarized in Figure 7.3.

Ion Transport by Red Cells

Red cells have been used extensively in the study of ion transport because they are very convenient to handle experimentally, rather than on account of any special physiological importance the process may have in erythrocyte physiology.

The concentration of potassium in red cells is much higher than in plasma whilst that of sodium is much lower (Table 7.1). This uneven distribution of ions was once thought to arise during red cell differentiation and to be maintained in mature red cells because their membranes were impermeable to both Na^+ and K^+. About 1940 this point of view had to be abandoned because experiments with ^{24}Na and ^{42}K indicated that red cell membranes were permeable to both Na^+ and K^+. It was also found that when blood was stored at $2°$, the red cells lost K^+ and gained Na^+; furthermore, when the temperature was raised to $37°$ the cells slowly regained their lost K^+ and expelled the excess Na^+. However, in the absence of glucose or in the presence of an inhibitor of glycolysis, such as fluoride or iodoacetate, cold-stored cells were unable to restore their ionic composition to

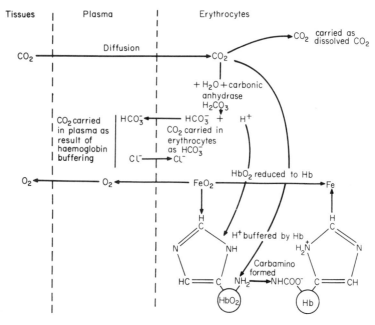

Figure 7.3 Schematic representation of the processes occurring when carbon dioxide passes from the tissues into the erythrocytes. The imidazole ring of histidine is shown as the reactive portion of the haemoglobin molecule. Modified from H. W. Davenport, *The ABC of Acid–Base Chemistry*, University of Chicago Press, Chicago, 1950

normal when incubated at 37°. These observations indicated that red cells are permeable to Na^+ and K^+ and have to maintain their ionic composition by expelling any Na^+ that enters and by taking up K^+ to replace that lost by leakage. Since such ion movements are against concentration gradients they require an input of energy in accordance with the formula:

$$\Delta G = RTn \ln \frac{[Na^+]_{out}}{[Na^+]_{in}} + RTm \ln \frac{[K^+]_{in}}{[K^+]_{out}} + (n - m) EF$$

where R is the gas constant, T is the absolute temperature, F the Faraday and E the membrane potential which is calculated to be about 10 mV. n is the number of sodium ions expelled and m the number of potassium ions taken up.

Table 7.1 Potassium and sodium concentrations in red cells and plasma

	Red cells	Plasma
$[K^+]$	150 mM	5·4 mM
$[Na^+]$	12–20 mM	144 mM

Evidence now exists that the net transport of Na^+ outwards, which is an active process, is dependent upon the inward movement of K^+. There is also a passive efflux of Na^+ that can be observed in the absence of glucose, or in the presence of glucose but absence of external K^+ (Figure 7.4). This efflux of Na^- represents internal Na^+ exchanging with external Na^+ and is abolished if Na^+ is omitted from the external medium. Such a passive exchange produces no net change in concentration so that it does not require an input of energy and can consequently be observed in the absence of glucose. Figure 7.4 shows that in the presence of

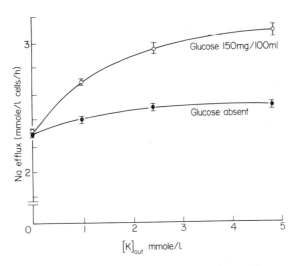

Figure 7.4 The effect of glucose on sodium efflux at different external potassium concentrations. Each point is the mean of three estimations. From Glynn, *J. Physiol. (London)*, 1956

glucose a potassium-dependent efflux of sodium is superimposed on the passive flux. This potassium-dependent sodium efflux is an active process driven by energy obtained from glucose breakdown and is stimulated half-maximally by 1·8 mM K^+.

Potassium influx also consists of two components, one independent of glucose metabolism but linearly dependent on the concentration of external potassium and another dependent upon glucose metabolism and showing Michaelis–Menten kinetics with regard to K^+ ions with a K_m of about 2 mM. The similarity between the K_m values of potassium of the K^+ influx and of K^+-dependent Na^+ efflux adds further support to the view that both processes are mediated by the same system.

Cardiac glycosides such as ouabain prevent the uptake of K^+ and expulsion of Na^+ by erythrocytes. This inhibition is by a direct action on the transport system rather than by interference with the glycolytic pathway and can be overcome by increasing the external concentrations of K^+ ions. In addition to inhibiting active transport the cardiac glycosides also decrease the size of the passive fluxes.

Ion Transport ATP-ase

Studies on the rate of K^+ influx as a function of the internal ATP concentration provided evidence that the energy required for active transport by red cells is supplied by ATP derived from glycolysis. Further clarification of the role of ATP came with the discovery of an ATP-ase in the membranes of haemolysed red cells that required the presence of Na^+, K^+ and Mg^{2+} for maximal activity. The ATP-ase and ion transport by red cells share many common features.

(1) Both systems are located in the red cell membrane.
(2) Both use ATP but not ITP.
(3) Both require the presence of Na^+ and K^+ together for maximal activity.
(4) The K_m value for Na^+ is similar for both systems.
(5) The K_m value for K^+ is similar for both systems.
(6) Cardiac glycosides inhibit both systems to the same extent.
(7) Excess potassium overcomes glycoside inhibition in both instances.
(8) The activation of both systems by K^+ is inhibited by high concentrations of Na^+.
(9) In both systems NH_4^+ can substitute for K^+ but not for Na^+.

On the basis of these similarities it was concluded that the ATP-ase activity was related to the ion transport system.

The connection between ion transport and ATP-ase activity has been examined by making use of a technique that enables the ionic composition within red cell 'ghosts' (ruptured and resealed red cells having lost most of their haemoglobin) to be defined. With such ghosts it has been found that the hydrolysis of ATP within the cells is dependent upon the external concentration of K^+ up to 10 mM and upon the internal concentration of Na^+ up to 100 mM but is independent of the external Na^+ concentration. At present it has not been possible to discover whether the ATP-ase activity is independent of the internal K^+ concentration because ghosts free from potassium have not been prepared. The asymmetric activation of the ATP-ase might be anticipated from the fact that ion transport is dependent on the external potassium and internal sodium concentrations. ATP added to the external medium is not degraded and phosphate liberated during the hydrolysis of internal ATP is released into the cells rather than into the medium. Ouabain added to the external medium inhibits activation of the membrane ATP-ase by both Na^+ and K^+.

By varying the internal concentration of Na^+ and the external concentration of K^+ the rate of ion transport can be altered. Such experiments demonstrated that the activity of the ATP-ase varies directly with the rate of ion transport so that the number of ions transported per molecule of ATP hydrolysed remains constant and is independent of the electrochemical gradient against which the ions are moved. The free energy made available by the hydrolysis of 1 mole of ATP in red cells has been calculated to be 13 kcal. During the time taken by the ouabain-sensitive ATP-ase to hydrolyse 1 mole of ATP, red cells can expend 9 kcal on ion transport. Thus the amount of energy liberated by the ATP-ase and the amount used for ion transport are similar. This is consistent with the view that ion transport and the ATP-ase form a coupled system.

The activity of Na^+/K^+-dependent ATP-ase exerts some control over the rate of glycolysis by red cells. Lactate production increases by up to 75 per cent above the basal level when ion transport is stimulated by raising the external K^+ and internal Na^+ concentrations. This extra lactate production is inhibited by ouabain. The control of glycolysis is thought to result from variations in the concentration of ADP formed from ATP by the Na^+/K^+ activated ATP-ase. During periods of increased ion transport the rate of ADP production rises and since ADP is a substrate for the phosphoglycerokinase and pyruvic kinase reactions, glycolysis is stimulated. This provides a further example of the way in which the activity of a pathway supplying energy is adjusted to meet the demands of the work being performed.

Mechanism of Ion Transport

A carrier, driven by ATP, which picked up a sodium ion at the internal surface of the membrane and transported it to the outer surface where it was exchanged for potassium which was then carried into the cell has been postulated by many workers in an attempt to account for active transport. More recently, carriers have tended to become unfashionable and current ideas are focused on enzymes that cause their cofactors to move in a vectorial fashion. A simple scheme of this type may be outlined as follows. ATP phosphorylates an enzyme in a reaction that requires Na^+ as cofactor at a site that can be approached only from the inside of the cell. During phosphorylation the conformation of the enzyme changes so that the only path available to the sodium leads to its emergence at the opposite side of the enzyme, that is, the side in contact with the extracellular solution. Dephosphorylation of the enzyme may then require the presence of potassium at a site accessible only from the outside of the cell and lead to the movement of potassium inwards.

CHAPTER 25

White Blood Cells

Phagocytosis

One of the special functions of leucocytes is to engulf and destroy bacteria that enter the body. Biochemical studies have shown that the respiration of isolated leucocytes is stimulated by the presence of dead bacteria or small particles and that the increased respiration is more or less proportional to the number of particles taken up. This work has been set on a quantitative basis by using polystyrene granules of known size. Polystyrene spherules are taken up rapidly at first (Figure 7.5) but after 20 min no more are engulfed. The total amount of polystyrene accumulated is independent of the size of the particles. Table 7.2 illustrates this feature of phagocytosis and also shows the three-fold stimulation of respiration that occurs in the presence of the spherules. Probably the most important factor contributing to the increased oxygen consumption is the oxidation of NADPH + H$^+$ by a particle-bound NADPH + H$^+$ oxidase. The NADPH + H$^+$ is formed by the preferential oxidation of glucose via the pentose phosphate pathway that accompanies phagocytosis. The physiological role of this

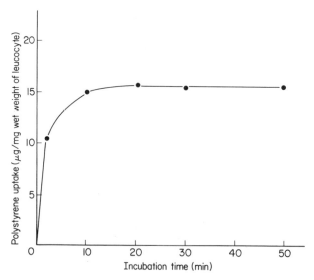

Figure 7.5 Uptake of polystyrene granules by leucocytes. From Roberts and Quastel, *Biochem. J.*, 1963

Table 7.2 Effect of polystyrene particles on phagocytosis and O_2 consumption by polymorphonuclear leucocytes from guinea pigs

Diameter of spherules (μ)	O_2 uptake in 40 min (μl)	Polystyrene uptake in 40 min (μgm/mg wet wt. of leucocyte)
0·088	17·3	7·4
0·264	44·7	30·1
1·305	51·2	34·3
3·04	50·6	35·9
7 to 14	14·1	0
No polystyrene	14·9	0

NADPH + H$^+$ oxidase is unknown. However, experiments suggest that the stimulated consumption of oxygen is subsequent to the accumulation of bacteria or particles by phagocytosis. These processes using oxygen play some part in destroying engulfed bacteria, because the bacteriocidal, but not the bacteria-accumulating, properties of leucocytes are impaired under anaerobic conditions. Furthermore, leucocytes that have been allowed to take up bacteria and have then been washed quickly to remove any extracellular bacteria still exhibit an elevated Q_{O_2} although phagocytosis is clearly impossible. The energy required for phagocytosis appears to be supplied by the glycolytic pathway rather than by oxidative phosphorylation since the production of lactate increases. Also the uptake of particles is inhibited by fluoride and iodoacetate but not by cyanide, dinitrophenol or anaerobic conditions.

The rate of incorporation of $^{32}PO_4$ into phosphatidic acids, phosphatidyl inositol and phosphatidyl serine is also increased greatly during phagocytosis. This increased turnover of phospholipids may be related to changes in the organization and amount of cell membrane that are required if particles are to be enveloped.

CHAPTER 26

Plasma

Plasma Proteins

The plasma is one of the most inactive parts of the body biochemically, yet it is the material that is most frequently analysed for diagnostic purposes. This is because its composition reflects the activity of other organs in the body and is not distorted by its own metabolic activity. Besides its six main classes of proteins, the plasma contains enzymes that have leaked from the tissues. The six classes of plasma proteins are albumin, α_1-globulin, α_2-globulin, β-globulin, γ-globulin and fibrinogen. Albumins are soluble in water whereas globulins and fibrinogen are insoluble in water but soluble in dilute salt solutions. The various protein fractions are usually separated by paper or gel electrophoresis for clinical examination (Figure 7.6). Most of the plasma proteins enter the blood from the liver where they are synthesized, although the γ-globulins, which are rich in antibodies, are derived from lymphoid tissue and plasma cells. The presence of plasma proteins results in the development of a colloid osmotic pressure across the capillary walls that opposes the hydrostatic pressure of the blood and so prevents the tissues from becoming oedematous. Under normal conditions the plasma proteins turn over rapidly and must be removed at rates equal to those of their synthesis since their concentration in the plasma remains constant. Proteins removed from the circulation, possibly by pinocytosis, are degraded by intracellular proteinases to their constituent amino acids which can then be incorporated into new proteins. Thus the plasma proteins act as a mobile store of amino acids which is used by the tissues. In prolonged starvation the concentration of plasma proteins in general, and of albumins in particular, falls, and, especially in children, hunger oedema develops. In some cases of extreme malnutrition, albumins are injected intravenously to alleviate the oedema and to provide a source of amino acids.

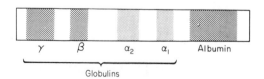

Figure 7.6 Paper electrophoretic pattern of human serum, showing the separation of albumin and globulins (α, β and γ)

Plasma Proteins and Disease

The concentration of individual proteins in the plasma changes in certain diseases. For example, cirrhosis of the liver leads to the impairment of albumin synthesis and a consequent fall in the level of albumin in the plasma, whilst in nephrosis albumins are lost in the urine. The β-globulins, precursors of β-lipoproteins, increase in quantity during hepatitis, as do γ-globulins in cirrhosis of the liver. Apart from changes in response to disease, there are some inherited abnormalities in the composition of the plasma resulting from an inability to synthesize particular proteins such as fibrinogen, antibodies, albumins and thromboplastic factors.

Besides the normal proteins already considered, the plasma contains small quantities of enzymes that are of growing importance in clinical studies. Damage to tissues frequently results in a leakage of enzymes into the plasma and there is a roughly quantitative relationship between the amount of tissue damaged and the amounts of enzymes liberated. Thus, pathological changes in some tissues are accompanied by marked increases in the activity of characteristic enzymes in the plasma whose measurement can aid diagnosis of the condition. In prostatic cancer, the activity of acid phosphatase in the plasma is elevated, whilst pancreatitis leads to an increase in the amount of circulating amylase. Muscular dystrophy is attended by a rise in the plasma concentration of such enzymes as aldolase and creatine phosphokinase. In myocardial infarction glutamate–oxaloacetate transaminase (GOT), lactic dehydrogenase (LDH) and malic dehydrogenase (MDH) enter the circulation from the damaged tissue.

Some enzymes occur in several distinct forms, termed isoenzymes, that can be separated from one another by electrophoresis. The pattern of LDH isoenzymes found in heart differs from that found in liver (see Figure 1.24). Thus by studying the LDH isoenzymes in plasma following myocardial infarction it is possible to determine whether any secondary damage has occurred to the liver as a result of congestion caused by a diminished venous return.

Not all increases in the activity of plasma enzymes are indicative of pathological changes. Following severe muscular exercise there is an elevated level of aldolase in the plasma and during pregnancy the activity of plasma histaminase increases, presumably to protect the mother from the toxic effects of histamine which is present in high concentration in the foetus.

In addition to aiding diagnosis, measurement of changes in the activity of plasma enzymes can be of use in assessing the progress of recovery. For example, the healing of a cardiac infarction is attended by a decline in the level of enzymes entering the blood from the damaged region.

Blood Coagulation

Blood has to remain fluid so that it can be circulated around the body, but at the same time it has to be able to coagulate rapidly if bleeding is to be brought under control. The need to fulfil both of these requirements has led to the evolution of a highly complex series of processes that allow the clot-forming protein, fibrin, to be liberated from its precursor, fibrinogen, only following damage to the blood vessels (Figure 7.7).

When blood is shed, changes occur which result in the formation of a platelet thrombus and the liberation of thromboplastins having proteolytic activity from both damaged tissues and plasma. These initial changes involve interactions between many clotting factors, protein and lipid in nature, that are understood in part only. Provided that Ca^{2+} is available, the thromboplastins bring about the conversion of prothrombin to thrombin, a proteolytic enzyme whose natural substrate is fibrinogen. Under the influence of thrombin, fibrinogen is hydrolysed to

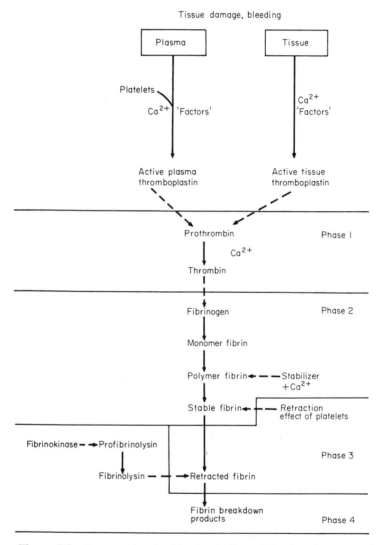

Figure 7.7 Processes involved in the clotting of blood; → is converted to, --→ acts on. Redrawn from *Documenta Geigy*, 6th ed., 1962

give 'monomer' fibrin and two fibrinopeptides. Fibrin monomers polymerize in the presence of Ca^{2+} and a stabilizer to form a network of proteins, cross-linked by disulphide bonds, called stable fibrin. Platelets cause this fibrin clot to retract, thereby expelling any occluded plasma. Clots are eventually broken down by the fibrinolytic system whose active agent is another proteolytic enzyme termed fibrinolysin.

Coagulation can be inhibited for clinical purposes by heparin which amongst other actions interferes with the prothrombin-to-thrombin conversion and by vitamin K antagonists such as dicoumarol since the formation of prothrombin depends upon the presence of vitamin K.

CHAPTER 27

The Metabolism of Iron

The red cells contain about 3 g of iron representing some 60 per cent of the total in the body; the greater part of the remainder is stored in an iron-containing protein called ferritin, especially in the liver and spleen. The structure of ferritin is shown in Figure 7.8. Myoglobin, cytochromes and enzymes in the tissues contain only about 200 mg or 4 per cent of the total.

In males the normal daily loss of iron from the tissues, chiefly as a result of cells shed from the skin and intestinal mucosa, is about 1 mg per day and any loss in excess of this is the result of bleeding. In adult women additional iron is lost during menstruation and to the foetus during pregnancy. Although most mixed diets contain between 15 and 40 mg of iron, only 1 to 2 mg are absorbed unless

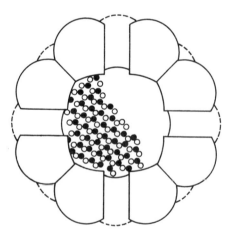

Figure 7.8 Schematic drawing of a ferritin molecule seen in section, containing about half its maximum complement of iron. The molecule consists of a multisubunit shell, in which there are spaces to allow the passage of iron in or out of the molecule, and a crystalline hydrated iron oxide core. The orientation of the latter with respect to the protein appears to be random. By kind permission of Dr P. M. Harrison

there is an increased need for iron following blood loss. If all the dietary iron were absorbed, a large, toxic excess of iron would rapidly build up as there is no method for excreting it. Hence control of iron balance is achieved at the level of absorption. Most of the dietary iron is in the ferric state which, unlike the ferrous, is not at all well absorbed. In this context it is interesting that infants may develop anaemia during vitamin C deficiency. The probable reason for this is that ascorbic acid plays a part in reducing ferric iron to ferrous and in forming more readily absorbed iron complexes. In the scorbutic infant, probably fed on pasteurized or dried milk, there is a combination of low vitamin C intake and low iron intake since cows milk contains not more than 0·2 mg iron and 1 mg vitamin C per 100 ml. Reduction of ferric iron to the ferrous state is favoured by the acid medium of the stomach, consequently iron absorption is impaired in cases of achlorhydria. The absorption of iron is also influenced by the other components of the diet; phosphates, present in wheat products, for example, impair iron absorption whilst sugars and amino acids enhance it. Haemoglobin provides a good source of absorbable iron and it appears that the iron is absorbed as intact haem.

The actual quantity of iron absorbed by the intestinal mucosa depends upon the size of iron pool in the body. Following iron loss by bleeding the percentage of dietary iron absorbed increases and remains elevated until the deficiency is made good. The manner in which this control is exercised is not entirely clear. Ferrous iron is taken up by mucosal cells and probably conveyed to the serosal pole of the cells in combination with a soluble protein similar to transferrin. The quantity of this carrier protein in the mucosal cells increases in iron deficiency, and it may be the factor whose level controls the amount of iron transferred across the gut wall. Iron subsequently becomes bound as Fe^{3+} to the plasma protein transferrin; the attachment to protein prevents iron being lost into the glomerular filtrate as the equilibrium binding constants for both atoms of bound iron are about 10^{30}. The binding of iron is accompanied by the binding of one mole of bicarbonate per mole of iron. Iron bound to transferrin is carried from the intestinal mucosa and from the sites of haemoglobin catabolism in the reticulo-endothelial system to the sites of erythropoiesis in the bone marrow. Within the marrow transferrin binds to sites on the red cell precursors before a direct transfer of iron takes place, possibly following removal of bicarbonate and a consequent fall in the affinity of transferrin for iron. Release of iron from transferrin probably also involves its reduction to the ferrous state.

Normally plasma transferrin is about 30 per cent saturated with iron; if the saturation exceeds 60 per cent, the excess iron is deposited in the liver as ferritin.

In the absence of iron deficiency, the transfer of iron through the mucosal cells is low (possibly owing to a low level of intracellular transfer protein) and iron becomes sequestered in the mucosa in the Fe^{3+} state as ferritin. This store of iron is mobilized only slowly and as a result much of it tends to be lost as the cells are exfoliated.

Iron liberated in the breakdown of haemoglobin is deposited as ferritin in the liver, spleen and in the reticulo-endothelial cells of bone marrow if it is not immediately used for haemoglobin synthesis. Using radioactive iron it has been found that about 27 mg of the body's store of 5 g of iron is used daily for haemoglobin synthesis and of this about 20 mg is supplied by the direct reutilization of iron from degraded red cells.

Since iron is not excreted to a significant extent, it follows that transfusions increase the total iron of the body in a relatively irreversible manner. Excess iron is deposited in the liver as haemosiderin granules. Accumulation of haemosiderin can cause severe liver damage in patients suffering from haemolytic anaemia who have received many transfusions over a period of years.

CHAPTER 28

Specific Immune Responses

Introduction

The body possesses two systems able to recognize specifically, and to destroy, any foreign macromolecules and cells that may gain access to it; they are the systems of humoral and cell mediated immunity. These systems are principally directed against infective agents, but they also play a part in finding and destroying endogenous tumour cells by combining with specific antigens on the cell surfaces. The immune system is thus the arbitrator of what constitutes 'self' and 'non-self' at the molecular level; if this system is inadvertently directed against self, there ensues what is termed an auto-immune disease.

The lymphoid system of cells responsible for generating specific immune responses passes through a period during embryological development in which it recognizes all available antigens as contributing to the 'self' of the organism. Thereafter any new antigens entering the system are recognized as foreign or 'non-self' and act as triggers for the initiation of an immune response.

Antibodies

Antibodies or immunoglobulins (Ig) are proteins, produced by plasma cells, which combine specifically with antigens to effect the humoral aspects of immunity. Humans have five classes of immunoglobulin IgG; IgA; IgM; IgD and IgE of which the first three are the best understood. Class IgG is the most abundant and is responsible for neutralizing antigens carried in the blood; most of the members of this class of immunoglobulin are able to fix complement (see below). Class IgA is also found in serum but its chief role is as an antibody secreted by lymphoid tissue in the lining of the gut and respiratory and urinogenital tracts. Before being released onto the tract surface, pairs of IgA antibodies combine with a component that protects them against proteolytic activity. IgA immunoglobulins do not fix complement. Immunoglobulins of class IgM, which are able to fix complement, are of high molecular weight and are confined to the intravascular space where they agglutinate organisms such as Gram negative bacteria.

Structure of Immunoglobulins

Immunoglobulins of the IgG class consist of two, identical, light and two identical, heavy, polypeptide chains linked together by disulphide bridges (Figure

7.9). The heavy chains of the IgG antibodies are always of a single type (γ) but the light chains may be of either of two different types (κ or λ). Each IgG molecule possesses two identical antigen binding sites, each consisting of the —NH₂ terminal portions of one heavy (γ) and one light (κ or λ) chain. Thus each IgG molecule is able to bind to two identical antigenic sites. The amino terminal ends of the heavy and light chains have very variable amino acid sequences for about a hundred residues and it is this feature that accounts for the existence of

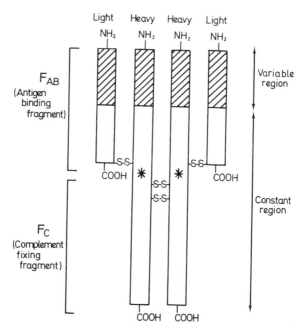

Figure 7.9 Structure of an immunoglobulin of class IgG. *Sites of proteolysis by papain to give two F_{AB} fragments and one F_C fragment per molecule

large numbers of different IgG antibodies with specificities for a wide range of antigens. Figure 7.10 shows the structure of IgG, IgA and IgM immunoglobins.

Any one antigen, be it pure protein or a whole microbe, carries a large number of different antigenic determinants, each of which will stimulate the production of a specific antibody. Thus for any one antigen there are several species of specific antibodies. This and the fact that antibodies have more than one valency allows a lattice of antigen and immunoglobulin to form with the result that invading bacteria are agglutinated and macromolecules are precipitated. Combination of antigen with antibody is merely the beginning of the counterattack against the foreign body; the next stage of the battle involves the serum complement system.

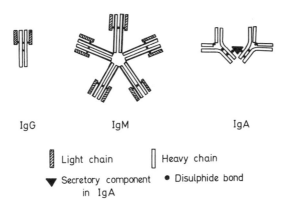

Figure 7.10 Structure of immunoglobulins

The Serum Complement System

The serum complement system, consisting of a number of plasma proteins, is called into action once antibodies and antigen have combined. A component of the complement system binds to the antigen/antibody complex and thereafter the system releases a chemotactic factor that attracts granulocytes to the site of the immune response, renders the foreign particle more susceptible to phaocytosis, increases vascular permeability and causes lysis of invading cells. The sequence of reactions occurring when the serum complement system is activated is shown in Figure 7.11. Even a single IgG antibody binding to the surface of a foreign cell can bring about its destruction, as activation of the complement system leads to a great amplification of the initial immune response.

Antibody Production

It is generally believed that the members of a clone of mature plasma cells are able to synthesize only one, unique, immunoglobulin. This is indicated by cases of multiple myeloma in which there is a vast overproduction of a monoclonal immunoglobulin; however, in every case investigated the nature of the immunoglobulin produced has been different. Presumably, in all cases, the neoplastic plasma cells are derived from a single parent cell. The reason for small lymphocytes (the precursors of plasma cells) synthesizing only a single immunoglobulin is unknown, as is the process for eliminating those clones of lymphocytes producing immunoglobulins directed towards endogenous or 'self' antigens. Such 'anti-self' clones are destroyed during embryological development if the lymphocytes come into contact with the corresponding antigen. Clones producing antibodies to sequestered antigens can however survive *post partum* and can be responsible for autoimmune diseases if a sequestered antigen becomes accessible to the appropriate lymphocytes.

Following the first introduction of a foreign antigen into an animal there is a latent period during which the antigen is processed by macrophages and trans-

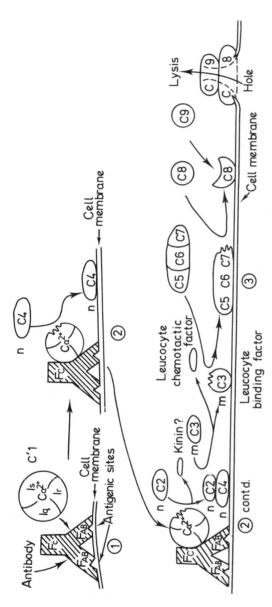

Figure 7-11 Activation of the serum complement system.

(1) Antibody binds to antigenic sites on the membrane of an invading Gram negative bacterium and complement component C'1 attaches to its Fc moiety via subunit 1q. The consequent conformational change converts subunit 1s, a proenzyme, into an active protease.

(2) 1s then modifies molecules of component C4 so that they can bind to the cell surface. It also hydrolyses molecules of component C2 to yield two products, one able to bind to C4 and the other, a kinin like peptide able to increase vascular permeability and contract smooth muscle.

(3) The C2/C4 complex on the surface of the cell then acts upon many molecules of component C3 so that they bind to the membrane. In the process a leucocyte chemotactic factor is released. The membrane-bound C3 units render the cells readily susceptible to phagocytosis by the leucocytes called up by the chemotactic factor. Membrane-bound C3 also converts a complex (C5/C6/C7) to an active, membrane-bound, form able to activate component C8. When C8 binds to the membrane and interacts with C9, holes appear in the membrane and lysis occurs.

If components C4 and C3 do not bind to the membrane as soon as they are formed they undergo a rapid conformational change that renders them inactive, and harmless to host cells

ferred to members of the appropriate clone of small lymphocytes. The appearance of specific IgM antibodies produced by plasma cells derived from the lymphocytes marks the end of the latent period, but these antibodies decline in abundance later. It is suggested that this coincides with a switch from IgM to IgG immunoglobulin production (Figure 7.12). In the absence of further immunological stimulation, the level of circulating antibodies falls to the resting value. However, subsequent exposure to the same antigen leads, without a long latent period intervening, to a rapid proliferation of small lymphocytes and their transformation into plasma cells producing chiefly IgG immunoglobulins.

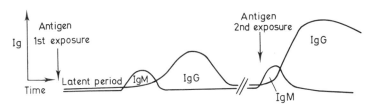

Figure 7.12 Primary and secondary antibody responses

Antibody-producing plasma cells are derived from thymus independent lymphocytes (or B-lymphocytes), so called because their proliferation in the lymph nodes is not dependent upon the thymus.

Immunological stimulation also gives rise to a population of lymphocytes whose proliferation and development in the lymph nodes is dependent upon influences from the thymus. These thymus dependent cells (or T-lymphocytes) are responsible for cell mediated immune responses (delayed hypersensitivity responses).

Cell Mediated Immunity

Cell mediated immunity is exercised by antigen sensitized lymphocytes (T-lymphocytes) that have followed a thymus-dependent course of development. Following combination of antigens with specific receptors (probably similar to antibodies) on the surface of T-lymphocytes, several events occur. The lymphocytes divide and release both a polypeptide transfer factor able to transfer reactivity to other uncommitted lymphocytes and a mitogenic factor that stimulates their division. They also release chemotactic factors for both macrophages and polymorphonuclear leucocytes and macrophage activation factors. The latter appear to increase phagocytosis of bacteria or foreign cells introduced by tissue transplantation.

Activated lymphocytes liberate interferon, a protein factor able to protect other cells against viral infection by inducing them to produce an anti-viral protein. Interferon is also formed and released by other cell types when they are invaded by viruses. Activated lymphocytes are also the source of lymphotoxins that destroy their target cells. Full development of the cell mediated response in sensitized individuals takes about 3 days in contrast to the much more rapid humoral response.

Blood Group Antigens and Histocompatibility Antigens

Individual human beings are characterized by their possession of a unique set of antigens termed homologous antigens or isoantigens. Examples of these are the blood group substances (responsible for the division of people into blood groups A, B, AB and O), the Rhesus factor and the histocompatibility antigens. It is the existence of these and other antigens that prevent tissues being freely interchanged between individuals whether it be by blood transfusion or heart transplantation. For example, a person with A blood group substance cannot donate blood to a B group recipient as the latter recognizes A group red cells as foreign and is, therefore, able to raise antibodies against them. Similarly, a Rhesus negative mother (one lacking the Rhesus antigen) of a Rhesus positive foetus raises antibodies against Rhesus antigens entering her circulation on any foetal cells. Maternal anti-Rhesus immunoglobulins entering the foetal circulation during that or any subsequent Rhesus positive pregnancy cause severe anaemia and leukopenia in the foetus.

The blood group antigens are complex polysaccharides; that of Group O individuals, the H-antigen, has the structure:

$$\beta\text{-galactose} \xrightarrow{\text{or } 1,3 / 1,4} N\text{-acetylglucosamine} ----$$
$$\Big| \alpha\text{-1,2}$$
$$\text{fucose}$$

as its non-reducing terminal. Group A individuals possess an enzyme able to add N-acetylgalactosamine to the terminal galactose residue

$$\alpha\text{-}N\text{-acetylgalactosamine} \xrightarrow{1,3} \beta\text{-galactose} \xrightarrow{\text{or } 1,3 / 1,4} N\text{-acetylglucosamine} ---$$
$$\Big| \alpha\text{-1,2}$$
$$\text{fucose}$$

whilst group B individuals possess an enzyme adding a galactose residue in the same position

$$\alpha\text{-galactose} \xrightarrow{1,3} \beta\text{-galactose} \xrightarrow{\text{or } 1,3 / 1,4} N\text{-acetylglucosamine} ---$$
$$\Big| \alpha\text{-1,2}$$
$$\text{fucose}$$

Group AB individuals possess both enzymes whilst those of Group O lack both.

References

The Mechanism of Blood Coagulation. Owen P. A. and Stormorken, H. *Ergebnisse des Physiologie/Reviews of Physiology,* **68**, 1–54, 1973.
Iron in Biochemistry and Medicine, Jacobs, A. and Worwood, M. (eds.), Academic Press, London, 1974.
The Immune System. Jerne, N. K. *Scientific American,* 1973, July, p. 52.
The Complement System. Mayer, M. M. *Scientific American,* 1973, November, p. 54.
Markers of Biological Individuality. Reisfeld, R. A. and Kahan, B. D. *Scientific American,* 1972, June, p. 28.
The Induction of Interferon. Hilleman M. P. and Tytell, A. A. *Scientific American,* 1971, July, p. 26.

SECTION 8

The Lungs

CHAPTER 29

The Biochemistry of the Lungs

The lungs weigh about 1 kg and their role, both physiological and biochemical, is complex and not only related to respiration. The lungs have a structural net of collagen to which are attached four cell types, all with separate functions and biochemistry. These are:

(1) Large alveolar epithelial cells through which the gaseous exchange occurs and which probably secrete the special phospholipid that lines the lung. The size of the small spherical alveoli is such that once deflated, surface tension forces of normal watery fluids would effectively prevent reinflation. The lowering of surface tension brought about by the phospholipid lining (surfactant) enables the reinflation to occur.

(2) Alveolar macrophages which contain large amounts of hydrolases including the bacteriolytic lysozyme.

(3) Endothelial cells of capillaries which may give rise to lipoprotein lipase and also to the activators of fibrinolysin.

(4) Mast cells which contain heparin and histamine and also proteolytic enzymes.

These cells acting together ensure that the lung carries out its functions of gas exchanger, scavenger of particulate matter, excretory organ, heat exchanger and endocrine gland. For these many roles oxidative energy is required and the lungs consume some 4 per cent of the body's oxygen intake. This is abstracted from the blood which flows through them at about 5 l/min; this is of course the entire output of the right ventricle. Some 10 per cent of the blood volume is contained within the lungs and the surface area of the alveoli (some 50 to 80 m^2; the body's largest membrane) is 25 to 40 times greater than the surface area of the body. Because of its anatomical position the lung has the first exposure to the chylomicra entering the circulation via the thoracic duct and has a rather special fat metabolism. Of the cells described, it is probable that the large alveolar cells by virtue of their mitochondrial content and large numbers (equivalent *in toto* to an organ the size of the spleen (0·15 kg)) are responsible for the major oxygen consumption of the lung.

Scavenging Function of the Lungs

Besides acting as an excretory organ for gases, the lung removes debris brought to it in the blood stream, and microorganisms and particles that are inhaled. In both cases the process is essentially one of hydrolysis by the action of the lungs' proteinases and lipases. Although the small blood clots, agglutinated cells, fat, etc. from the blood stream that are hydrolysed are all formed by normal mechanisms, they may nevertheless give rise to pathological emboli. Thus the lungs act as a sort of living sieve. The alveolar macrophages are particularly rich in lysozyme (see Figure 1.40) and are responsible for removing microorganisms. Apart from the respiratory function of removing excess carbon dioxide, the lungs cleanse the blood of volatile materials which may arise from faulty metabolism or from the breakdown of ingested material. For example the concentration of acetone in the alveolar air reflects the concentration of ketone bodies (see p. 253) in the blood and is particularly noticeable with uncontrolled diabetics. The characteristic odour after eating garlic is due to the formation of the volatile allicin from the sulphur-containing amino acid alliin. Alcohol also appears in the breath reflecting the blood concentration and is the basis of the breathalyser system of detecting an excessive blood level (over 80 mg per cent) of ethanol.

Maintenance of Blood Fluidity

The fluidity of the blood is determined firstly by the opposing effects of heparin (which prevents blood clotting) and prothrombin (which, on conversion to thrombin, promotes it). The amount of fibrin that circulates in the blood is determined by the relative rates of formation and hydrolysis of the fibrin clot. The proteolytic enzyme fibrinolysin responsible for the hydrolysis is found in the blood as an inactive form, profibrinolysin, and the lung is a very good source of the activator that converts profibrinolysin to fibrinolysin. Lung is also a good source of the phosphatide–protein complex called thromboplastin which converts prothrombin to thrombin and of heparin which prevents the reaction. The processes are summarized in Figure 7.7.

The Mast Cells

Mast cells, whose role remains obscure, contain the sulphated mucopolysaccharide heparin, neutralized by histamine. The mast cells contain the decarboxylases for producing the amine from the amino acid and probably those for synthesizing heparin, although both substances could be made elsewhere and stored in the mast cells. The mast cells also contain several proteases. Disruption of the mast cells will liberate the stored substances and allow the histamine to have its effect, for example on the permeability of the small blood vessels.

The lungs can contain about 400 mg of heparin. It is destroyed by an enzyme heparinase which is found only in the liver but heparin is also excreted in the urine. Heparin acts as an antithromboplastin and an antithrombin and thus inhibits blood clotting. It also has an antihistamine effect. On the positive side, heparin stimulates lipoprotein lipase which is responsible for the breakdown of the chylomicra and the lipolysis of lipoproteins (see p. 249).

Lipid Metabolism of the Lung

The chylomicra from the thoracic duct pass with venous blood into the pulmonary circulation and are there exposed to the action of the lipoprotein lipase located in the endothelial membrane. Thus substantial amounts of free fatty acid become liberated for oxidation by the lungs or for synthesis into phospholipid. Lung oxidizes palmitate at half the rate of kidney and three times the rate of the heart, but it also uses a great deal of the palmitate presented to it to synthesize the special phospholipid (dipalmityl lecithin) that lines the alveoli to reduce surface tension. The alveoli are coated with several layers of phospholipids and as the lungs expand the molecules slide over each other exposing the lower layers of phospholipid but ensuring that the film remains intact at all lung volumes. The synthesis of the lung surfactant starts late in foetal life and the content in the immature lung is small. It is therefore possible that the respiratory distress syndrome of the premature infant is caused by the high surface tension in the alveoli due to lack of the phospholipid layer.

Reference

Non-respiratory Function of Mammalian Lung. Hememann, H. O. and Fishman, A. P. *Physiological Reviews,* **49**, 1–45, 1969.

SECTION 9
The Kidneys

CHAPTER 30

The Biochemistry of the Kidneys

The kidneys are the final arbiter of the solute and water balance of the body. This is achieved by a constant filtration of the blood plasma (blood flow to the kidney about 1200 ml/min) coupled with a restoration of some proportion of the solutes to the blood by active transport, a counter-current diffusion process to reabsorb water, an active secretion of some substances from the blood to the urine and various ion exchange processes. The semipermeable membrane of the glomerulus in general allows the passage of molecules with molecular weights below 10,000 and virtually excludes all molecules with molecular weight above 50,000. In between these two extremes there is a progressively decreasing filtration rate with increase in the molecular weight. The volume of fluid that passes the glomerular filter is about 125 ml/min. The driving filtration pressure is the blood pressure in the glomerular capillaries which is some 70 per cent of the arterial blood pressure. About one-fifth of the plasma flowing through the capillaries is filtered at the glomerulus. Filtration stops when the arterial pressure falls below about 40 mm Hg. The colloid osmotic pressure of the blood (see p. 290) is about 25 mm Hg (about 1/200 of the total osmotic pressure of the blood) and it is this colloid osmotic pressure that sets the physical limits on the glomerular filtration. The result of the filtration and readjustment processes is to produce urine whose composition varies widely but is approximately that given in Table 9.1 which also compares the concentrations of the major constituents of the plasma. The volume of the 'normal' urine ranges from about 600 ml to 1600 ml daily and the amounts of the constituents excreted also varies considerably, particularly the nitrogenous constituents which largely reflect the protein content of the diet. However, an individual's output of creatinine is comparatively constant and is sometimes used as a standard to relate to other excreted substances. Also, since creatinine is completely filtered at the glomerulus but is not reabsorbed, its measurement in the urine and the blood gives some measure of the glomerular filtration rate.

Processes Involved in Restoring Filtered Substances to the Blood

The transport of glucose back from the tubule lumen is an active process and is of the type described in Chapter 13 in connection with intestinal absorption. Sodium entering the cells of the kidney tubules from the lumen is actively pumped into the extracellular fluid and hence into the blood by a Na^+/K^+ activated ATP-ase (see p. 286). The reabsorption of amino acids is thought to be by

Table 9.1 Comparison of constituents of blood and urine

Substance	Concentration in arterial blood (mM)	Concentration in urine of nominal daily volume of one litre (mM)	Percentage of substance filtered at glomerulus that is absorbed later	Notes
Glucose	5.4	0	100	The bracketed values are those of the plasma. The osmolarity of the plasma is calculated from the osmotic pressure. The percentage of filtered solute restored to the blood is calculated from the plasma concentration and the assumption that 120 volumes of glomerular filtrate gives rise to 1 volume of urine
Urea	3.1–5.4	330–580	Variable but up to 40 decreasing with increased urine flow	
Uric acid	0.06–0.11	0.4–5.8	Variable round 50	
Amino nitrogen	3.3–4.8	19–31	95	
Ammonia	0.07–0.2	29–72	0—actively excreted by kidney	
Creatinine	0.1–0.13	9.5–28.5	0—some active excretion	
Na^+	79.3–91 (134–154)	190	99	
K^+	32.2–53.8 (3.3–5.3)	70	87	
Cl^-	77–90.6 (100–107)	200	98	
PO_4	0.7–1.2	35	70	
SO_4	0.09–0.2	21–34	60	
pH	7.33–7.45	5.5–7		
Osmolarity	(300)	800–1200		

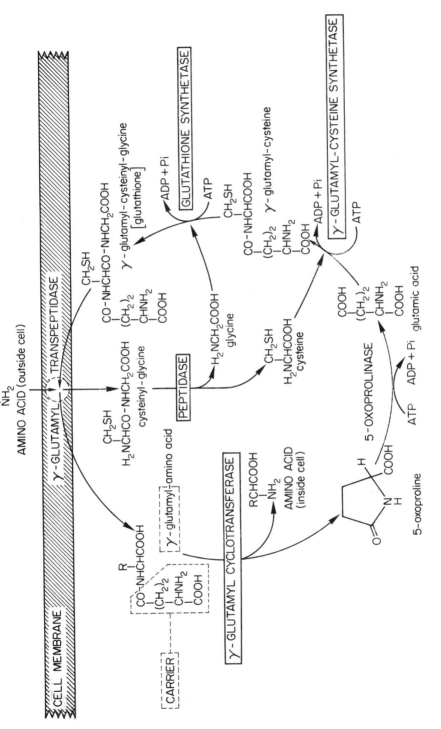

Figure 9.1 The γ-glutamyl cycle

an active process in which glutathione plays a large part. The process is a cycle, each turn of which consumes three molecules of ATP (Figure 9.1).

The average man restores to his blood from his glomerular filtrate 0·4–0·5 moles of amino acid daily, requiring 1·5 moles of ATP. Such a high consumption of energy ensures that the process is driven quantitatively in the required direction. The existence of the cycle is supported by the observations that the key enzyme which reacts with the amino acid (γ-glutamyl transpeptidase) is membrane bound; furthermore the mechanism provides a role for the ubiquitous glutathione. Also the operation of the cycle is supported by the findings that there are rare clinical conditions in which there is an inability to convert 5-oxoproline to glutamate, 5-oxoproline is then excreted in the urine and the amount excreted rises as the level of blood amino acids increase. If the level of blood amino acids exceeds the capacity of the system to make 5-oxoproline there is an aminoaciduria. Similarly, cases have been found where an apparent deficiency of the ability to synthesize glutathione has been accompanied by an aminoaciduria. All the enzymes of the cycle are found in the kidney at activities easily able to cope with the measured transport of the amino acids from the glomerular filtrate. Such a scheme, although it may well function, still leaves unexplained the known dependence of amino acid transport in kidney slices on the sodium concentration in the medium.

The passage of urea into the blood from the glomerular filtrate is essentially a passive diffusion process following the reabsorption of salts and water by the kidney tubules. The amount reabsorbed depends partly on the urine flow. The faster the flow the less urea is absorbed.

Chloride, phosphate and sulphate and other anions are in general reabsorbed in conjunction with sodium transport to ensure neutrality.

pH Adjustment of the Urine and Acid–Base Balance

The minute-to-minute adjustment of the body's pH is by altering the bicarbonate content of the blood which, in turn, depends on the CO_2 pressure in the alveolus. This pressure is automatically adjusted by the respiratory cells in the brain to give the rate of ventilation that will ensure CO_2 loss at the appropriate rate. In some respiratory disorders or experimental systems, the CO_2 adjustment cannot be made and the blood may then become more acid (excess CO_2) or more alkaline (deficiency in CO_2). CO_2 itself is a neutral substance but in contact with water will produce carbonic acid. This combination with water is a slow reaction but it is catalysed by the enzyme carbonic anhydrase found in the red cells, pancreas, gastric mucosa, salivary glands and kidney. Such tissues will therefore lower the concentration of CO_2 and produce an accumulation of carbonic acid which dissociates to give largely HCO_3^- and H^+. In these tissues possessing carbonic anhydrase, therefore, there is always an equilibrium between bicarbonate and CO_2 and their pH is determined by the partial pressure of CO_2 in the system. Since CO_2 can always readily enter such cells by diffusion they have a constant need to dispose of the excess acid (H^+). The kidney deals with this by secreting acid (H^+) to the outside of the body. In the cells of most tissues the CO_2 produced in respiration is not in the short term in equilibrium with the cells' bicarbonate as they do not contain carbonic anhydrase. Thus a burst of CO_2 production from any decarboxylation will increase the CO_2 tension and enable the CO_2 to diffuse naturally down a concentration gradient to the outside of the cell. Any rapid equilibration of CO_2 with cellular bicarbonate would cause buffering difficulties and would hinder the easy loss of the CO_2 produced. Apart from the general problem of bicarbonate and CO_2 adjustment, which is largely a respiratory one (see Section 8), there are other ways in which the pH of the body can be altered during normal metabolism. The ingestion of any neutral substance which can give rise to the net formation of non-metabolizable acid or base will produce a pH problem that must be adjusted by the kidney. For example if sodium citrate or sodium glutamate is eaten the anion can be completely metabolized in one case to CO_2 and water and in the other case to CO_2, H_2O and urea. In both cases there remains an excess of fixed base (NaOH) to be disposed of. On the other hand, if a neutral substance such as ammonium chloride is eaten the NH_3 is rapidly metabolized to urea leaving HCl to be disposed of. Similarly the oxidation of large amounts of sulphur-containing amino acids results in the formation of sulphuric acid. There also are circumstances such as prolonged starvation or the consumption of a diet very high in fat where acetoacetate and β-hydroxybutyrate are made so fast that the blood level of these two acids rises and in both cases the body is faced with an excess of acidity. The kidney finally disposes of this excess acidity as it does also when excess phosphate occurs possibly from the oxidation of phosphoproteins and nucleic acids.

The adjustment of excess alkalinity in the plasma is by the tubule failing to completely reabsorb $NaHCO_3$ filtered by the glomerulus; this continues until the plasma pH approaches normal. Thus the urine excreted under these conditions is alkaline and under some extreme circumstances may approach pH 8·0. Some further adjustments are made as the urine passes down the tubule such as exchanging rather more potassium for sodium than at neutral pH and also allowing two cations to be associated with the phosphate excreted.

The adjustment to excess of acid other than carbonic is rather more complex but is summarized by Figure 9.2. A is a non-metabolizable anion and the glomerular filtrate is a mixture of NaA, $NaHCO_3$ and HA. In the proximal tubule sodium ions pass into the cells in exchange for hydrogen ions which, together with the bicarbonate ions in the filtrate, gives rise to H_2CO_3. The proximal tubules have the enzyme carbonic anhydrase attached to their walls. Thus the H_2CO_3 is virtually in equilibrium with H_2O and CO_2. Since CO_2 passes all cell membranes easily, the CO_2 will move down the concentration gradient into the cell and hence into the blood. The cell itself contains carbonic anhydrase and the CO_2 in the cell is therefore also in equilibrium with cell bicarbonate. The pH of the filtrate at this time will be determined by how much Na^+ has been restored to the cell in exchange for H^+ and how much of that H^+ is associated as HA and how much associated as H_2CO_3, the latter being determined by the CO_2 concentration. In the distal tubule the membrane is not lined by carbonic anhydrase

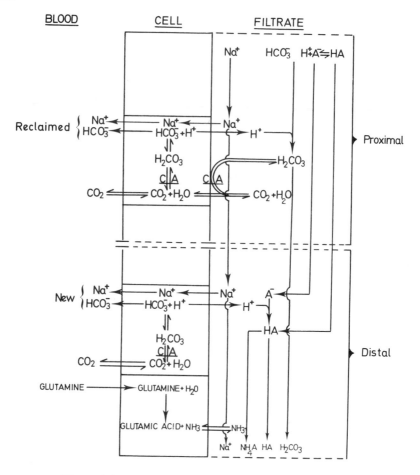

Figure 9.2 Mechanism of pH adjustment by the kidney

and therefore there is not the requirement that CO_2 must be in equilibrium with H_2CO_3. Therefore CO_2 is not rapidly formed from the carbonic acid and the partial pressure of CO_2 is unlikely to become high enough to drive CO_2 into the cells. If anything CO_2 will tend to move out from the cells. At this point the exchange of Na^+ for H^+ is still occurring, thus the filtrate is becoming progressively more acid with no possibility of transporting the acidity back into the cell in the form of CO_2. For some reason, which is not entirely clear, there appears a need to reduce the concentration of H^+ ions in the filtrate and this occurs by a direct neutralization by NH_3. NH_3 is liberated by the enzyme glutaminase from glutamine with the whole process being controlled to work faster as the cell becomes more acid.

The ammonia diffuses freely through the cell membrane and combines with H^+ to give NH_4^+ and thus a good deal of the fixed acids are finally excreted as their ammonium salts. The major source of urinary ammonia is blood glutamine which reacts with kidney glutaminase to give glutamic acid and ammonia. The rate of the reaction is controlled partly by the concentration of glutamate. The glutaminase of the kidney is inhibited about 50 per cent by 2 mM glutamate (within the physiological range) and it is known that glutamate concentration is lower than normal in acidosis and higher than normal in alkalosis.

Correlation of Gluconeogenesis and Ammonia Production in the Kidney

The kidneys carry out possibly up to one-fifth of the body's gluconeogenesis using blood lactate and restoring the synthesized glucose directly to the blood. The kidney also uses gluconeogenic amino acids for the same purpose. A fall in pH has been shown to stimulate gluconeogenesis in the kidney and thus the usage of glutamate for this purpose will increase. The necessary first step of removing ammonia from the glutamate by glutamic dehydrogenase will liberate ammonia for urine neutralization and the reduction in the concentration of the glutamate will allow the glutamine to be more rapidly hydrolysed by glutaminase and add to the ammonia available whilst at the same time topping up the supply of glutamate for gluconeogenesis. The synthesis of glutamine from glutamate and ammonia brought about by glutamine synthetase is inhibited by increase in acidity and therefore this reaction will not compete for the ammonia required for the urine neutralization (Figure 9.3).

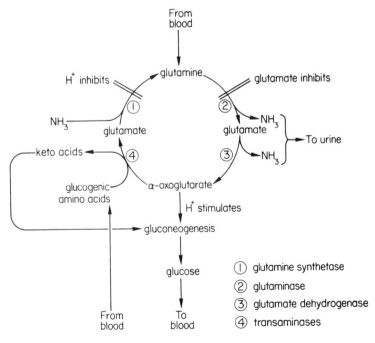

Figure 9.3 Relationship between gluconeogenesis and ammonia production in the kidney

Diuretics

The kidney tubules normally reabsorb a roughly balanced mixture of anions and cations from the glomerular filtrate and return them to the blood. The water that moves with the ions is that required to stabilize the osmolarity of the blood. Thus in any situation where there is any inhibition of solute reabsorption by the tubules (see Figure 9.2) there will also be a smaller amount of water reabsorbed and a diuresis will occur. Diuretics are substances given to bring about this phenomenon. In principle it does not matter whether it is cation or anion movement (or both) that is inhibited since a reasonably neutral fluid must be reabsorbed. The principal cation involved is sodium and the principal anion is bicarbonate. Therefore inhibitors of either reabsorption will ensure a greater volume of urine and a greater loss of cation and anion. Organic mercurials, presumably by reaction with SH groups in the membrane of the tubule, inhibit Na^+ uptake and therefore cause a diuresis. Inhibitors of carbonic anhydrase (usually sulphonamide derivatives) also promote a diuresis. This occurs because the filtered bicarbonate and the carbonic acid formed from it is no longer able to decompose rapidly to CO_2 and H_2O. Thus the osmotic pressure of the filtrate fails to fall in the normal manner and instead of the free passage of CO_2 back into the cells, bicarbonate ions, which are not readily reabsorbed, tend to remain in the filtrate. Also within the cells, the formation of carbonic acid from CO_2 and H_2O is inhibited and thus the formation of H^+ ions for exchange with sodium is prevented. Under these conditions there is a diuresis of an alkaline urine.

The efficiency of sodium uptake by the tubules is increased by the presence of aldosterone secreted by the adrenal cortex. Aldosterone is secreted in response to lack of sodium and reduced when the sodium content of the diet is high and it also promotes the excretion of potassium and hydrogen ions. Antagonists of aldosterone such as spironolactone (see p. 408) prevent the stimulation of uptake of sodium by aldosterone and therefore promote a diuresis.

The secretion of aldosterone may in part be a direct response to the sodium concentration of the blood but it is also influenced by a hormonal mechanism. It has been found that during sodium deprivation in normal subjects there is a rise in the blood level of a proteolytic enzyme named renin which is released from the kidney. The substrate of the renin appears to be an α-globulin found in the plasma which, on hydrolysis, gives rise to a decapeptide Asp-Arg-Val-Tyr-Ile-His-Pro-Phe-His-Leu called angiotensin I. Further hydrolysis by an unspecified enzyme removes the last two amino acids to give angiotensin II. Both these two substances may be broken down further by angiotensinase found in the kidney and the lung. The angiotensins stimulate the output of aldosterone (and possibly its synthesis) by the adrenal cortex. There is thus a relationship between the level of renin and angiotensin and between the level of angiotensin and that of aldosterone; the last determines the efficiency of sodium reabsorption in the kidney. Thus the uptake of sodium is decided by the balance of a hydrolytic activity of a proteinase (renin) and a peptidase (angiotensinase) neither of which is directly involved as the modifier of sodium transport.

So far we have been considering circumstances where the ionic content of the glomerular filtrate and the efficiency of sodium uptake determines the urine volume. There is also the separate mechanism where the control of urine volume is largely by modulating the water returned to the blood. Water in the distal tubules and the small collecting tubules passes through their walls into the extracellular fluid because the osmotic pressure of the filtrate is less than that of the extracellular fluid since some salts have already been restored to the extracellular fluid by active transport. The rate at which the water enters the extracellular fluid is dependent on the osmotic gradient between tubular contents and extracellular fluid and on the permeability of the tubules. The permeability of the tubules is greatly increased by the pituitary peptide hormone vasopressin (or antidiuretic hormone). Thus in times of water shortage, the uptake of water by the tubules follows the osmotic gradient very closely. After excess drinking the secretion of vasopressin is reduced or stopped and the resulting impermeability of the tubules to water causes the water to move down the tubule rather than to percolate through its walls into the blood. Therefore a dilute urine is excreted.

Reference

The Kidney, Morphology, Biochemistry and Physiology. Rouiller, C. and Muller, A. (eds.). Vols. 1–4. Academic Press Inc., New York, 1969, 1971.

SECTION 10

The Nervous System

CHAPTER 31

General Metabolism of the Brain

The gross metabolism of a human subject can be described by his oxygen consumption, carbon dioxide evolution, calorie intake and heat output under a variety of physiological conditions. The overall picture obtained, although useful and interesting, indicates nothing of the marked differences in metabolism that exist between the various tissues that comprise his body. Similarly, although an overall study of the metabolism of the brain illustrates some gross features of cerebral activity, a deeper understanding can only be achieved when the subtle differences in the structure, metabolism and function of the component parts are appreciated.

A 70 kg man has a brain weighing about 1·5 kg; this small mass of tissue receives some 15 per cent of the cardiac output and accounts for about 20 per cent of the total oxygen consumption of the body at rest. Under normal conditions of nutrition, the *RQ* (moles CO_2 liberated/moles O_2 consumed) calculated from the concentrations of carbon dioxide and oxygen in the blood leaving and entering the brain is about unity, indicating that carbohydrate provides the fuel for respiration. Furthermore, glucose is the only carbon source removed from the cerebral blood supply in substantial quantities under these conditions. Despite some suggestions to the contrary, there is little firm evidence that lactate and pyruvate accumulation accounts for more than 10 per cent of the glucose consumed. This view is confirmed by the oxygen consumption being sufficient to account for the complete oxidation to CO_2 and H_2O of over 90 per cent of the glucose consumed. Such a large flow of glucose through the glycolytic pathway produces considerable quantities of cytoplasmic $NADH + H^+$ which must be reoxidized. In the brain this is brought about by the activity of two L-α-glycerophosphate dehydrogenases. One in the cytoplasm converts dihydroxyacetonephosphate to L-α-glycerophosphate which can pass into the mitochondria to be oxidized by the other.

Cytoplasm $NADH + H^+ + $ dihydroxyacetone $PO_4 \longrightarrow NAD^+ + $ L-α-glycero PO_4

Mitochondria L-α-glycero $PO_4 + NAD^+ \longrightarrow $ dihydroxyacetone $PO_4 + NADH + H^+$

Since it has stores of glucose and glycogen sufficient for only a few minutes normal activity, the brain is dependent upon the glucose supplied by the blood. A small fall in the concentration of blood glucose can be compensated by increases in cerebral blood flow but a more severe fall (occasioned, for example, by an overdose of insulin) leads to a progressive deterioration of brain function until, at about 9 mg glucose/100 ml, coma is induced and brain respiration falls to 44 per cent of the normal value. The very rapid loss of brain function that follows circulatory arrest is almost certainly a consequence of the interruption of the oxygen supply to the brain as the cerebral content of oxygen is only sufficient for about 8 seconds activity. Although it was once thought that glucose was the sole fuel used by the brain under physiological conditions, this is now known not to be the case. Adult brain has the enzymic capacity to provide all the acetyl-CoA needed to fuel the citric acid cycle from the ketone bodies, D-(—)-β-hydroxybutyrate and acetoacetate. However, under normal conditions the concentration of ketone bodies in the blood is so low (Table 10.1) that they fail to make a significant contribution to the brain's fuel supplies. During periods of extended

Table 10.1 Arterial concentrations of energy substrates in fed and starved humans

	Fed	Starved 5 to 6 weeks
Glucose	5·0 mM	4·49 mM
D-β-Hydroxybutyrate	0·02 mM	6·67 mM
Acetoacetate		1·17 mM

starvation, such as those used to induce weight loss by obese patients, the level of circulating ketone bodies rises dramatically whilst the level of blood glucose falls by about 10 per cent. Under these conditions oxidation of ketone bodies accounts for about 60 per cent of the oxygen consumption of the brain and glucose for only some 30 per cent. In these periods of prolonged starvation, glucose consumption by brain is reduced to about half the normal value despite the almost normal level of blood glucose. Moreover, roughly half of the glucose taken up is degraded only as far as lactate and pyruvate, both of which can be reconverted to glucose in the liver and kidneys.

During prolonged starvation the RQ of brain falls as low as 0·63 which is much lower than can be accounted for by the complete oxidation of its major metabolic fuels (RQs for the complete oxidation of D-(—)-β-hydroxybutyrate, acetoacetate and glucose are 0·89, 1·0 and 1·0 respectively). This suggests that during starvation CO_2 fixation is occurring on a larger scale than usual.

In infancy, when the activities of the enzymes involved in converting ketone bodies to acetyl-CoA are higher than in adulthood, ketone bodies provide a significant part of the oxidizable substrate of the brain. At birth there is a transient hypoglycaemia coupled with a low level of ketone bodies but once suckling begins the infant is transferred to a high-fat diet and the concentration of ketone bodies rises sufficiently for them to act as an energy source.

There is now evidence that ketone bodies are used by the brain during diabetic ketoacidosis and that glucose uptake decreases. In diabetic coma, oxygen uptake by the brain is reduced by 20 to 50 per cent although glucose is present in great excess and the cerebral blood flow is not much diminished. It would appear that the level of ketoacids is the chief factor responsible for the coma but whether it is entirely accounted for by the accompanying acidosis is uncertain.

Although the nature of the metabolic fuel consumed by the brain undergoes a radical change during starvation, the overall size and composition of the brain remains remarkably constant. This feature is in strong contrast to the behaviour of the remainder of the body which may lose half its original weight before death ensues (Table 10.2). Other work has shown that the protein and RNA contents of brain are unaffected by starvation.

Table 10.2 The body weights, brain weights and brain composition of fed rats and rats maintained on a diet of sucrose and water for 5 weeks

	Control (normal diet)	Starved (sucrose and water)
Body weight (g)	289	160
Brain weight (g)	1·71	1·67
Brain cholesterol (mmoles/kg)	49·7	49·7
Brain total phospholipids (mmoles/kg)	65·1	66·0
Brain DNA–phosphorus	4·01	4·24

The Metabolism of Ammonia and Glutamate

The nervous tissue contains several enzymes such as adenylate deaminase (see p. 219) capable of producing ammonia at a high rate. However, ammonia is a toxic substance and its steady state concentration within the brain is maintained at about 0·3 mM. Excess ammonia is fixed into glutamic acid by the action of glutamine synthetase and this is reflected in the difference between the inflow of glutamate in the blood (0·79 mg/100 ml) and the outflow of glutamine (1·86 mg/100 ml). The equilibrium and activity of the glutamate dehydrogenase system

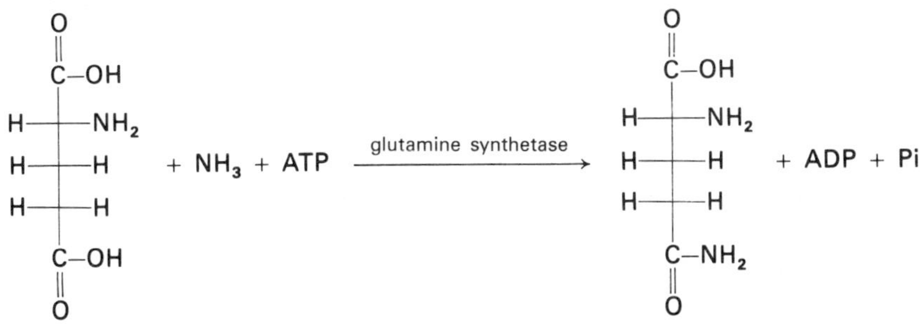

$$\begin{array}{c} \text{COOH} \\ | \\ \text{H—C—NH}_2 \\ | \\ \text{H—C—H} \\ | \\ \text{H—C—H} \\ | \\ \text{COOH} \end{array} + NH_3 + ATP \xrightarrow{\text{glutamine synthetase}} \begin{array}{c} \text{COOH} \\ | \\ \text{H—C—NH}_2 \\ | \\ \text{H—C—H} \\ | \\ \text{H—C—H} \\ | \\ \text{CONH}_2 \end{array} + ADP + P_i$$

ensures that ammonia concentrations will be kept low provided that there is a supply of α-oxoglutarate and reducing power. Any extensive conversion of α-oxoglutarate to glutamate drains intermediates from the citric acid cycle; such depletions must be made good by carbon dioxide fixation. It has often been assumed that the enzyme responsible for this is the malic enzyme; however, when

$$\begin{array}{c}\text{O}\\\|\\\text{C–OH}\\|\\\text{C=O}\\|\\\text{H–C–H}\\|\\\text{H–C–H}\\|\\\text{C–OH}\\\|\\\text{O}\end{array} + NH_3 + NADH + H^+ \xrightarrow{\text{glutamic dehydrogenase}} \begin{array}{c}\text{O}\\\|\\\text{C–OH}\\|\\\text{H–C–NH_2}\\|\\\text{H–C–H}\\|\\\text{H–C–H}\\|\\\text{C–OH}\\\|\\\text{O}\end{array} + NAD^+ + H_2O$$

consideration is given both to the availability of its substrates and to the kinetic properties of its two molecular forms, it would appear that it is less likely to be important for CO_2 fixation than the smaller amount of pyruvic carboxylase which is present in brain.

On balance about 63 per cent of the glutamate converted to glutamine can be considered as coming as such from the blood whilst the remainder is chiefly derived from blood glucose (Figure 10.1). The net loss of glutamine provides a mechanism whereby excess ammonia produced in the brain can be transported to the liver for conversion to urea. One especially important series of reactions for the brain which produces ammonia is the formation and removal of the inhibitory neurotransmitter γ-aminobutyric acid (GABA). This series of reactions starts with the decarboxylation of glutamate in a reaction requiring pyridoxal phosphate as a coenzyme (Figure 10.2). γ-Aminobutyric acid is deaminated to form succinic semi-aldehyde and ammonia. The succinic semi-aldehyde can be oxidized to succinate and thence to oxaloacetate, which, by condensation with acetyl-CoA, can eventually give α-oxoglutarate. Ammonia can combine with α-oxoglutarate to give glutamate and hence γ-aminobutyric acid again. Provided that any dicarboxylic acids drained out of the cycle can be replaced by CO_2 fixation, the γ-aminobutyric acid cycle will operate as long as C_2 units are supplied.

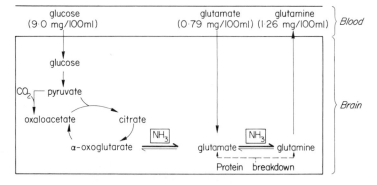

Figure 10.1 Reactions involved in the removal of free ammonia. The figures indicate the net amounts of materials entering or leaving a human brain in each 100 ml of blood passing through

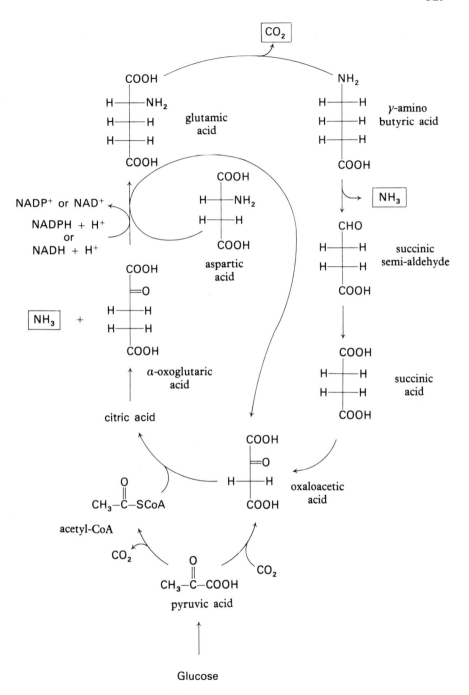

Figure 10.2 The γ-aminobutyric acid cycle and associated reactions

It appears that about 10 per cent of the total glucose turnover of the brain is metabolized by the γ-aminobutyric acid 'shunt' of the citric acid cycle.

The origin of the large quantities of ammonia produced under postmortem conditions or when brain slices or homogenates are incubated without substrate is not clear. The available evidence suggests that neither the hydrolysis of glutamine by glutaminase nor the oxidative deamination of glutamate are involved. AMP is readily deaminated by adenylic acid deaminase but this reaction alone could not account for the large quantities of ammonia produced by brain slices incubated without substrates unless AMP was regenerated from IMP by a plentiful supply of —NH$_2$ groups. It is possible that the necessary regeneration could be accomplished by an adenine nucleotide cycle of the type thought to exist in muscle (see p. 219). Whether the flow of amino acids through such a cycle is

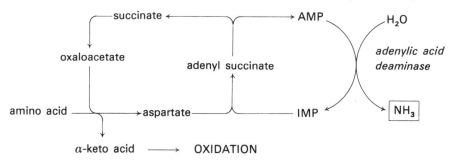

related to the turnover of brain proteins remains uncertain. A further possibility is that about a quarter of the ammonia produced in brain originates from the hydrolysis of amide groups on proteins.

Vitamins and Brain Metabolism

Dietary deficiencies of those vitamins that provide coenzymes for reactions supplying energy can lead to the development of nervous disorders. This is hardly surprising when one considers the high rate of metabolism of the brain and the rapid deterioration of brain function when the supply of glucose is restricted.

Thiamine (B_1)

Thiamine pyrophosphate is required for the decarboxylation of pyruvate during its conversion to acetyl-CoA (see p. 122). Thus, in thiamine deficiency, the citric acid cycle (normally the main source of energy for cerebral activity) is inhibited and pyruvate and lactate accumulate. The human brain contains about 5 nmoles of thiamine/g of tissue and it retains this more tenaciously than any other organ. Thus, during the first 15 days of a diet deficient in thiamine, the brain retains its normal content of the vitamin whilst other tissues keep only some 30 per cent of their original content. After 15 days, when the brain thiamine begins to fall, nervous symptoms develop. Beriberi, a disease resulting from a deficiency of thiamine, is endemic in those parts of the world where polished rice provides the staple diet. Two forms of the disease, dry and wet beriberi, are known. The former is characterized by peripheral neuritis, muscular wasting, impaired cardiac function and mental confusion; in the latter a general oedema obscures the wasting. Thiamine deficiency is encountered more commonly in the West as Wernicke's syndrome in alcoholics who have a low intake of normal foodstuffs. Although some of the milder symptoms of this disease may be relieved by giving thiamine, some cerebral lesions cannot be reversed and leave a clinical condition termed Korsakoff's syndrome, which comprises mental confusion and a failure to remember recent events.

Pyridoxal (B_6)

A phosphorylated form of vitamin B_6, pyridoxal phosphate (see p. 214), is involved in several important reactions in brain tissue. The transamination occurring between glutamate and oxaloacetate and decarboxylations of glutamate to γ-aminobutyric acid, of 5-hydroxytryptophan to 5-hydroxytryptamine and of dopa to dopamine all require pyridoxal phosphate as a coenzyme. Vitamin B_6 deficiency is characterized by convulsions which may result from a diminished ability to form γ-aminobutyric acid, which has an inhibitory effect on neuronal activity.

Nicotinic Acid (B_3)

Nicotinic acid is a component of NAD^+ and $NADP^+$ (see Figure 2.13) and is hence linked intimately with reactions supplying energy for cerebral activity. Mental disturbances, such as depression, are amongst the first symptoms of pellagra and may develop into dementia as a result of a severe niacin deficiency. Other symptoms of pellagra are dermatitis, diarrhoea and magenta tongue.

Protein Turnover

The protein content of the adult brain remains constant although studies involving labelling brain proteins with ^{14}C dietary lysine have shown that about 90 per cent of the cerebral proteins have a turnover time of less than 20 days. This implies that the brain is able to synthesize proteins to make good the loss. In agreement with this expectation, it has been found that brain slices are able to incorporate amino acids into protein at a rate comparable with liver slices. The mechanism of protein synthesis appears to be no different from that encountered in other mammalian tissues. The heavy commitment of neurones to protein synthesis is indicated by the dense accumulation of polysomes in their perikaryia known to light microscopists as Nissl substance.

Lipids

Chemically the brain is characterized by its high content of lipid which accounts for 56 per cent of the total dry weight of white matter and 32 per cent of that of grey matter. The lipids themselves are of a special nature, containing a large proportion of cerebrosides, sphingomyelins, plasmalogens and sulphatides as well as the more usual lecithins and phosphatidyl ethanolamines. However, the most abundant of all the lipids is cholesterol which constitutes 4 per cent of the fresh weight of white matter. Most of the lipids are metabolically inert, especially those found in myelin, and have a structural rather than a metabolic role. However, there is a small fraction of phospholipids, comprising phosphoinositides and phosphatidic acids, which turns over rapidly. This high rate of turnover is shared by brain phosphoproteins and is related to the level of neuronal activity. The gangliosides (see Figure 12.12) are another interesting group of brain lipids which are more concentrated in grey matter than in white. They are the most hydrophilic lipids present in brain and are characterized by containing *N*-acetylneuraminic acid.

Glial Cells

In considering the metabolism and function of the nervous system attention is naturally directed towards the neurones. However, it must not be forgotten that half of the mass of the brain is accounted for by the various non-neuronal supporting cells. The role of these cells is still largely unknown as is their contribution to the overall metabolism of the nervous system.

CHAPTER 32

Membrane Potentials and Ion Transport

Resting Potential

When two solutions of KCl of different concentrations are separated by a membrane impermeable to Cl^- ions, a potential is developed across the membrane so that the more concentrated solution (solution 1) is at a potential E volts more negative than the more dilute solution (solution 2). This potential arises as a result of the charge separation occurring when K^+ ions diffuse from the more concentrated solution unaccompanied by Cl^- ions. Equilibrium is established when the potential exactly counteracts the tendency of K^+ ions to diffuse from the more concentrated solution to the more dilute. At equilibrium, the free energy required to move one gram ion of potassium against the concentration gradient ($RT \ln [K^+]_2/[K^+]_1$) is equal to the free energy (EF) released as it moves down its electropotential gradient; i.e.

$$RT \ln \frac{[K^+]_2}{[K^+]_1} = EF$$

Therefore

$$E = \frac{RT}{F} \ln \frac{[K^+]_2}{[K^+]_1}$$

This equation relating the transmembrane potential to the potassium concentration gradient is called the Nernst equation.

Now we must consider a more complex situation in which a semipermeable membrane (one impermeable to macromolecules) separates a solution containing Na^+, K^+, Cl^- and non-diffusible anions such as proteins (solution 1) from one containing only Na^+, K^+ and Cl^- ions (solution 2). Such a system will eventually reach equilibrium with the solution containing the non-diffusible anion at a negative potential with respect to the other one. At equilibrium, the distribution of diffusible ions is defined by the relationship

$$E = \frac{RT}{F} \ln \frac{[Na^+]_2}{[Na^+]_1} = \frac{RT}{F} \ln \frac{[K^+]_2}{[K^+]_1} = \frac{RT}{F} \ln \frac{[Cl^-]_1}{[Cl^-]_2}$$

Therefore when a Donnan equilibrium of this sort is established

$$\frac{[Na^+]_2}{[Na^+]_1} = \frac{[K^+]_2}{[K^+]_1} = \frac{[Cl^-]_1}{[Cl^-]_2}$$

Neurones and muscle fibres have a high intracellular concentration of potassium and a low intracellular concentration of sodium compared with the extracellular fluid. Furthermore, the inside of these cells is at a negative potential with respect to the outside. Such an arrangement cannot be a Donnan equilibrium because

$$\frac{[Na^+]_{out}}{[Na^+]_{in}} \neq \frac{[K^+]_{out}}{[K^+]_{in}}$$

This conclusion led to the suggestion that cell membranes are impermeable to sodium ions and that the resting potential is an equilibrium potential determined by the transmembrane distribution of K^+ ions, i.e.

$$E = \frac{RT}{F} \ln \frac{[K^+]_{out}}{[K^+]_{in}}$$

However, with the realization that cell membranes are permeable to Na^+ ions, that resting cells continually lose K^+ and gain Na^+ and that a continuous expenditure of energy is required to maintain ion gradients, it became evident that the membrane potential is a consequence of the steady state distribution of ions across the membrane rather than an equilibrium potential. In other words, the concentration gradient of potassium ions is not exactly balanced by the transmembrane potential. Hence, the Nernst equation gives only an approximate description of the actual state of affairs. An equation giving a satisfactory description of the membrane potential under non-equilibrium conditions has been provided by Hodgkin, namely:

$$E = \frac{RT}{F} \ln \frac{[K^+]_{out} + \beta[Na^+]_{out}}{[K^+]_{in} + \beta[Na^+]_{in}}$$

where β is the ratio: membrane permeability to sodium (P_{Na})/membrane permeability to potassium (P_K). Permeabilities of frog muscle fibre membranes to Na^+ and K^+ ions determined from studies using radioactive tracers are given below

P_K efflux	6.2×10^{-7} cm/sec
P_K influx	5.7×10^{-7} cm/sec
P_{Na} influx (moving down electrochemical gradient)	7.9×10^{-9} cm/sec
P_{Na} efflux (moving against electrochemical gradient)	3.7×10^{-6} cm/sec

The permeability for sodium efflux is anomalous because it is much greater than for sodium influx despite the fact that efflux occurs against the electrochemical gradient. The permeability for sodium efflux therefore refers to an active process. If we consider only the passive fluxes, β has a value of 0·013. For the muscle fibres studied it was found that $[K^+]_{in} = 140$ mM; $[Na^+]_{in} = 9\cdot2$ mM; $[K^+]_{out} = 2\cdot5$ mM and $[Na^+]_{out} = 120$ mM.

Hence using Hodgkin's equation

$$E = \frac{RT}{F} \ln \frac{2\cdot5 + 0\cdot013 \times 120}{140 + 0\cdot013 \times 9\cdot2} = -89 \text{ mV}$$

the membrane potential is calculated to be -89 mV compared with an experimentally determined value of -90 mV. If the contribution of sodium is ignored and the potential is calculated using the Nernst equation

$$E = \frac{RT}{F} \ln \frac{2\cdot5}{140} = -101 \text{ mV}$$

a value of -101 mV is obtained. Clearly Hodgkin's equation, by allowing for the contribution made by sodium ions, provides a better description of the resting potential than the Nernst equation which ignores it. However, the contribution of sodium ions to the resting potential is small on account of the low value of β. Thus, whilst altering $[K^+]_{out}$ or $[K]_{in}$ in giant axons from squids markedly alters the value of E, varying $[Na]_{out}$ or $[Na]_{in}$ has very little effect.

Action Potential

During the passage of an action potential, the membrane potential swings over a period of about 4 msec, from the resting value of some -70 mV to around $+50$ mV and then back to about -80 mV before returning to the resting value (Figure 10.3). It has been shown that these changes in potential are consequences of the permeabilities towards sodium and potassium increasing and decreasing in an ordered sequence. During the rising phase of the action potential, the permeability of the membrane towards sodium increases rapidly so that at the peak of the potential it is very much greater than that towards potassium (Figure 10.3). Hence, at this peak, the recorded membrane potential approximates closely to that of the sodium equilibrium potential. Shortly before the action potential reaches its peak, the permeability towards potassium begins to rise above its resting level and continues to do so after the peak has passed and the sodium permeability begins to fall. When the permeability towards sodium has returned to its basal level, the permeability towards potassium is still much above its resting value so that a period of hyperpolarization is produced. The resting potential is restored when the membrane permeabilities towards both cations have returned to their original values. The net effect of these changes is that the axon loses a little potassium and gains a little sodium. Unlike that of the resting potential, the magnitude of the action potential depends upon the concentrations of sodium inside and outside the axon; thus increasing the external sodium concentration increases the height of the spike potential.

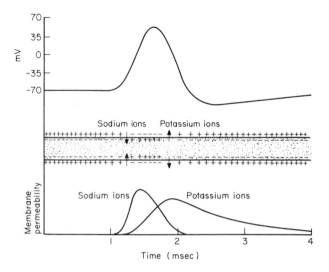

Figure 10.3 Nerve impulse travels along a nerve fibre as a self-propagating wave of electrical activity. The potential across the outer membrane of the fibre reverses and then returns to normal again. These changes are caused by rises in ionic permeability, which permit rapid movement of sodium ions into the fibre, followed by an egress of potassium ions. From Keynes, *Sci. Am.*, 1958

Active Ion-transport

As the axon is continually losing K^+ and gaining Na^+ (slowly at rest and more rapidly during the propagation of an action potential) it is necessary to restore the ionic gradients by the expenditure of metabolic energy. This is achieved by pumping out sodium ions and allowing potassium ions to enter the axon passively. The reason for supposing that sodium ions rather than potassium ions are transported actively is as follows. The resting potential (E) across the membrane is about -70 mV, which is approximately equal to

$$\frac{RT}{F} \ln \frac{[K^+]_{out}}{[K^+]_{in}}$$

that is to say, the energy required to move one gram equivalent of K^+ ions into the axon against the concentration gradient is

$$RT \ln \frac{[K^+]_{in}}{[K^+]_{out}}$$

which is approximately equal to EF, the energy gained when a gram equivalent of K^+ moves down its electropotential gradient into the axon. In other words, there is virtually no net change in free energy when a K^+ ion traverses the membrane, because the electrical 'driving force' in one direction is almost exactly balanced by the concentration 'driving force' in the other. However, for sodium

$$\frac{RT}{F} \ln \frac{[Na^+]_{out}}{[Na^+]_{in}}$$

does not equal -70 mV, the observed membrane potential, but is about $+55$ mV. Therefore in order to remove one gram equivalent of sodium ions from the axon, work equal to

$$RT \ln \frac{[Na^+]_{out}}{[Na^+]_{in}} + EF$$

has to be done against both the chemical concentration *and* electropotential gradients.

Since sodium is actively extruded from nerves in order to maintain the gradients of sodium and potassium across the cell membrane, the properties of the sodium pump will now be considered.

If a giant nerve fibre from a squid is loaded with $^{24}Na^+$ and then bathed by a continuous stream of a solution containing unlabelled sodium, the amount of $^{24}Na^+$ leaving the fibre can be measured at regular intervals (Figure 10.4). The rate of $^{24}Na^+$ loss falls as the concentration of the isotope within the nerve diminishes (Figure 10.5). If 0·2 mM dinitrophenol (DNP) is added to the bathing solution the efflux decreases dramatically. When the DNP is washed away, the rate of $^{24}Na^+$ efflux is restored to the control value. These observations suggest that sodium efflux depends upon a supply of ATP derived from oxidative phosphorylation. Similar effects on the rate of $^{24}Na^+$ expulsion can be obtained

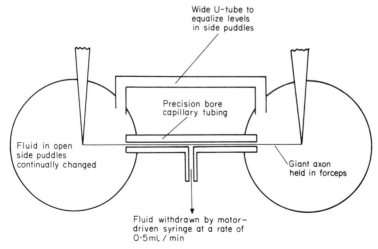

Figure 10.4 The experimental arrangement for studying ionic fluxes across the giant axon. From Hodgkin and Keynes, *Symp. Soc. Exp. Biol.*, 1954

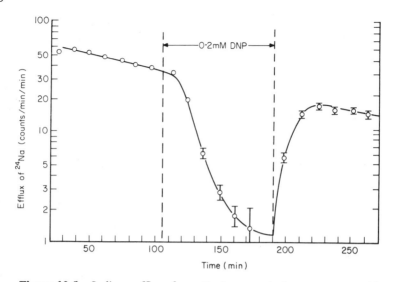

Figure 10.5 Sodium efflux from *Sepia* axon during treatment with dinitrophenol. The internal concentration of sodium had been raised initially by stimulation for 4 minutes at 156 pulses/sec in a Ringer solution containing ^{24}Na-labelled sodium. Poisoning the axon reduces the efflux. From Hodgkin and Keynes, *J. Physiol.*, 1953

by adding cyanide (CN^-) or azide (N_3^-) to the external medium. Axons are still able to conduct action potentials of undiminished size when the active transport of sodium is inhibited. This is to be expected because the ion fluxes forming the action potential are spontaneous or downhill processes, whilst maintenance of the ionic gradients is an active process. This view is confirmed by experiments in which all the axoplasm is removed from a giant axon and replaced by a solution of KCl. If such an axon is suspended in a solution of NaCl, it is still able to propagate action potentials despite the absence of a metabolic energy source. From these considerations it can be concluded that two systems concerned with ion movements exist in the axon membrane, one driven by metabolism and responsible for building up ionic gradients and another, independent of metabolism, responsible for controlling the downhill movement of sodium and potassium ions at rest and when the membrane is depolarized. This is illustrated in Figure 10.6 which also summarizes differences between the two systems. Sodium efflux from axons poisoned with cyanide can be restored by injecting ATP into the axoplasm but not by adding ATP to the external solution. This recalls the observation that the ion-transport ATP-ase of red cells cannot degrade ATP external to the erythrocytes and that ion transport is coupled to ATP breakdown inside the cells. As with red cells, the rate of sodium expulsion from nerves is dependent upon the external concentration of potassium, suggesting that active transport of sodium is coupled to the inward movement of potassium. An ATP-ase that requires Na^+, K^+ and Mg^{2+} for maximum activity has been located in the membrane of crab nerve; it is inhibited by ouabain and behaves

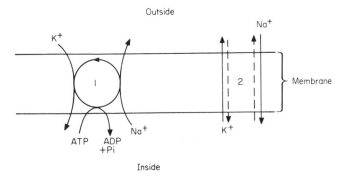

Figure 10.6 Systems regulating the passage of ions across the axon membrane. (1) Processes maintaining ionic gradients. Inhibited by DNP: CN^-: N_3^-; inhibited by ouabain; lithium not readily expelled; maximum rate of ion movement 50 pmole/cm²/sec. (2) Depolarization fluxes. Not affected by DNP: CN^-: N_3^-; not affected by ouabain; lithium can replace Na^+ in the action potential fluxes; maximum rate of ion movement 10,000 pmole/cm²/sec

similarly to 'ion-transport' ATP-ases found in other tissues including brain. Such an ATP-ase is probably involved in the active transport of sodium in nerve tissue in a manner similar to that proposed for the ATP-ase of red cells (see p. 286). Indeed, it has been found that the phosphorylation of brain microsomes by ATP requires Na^+ ions whilst dephosphorylation requires K^+ ions.

Stimulation of frog nerves causes a marked increase in their rate of oxygen consumption and carbon dioxide evolution. It can be calculated that about 2·6 out of every 3 molecules of ATP produced as a result of this extra consumption must be used to pump sodium out of the axons. A feedback system ensures that the supply of ATP increases to meet the demand of the pump at any instant. The faster the pump works, the more rapidly ADP is produced and can act as a phosphate acceptor in a tightly coupled system for oxidative phosphorylation.

CHAPTER 33

Secretion of Neurotransmitters

Acetylcholine

Electrophysiological studies have shown that neurones innervating skeletal muscles release acetylcholine in discrete quanta at rest and that the frequency of quantal release increases greatly when the nerve terminal is invaded by an action potential. The discovery of quantal release occurred at about the time that synaptic vesicles were first seen in electron micrographs of nerve endings; not surprisingly the idea developed that a quantum of acetylcholine may correspond with the content of a single synaptic vesicle. More recently, with their isolation from nerve endings in brain and electric organs, it has been shown that synaptic vesicles do indeed store acetylcholine, probably together with ATP and a water soluble protein, and that they are bounded by a lipoprotein membrane. Furthermore, treating neuromuscular junctions with venom from Black Widow Spiders causes a massive increase in the frequency of quantal acetylcholine release and depletes the nerve ends of synaptic vesicles.

Acetylcholine, released by a Ca^{2+} dependent process in response to the arrival of an action potential, interacts with receptors on the surface of the muscle fibres and thereby initiates the changes in ion permeabilities that underlie the endplate potential and action potential. This physiological activity is terminated by acetylcholine esterase hydrolysing acetylcholine to acetate and choline. Choline is

$$CH_3\text{-}\overset{O}{\overset{\|}{C}}\text{-}OCH_2\text{-}CH_2\text{-}N^+(CH_3)_3 + H_2O \longrightarrow CH_3\text{-}\overset{O}{\overset{\diagup}{C}}\diagdown_{OH} + HOCH_2CH_2N^+CH_3$$

acetylcholine acetic acid choline

retrieved by the nerve endings and used for the synthesis of more acetylcholine. Choline uptake is driven by being coupled to the downhill movement of sodium ions into the nerve endings, whilst the sodium gradient across the membrane is maintained by the activity of the ATP-dependent sodium pump. Within the nerve ending choline is reacetylated to form acetylcholine in a reaction catalysed by

$$CH_3\text{-}\overset{O}{\overset{\|}{C}}\text{-}SCoA + HO\text{-}CH_2CH_2N^+(CH_3)_3 \longrightarrow CH_3\text{-}\overset{O}{\overset{\|}{C}}\text{-}OCH_2CH_2N^+(CH_3)_3 + CoA\text{-}SH$$

acetyl-CoA choline acetylcholine coenzyme A

choline acetylase. The acetylcholine is then stored within the synaptic vesicles. The rate of acetylcholine synthesis is governed by the electrical activity of the neurones. For example, the rate of acetylcholine synthesis in a cat superior cervical ganglion at rest is about 4 ng/hr but rises to 28 ng/hr when the preganglionic fibres are stimulated maximally. The way in which this control is exercised is unknown. Figure 10.7 summarizes some of the events occurring at a neuromuscular junction.

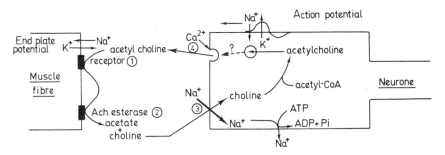

Figure 10.7 Events occurring at a neuromuscular junction. (1) Receptor blocked by D-tubocurare. (2) Acetylcholine esterase blocked by eserine. (3) Choline uptake blocked by hemicholinium. (4) Acetylcholine release blocked by tetrodotoxin. (?) Indicates uncertainty about the mechanism of acetylcholine secretion

Noradrenaline

In recent years biochemical understanding of events occurring at the endings of postganglionic, noradrenergic neurones has outstripped that of cholinergic endings on account of the availability of sensitive fluorimetric methods for estimating noradrenaline and for locating noradrenergic neurones histochemically.

Noradrenaline is stored in two types of particle in the varicosities of noradrenergic neurones which engage in *en passant* contacts with effector cells. The most abundant particle is the smaller which has a diameter of about 45 nm, whilst the larger particle, which is much less abundant, has a diameter of around 80 nm. Both types of particle are bounded by a lipoprotein membrane and the large variety (and possibly the small) contains dopamine-β-hydroxylase, the enzyme catalysing the conversion of dopamine to noradrenaline. These storage particles can take up both dopamine and noradrenaline from the cytoplasm by an ATP-dependent process that can be inhibited by reserpine. When animals are treated with reserpine the stores of noradrenaline in sympathetic nerve endings are depleted because noradrenaline leaking from the storage vesicles and that formed *de novo* cannot gain access to the storage sites. Instead it is oxidized to 3,4-dihydroxymandelic aldehyde by monoamine oxidase (MAO) present in the outer membrane of the mitochondria. However, if animals are given monoamine oxidase inhibitors (such as iproniazid) after reserpine treatment, noradrenaline accumulates in the cytoplasm. This cytoplasmic noradrenaline cannot be secreted when the neurones are stimulated, despite the fact that metaraminol, a com-

$$\underset{\text{noradrenaline}}{\begin{array}{c}\text{H} \\ | \\ \text{H}-\text{C}-\text{NH}_2 \\ | \\ \text{H}-\text{C}-\text{OH} \\ | \\ \text{C}_6\text{H}_3(\text{OH})_2 \end{array}} + \text{H}_2\text{O} + \text{O}_2 \xrightarrow{\text{MAO}} \underset{\text{3,4-dihydroxymandelic aldehyde}}{\begin{array}{c} \text{H}-\text{C}=\text{O} \\ | \\ \text{H}-\text{C}-\text{OH} \\ | \\ \text{C}_6\text{H}_3(\text{OH})_2 \end{array}} + \text{H}_2\text{O}_2 + \text{NH}_3$$

pound able to gain access to the vesicles by a reserpine resistant pathway, is secreted under these conditions as a false transmitter. This sort of study indicates that the noradrenergic storage vesicles are directly involved in the secretion, as well as in the storage, of noradrenaline. Following the discovery that

$$\underset{\text{metaraminol}}{\begin{array}{c} \text{H}-\text{N}-\text{H} \\ | \\ \text{H}-\text{C}-\text{CH}_3 \\ | \\ \text{H}-\text{C}-\text{OH} \\ | \\ \text{C}_6\text{H}_4\text{OH} \end{array}}$$

catecholamine secretion from the adrenal medulla occurs by a process of exocytosis (in which chromaffin granule membranes fuse with the cell membrane to allow direct egress of amines and proteins from the granules to the cell exterior (see Figure 13.6)), it was shown that dopamine β-hydroxylase was released from sympathetic neurones together with noradrenaline. It is therefore possible that the secretion of noradrenaline from the endings of sympathetic neurones also occurs by exocytosis. In support of this view it has been shown that, in common with secretion from other tissues where exocytosis is thought to be involved, noradrenaline release from noradrenergic neurones in response to nerve stimulation is Ca^{2+}-dependent.

Once noradrenaline has been released, it can interact with receptors on the surface of its target cell and induce changes in the membrane potential (see p. 333). The nature of the change depends upon the nature of the cell; in some instances hyperpolarization is induced, in others depolarization follows. There is some evidence that noradrenaline causes some target cells to synthesize and release prostaglandins which, on interacting with the presynaptic nerve ending, can attenuate noradrenaline release. This inhibitory feedback process can be blocked by inhibitors of prostaglandin synthesis such as eicosa-5,8,11,14-tetraynoic acid and aspirin (the latter however has other actions in addition to inhibiting prostaglandin synthesis).

Despite the existence of substantial stores of noradrenaline at nerve endings its rate of synthesis from tyrosine is greatly increased during periods of nerve activity.

arachidonic acid → prostaglandin E$_2$

The feedback control may be mediated through changes in the level of cytoplasmic noradrenaline or dopamine, both of which inhibit tyrosine hydroxylase, the first enzyme in their biosynthetic pathway.

Noradrenaline has its physiological activity terminated by being pumped back into the nerve endings and then restored in the vesicles. As with choline uptake, noradrenaline uptake into the nerve endings is a sodium-dependent process which can be indirectly inhibited by ouabain. The uptake process itself is inhibited by tricyclic antidepressants such as imiprimine. In contrast to acetylcholine, noradrenaline is not inactivated enzymically to any significant extent although any overflowing from the junction region is methylated by catechol-O-methyl transferase (see Figure 13.7). The reason for this difference

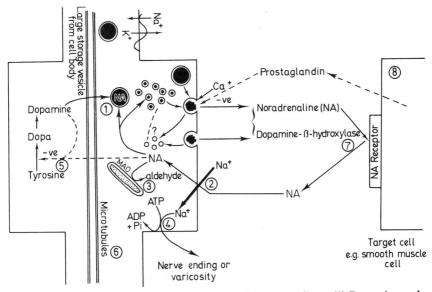

Figure 10.8 Processes occurring at noradrenergic nerve endings. (1) Dopamine and noradrenaline uptake into vesicles is blocked by reserpine. N.B. Dopamine taken into the vesicles is converted to noradrenaline by dopamine-β-hydroxylase. (2) Noradrenaline uptake blocked by imiprimine. (3) Monoamine oxidase inhibited by iproniazid. (4) Sodium pump inhibited by ouabain. (5) Tyrosine hydroxylase inhibited by noradrenaline and dopamine. (6) Microtubules dispersed by colchicine which prevent axonal transport of large noradrenaline storage vesicles from cell body. (7) Noradrenaline receptors blocked by sympatholytic drugs such as phentolamine. (8) Prostaglandin biosynthesis blocked by eicosa 5,8,11,14-tetraynoic acid

probably lies in the fact that, as noradrenergic neurones fire at very low rates, uptake of noradrenaline into the nerve endings is sufficiently fast to remove all the transmitter before the next nerve impulse arrives.

The source of the storage vesicles is not entirely understood. The larger variety (diameter 80 nm) appear to be synthesized in the cell bodies of the neurones and to pass down the axons in a manner that is dependent upon the integrity of the intra-axonal microtubules. The perikaryal origin of the storage vesicles appears

Table 10.3 Regional variations in some neurotransmitter concentrations in human brain

	Noradrenaline	Dopamine	5-Hydroxytryptamine	Choline acetylase[a]
Cerebral cortex	0.3	0.5	0.2	2
Caudate nucleus	0.45	24	1.8	13.1
Thalamus	0.65	1.6	1.3	3.1
Hypothalamus	6.0	4.8	2.6	3.0

The values for noradrenaline, dopamine and 5-HT are in nmoles/g.

[a] Choline acetylase values are for dogs and are given in μmoles acetylcholine formed/g/hr. Choline acetylase distribution reflects acetylcholine distribution.

to be related to the small capacity for protein synthesis exhibited by nerve endings compared with the abundant machinery present in the perinuclear region and evidenced by the dense arrays of rough endoplasmic reticulum forming the Nissl substance. Figure 10.8 summarizes some of the processes occurring at noradrenergic nerve endings.

Many types of neurone (i.e. noradrenergic and dopaminergic) are found within the central nervous system and their distribution is not uniform, hence some areas are richer in a particular kind of neurone than other areas. As a result the concentrations of neurotransmitters in different regions show considerable variation (see Table 10.3).

Despite the wide range of neurones and neurotransmitters in the central nervous system, it seems probable that the mechanisms involved in transmitter release are basically the same as those described above for peripheral cholinergic and noradrenergic neurones.

CHAPTER 34

Functional Aspects of Neuronal Biochemistry

Cyclic-3',5'-AMP in the Superior Cervical Ganglion

Neural tissue is rich in adenyl cyclase, the enzyme that converts ATP to cyclic-3',5'-AMP (c-AMP); moreover, when brain slices are stimulated electrically or treated with the neurotransmitters noradrenaline or 5-hydroxytryptamine, the tissue content of c-AMP increases dramatically. Noradrenaline also inhibits the spontaneous electrical activity of Purkinje cells in the cerebellar cortex. This effect can be potentiated by inhibitors of phosphodiesterase (the enzyme responsible for degrading c-AMP) and mimicked by c-AMP. In order to begin to investigate the role of c-AMP in neuronal activity it is necessary to find an experimental tissue that is much less complex than the central nervous system. The superior cervical ganglion is a suitable object for study because, despite its relatively simple structure, it possesses the ability to engage in integrative activities similar to those that characterize neural tissue as a whole. Furthermore it is amenable to study by electrophysiological, pharmacological, biochemical and histochemical techniques.

In the ganglion, postganglionic–noradrenergic motor neurones (1) (Figure 10.9) receive excitatory preganglionic fibres (a) which release acetylcholine; this excitatory effect of acetylcholine is blocked by hexamethonium but not by atropine. The motor neurones (1) also receive inhibitory fibres from interneurones (2) releasing dopamine whose effect is blocked by phentolamine. The interneurones are activated by preganglionic fibres (b) releasing acetylcholine whose action is inhibited by atropine but not by hexamethonium. The physiological effect of activating the inhibitory pathway (interneurones, 2) is to hyperpolarize the noradrenergic motor neurones (1) and thereby to diminish their response to direct preganglionic stimulation (a).

On stimulating the preganglionic nerve (a and b) experimentally, the c-AMP content of the ganglion rises whether or not hexamethonium is present. On the other hand, in the presence of atropine or phentolamine, preganglionic stimulation does not change the c-AMP concentration. This suggests that the c-AMP content of the postganglionic–noradrenergic motor neurones (1) increases in response to dopamine released by the interneurones (2). In agreement with this is the finding that the addition of dopamine to incubated ganglia increases their content of c-AMP and that this effect, like that of preganglionic stimulation, can be potentiated by theophylline (a phosphodiesterase inhibitor) and blocked by phentolamine.

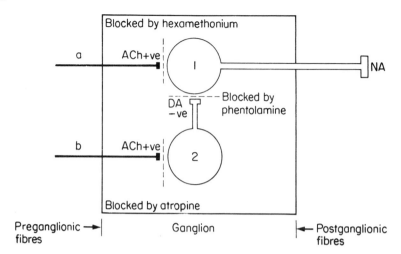

Figure 10.9 Diagrammatic representation of the principal connections in the superior cervical ganglion. Neurotransmitters: ACh = acetylcholine; DA = dopamine; NA = noradrenaline. Effect +ve = activates (depolarizes); −ve = depressive (hyperpolarizes)

Studies with microelectrodes show that both dopamine and lipid soluble derivatives of c-AMP (dibutyryl c-AMP) able to penetrate cell membranes hyperpolarize noradrenergic motor neurones. Furthermore, theophylline enhances dopamine induced hyperpolarization and impairs transmission through the ganglion. It is probable, therefore, that the hyperpolarization induced by dopamine is mediated via an increase in the intracellular concentrations of c-AMP. Evidence drawn from many sources suggests that c-AMP may activate a protein kinase (see p. 437) in the membrane of postganglionic motor neurones and that hyperpolarization may follow from the phosphorylation of a membrane protein by ATP. The decay of hyperpolarization might be brought about by the combined actions of phosphodiesterase degrading c-AMP and a membrane-bound phosphoprotein phosphatase dephosphorylating components of the membrane Figure 10.10).

This brief survey illustrates the methods and problems encountered in studying the integrative activity of the nervous system at the biochemical level; clearly much remains to be done but it is encouraging that a start has been made.

Pineal Gland and Melatonin

The existence of memory may well require the synthesis of macromolecules in response to a particular pattern of neuronal stimulation. At a later date, such molecules might be able to modulate neuronal activity in a manner that would allow the memory store to be tapped. An understanding of the complex processes underlying memory has hardly begun to emerge, but one example of how nervous activity can cause profound biochemical changes has been described for the

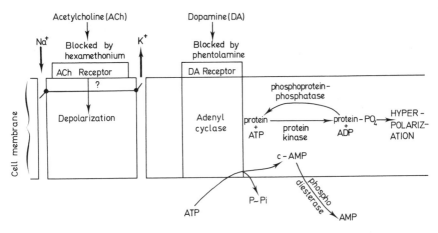

Figure 10.10 Diagram of the possible relationship between dopamine, c-AMP and hyperpolarization in noradrenergic motor neurones in the superior cervical ganglion

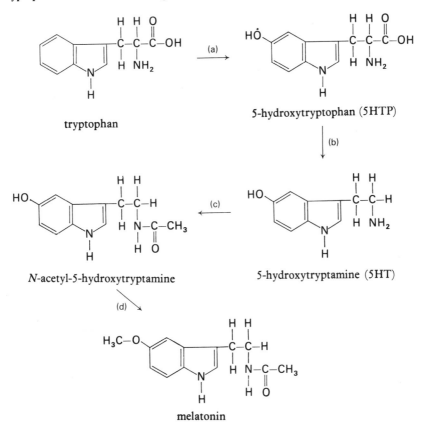

Figure 10.11 The biosynthetic pathway for melatonin. (a) Tryptophan hydroxylase. (b) 5HTP decarboxylase. (c) 5HT N-acetyltransferase. (d) Hydroxyindole-O-methyltransferase

pineal gland. The cells of the pineal synthesize the hormone melatonin (5-methoxy-N-acetyltryptamine) which, when released into the circulation, inhibits the onset of oestrus in females. Melatonin is synthesized from tryptophan via the series of reactions shown in Figure 10.11.

The pineal content of 5-hydroxytryptamine varies diurnally, falling at night as the production of melatonin rises and rising during the day as hormone production falls. Increased melatonin synthesis is associated with an increase in the activity of 5-hydroxytryptamine N-acetyl transferase as a result of *de novo* synthesis of the enzyme. The rate of enzyme synthesis is controlled by noradrenaline released from nerve fibres having their cell bodies in the superior cervical ganglion. The available evidence suggests that more noradrenaline is released from the neurones during the night than during the day and that it increases tissue levels of c-AMP by stimulating adenyl cyclase. Increased synthesis of the acetyl transferase at night is probably a consequence of the elevated c-AMP levels stimulating protein synthesis.

The night-time increase of noradrenaline secretion, and hence of acetyl transferase synthesis and of melatonin production, is controlled by an endogenous clock located in or near the suprachiasmatic nucleus of the hypothalamus and relaying to the superior cervical ganglion. These diurnal variations persist in blinded animals and those kept continuously in the dark. However, exposure of intact animals to light at night-time abolishes the response if it has developed. Furthermore, extension of the daylight period beyond the time when noradrenaline secretion normally rises prevents the onset of the rise. Thus the endogenous circadian rhythm is modulated by variations in environmental lighting.

The production of melatonin is therefore regulated by variations in neural activity which, in turn, control enzyme synthesis.

References

Brain Biochemistry, Bachelard, H. S. Chapman and Hall, London, 1974.
Neurotransmitters and Metabolic Regulation. *Biochemical Society Symposium 36.*
Biochemistry and the Central Nervous System. McIlwain, H. and Bachelard, H. S. Churchill-Livingstone, London, 1971.

SECTION 11

The Eye

CHAPTER 35

The Biochemistry of the Eye

The eye has a unique job to do in the body and, as might be expected, this is reflected in the biochemistry of its various tissues. Firstly, light falling upon the retina has to be absorbed by special pigments to initiate a series of chemical changes which finally yield an action potential in the optic nerve. Secondly, the lens, which focuses light on to the retina, must be constructed from materials which are transparent to light of a particular range of wavelengths. Normally the skin ensures that the tissues of the body are not penetrated by light and that the enzyme systems of cells are protected from the harmful effects of radiant energy.

The architecture of the eye presents two metabolic problems. If the cornea, lens and humours are to be transparent they cannot be supplied with blood vessels, therefore a strongly aerobic system of metabolism is not possible in those parts of the eye. On the other hand, since the retina is an extension of the brain, with the brain's sensitivity to oxygen lack, it must have a blood supply adequate to provide its high requirements for glucose and oxygen. Even a brief interruption of the blood supply to the retina can cause lasting damage.

The Lens

As the lens has no blood supply, it has to rely on the slowly circulating aqueous humour for its oxygen and raw materials and for the removal of its metabolic waste products. Since the amount of oxygen in the aqueous humour is limited to that held in physical solution, the oxygen supply to the lens is restricted. The small amount of oxygen that is consumed by the lens is probably used by the epithelial cells (which contain mitochondria) although some may penetrate slowly into the lens tissue itself and contribute to the oxidation of the SH groups of glutathione (see p. 280). In contrast to many animals the human lens does not have an initial rapid growth rate followed by an adult period of very little growth. Instead it grows throughout life at a slowly decreasing rate. It also differs from many other animals in maintaining a relatively constant water content of about 65 per cent throughout life and thus does not have the compressed dehydrated nucleus characteristic of the rat lens. Experiments carried out *in vitro* indicate that 90 per cent of the glucose used by the lens is glycolysed to lactate. The remaining 10 per cent is degraded via the pentose phosphate pathway and accounts for the carbon dioxide production by the isolated lens. The NADPH + H^+ produced by the pentose phosphate pathway can be used to reduce the oxidized dimer of glutathione to the reduced monomer GSH which is maintained at a high concentration in the lens presumably to keep the SH groups of the lens in the reduced state. (see Cataracts p. 354.)

Proteins of the Lens

The lens contains somewhat more than 30 per cent protein which falls as low as 20 per cent in the outer layers and is proportionally higher in the inner layers. Under normal circumstances about 90 per cent of the protein is soluble and about 85 per cent of the lens protein consists of the α, β and γ crystallins. These decrease in molecular weight from α to γ but increase in solubility. In the new born the proportions of the proteins are α-crystallin 44 per cent, β-crystallin 26 per cent, γ-crystallin 16 per cent and other proteins 14 per cent. With age all the soluble crystallins decrease with the γ-crystallin dropping to about 2 per cent whilst the other two decline by about 0·2 per cent per year. The protein mixture tends to have an enhanced solubility and greater transparency than the individual proteins.

Cataracts

Diabetic and Galactosaemic Cataracts

Certain tissues such as the lens contain the enzymes aldose reductase, and sorbitol dehydrogenase which together transform glucose to fructose and $NADPH + H^+$ to $NADH + H^+$ as shown

$$\text{D-glucose} + NADPH + H^+ \rightarrow \text{sorbitol} + NADP^+$$

$$\text{sorbitol} + NAD^+ \rightarrow \text{fructose} + NADH + H^+$$

(see p. 367). Aldose reductase has a low substrate specificity and can use galactose as well as glucose and in that case produces dulcitol as the sugar alcohol. Because the Km's for the sugars are comparatively high the sugar alcohol is not formed at a high rate unless the blood and tissue level is high in the relevant sugar. This occurs in the two diseases diabetes and galactosaemia. The sugar alcohols that are formed (polyols) do not easily pass membranes and since the polyol dehydrogenase is not sufficiently active to cope with the rate of formation there is an intracellular accumulation of the polyol and a consequent increase in the osmotic pressure within the cell. Water therefore passes into the cells and the lens swells. The cells finally disrupt and an opacity results. Inhibitors of aldose reductase are now used to prevent the formation of this type of cataract. In the process of forming the sugar alcohol $NADPH + H^+$ is consumed and this substance is also used by the enzyme glutathione reductase to produce reduced glutathione from the oxidized disulphide dimer. The reduced glutathione is itself used to maintain the SH side chains of the lens proteins which if they are not constantly re-reduced will form disulphide with each other and render the protein insoluble. It is of some interest that the amount of reduced glutathione in

Table 11.1 Nuclear cataracts

		Colour of lens		
Class of cataract	Pale yellow Normal	Pale even yellow brown I	Pale yellow brown cortex visible nucleus II	Hazel brown nucleus III	Deep brown nucleus IV
Per cent of soluble protein	80	63	53	32	17
Per cent of cross-linked protein	0	0.4	3	18	28
Per cent glutathione	100	37	12	7	10
Per cent oxidation of methionine	0	0	3	24	45
Per cent protein SH group	100	90	29	8	3
G-3-PDH activity per cent of normal	100	100	—	—	16

the human lens decreases with age from a value approaching 4·0 μmoles/g lens at birth to rather less than half this value at the age of 60. Thus the pool of reduced glutathione available for retaining the lens proteins in the reduced form is less and less with age and consequently any metabolic strain thrown on the system because of other reactions competing for the supply of NADPH + H$^+$ becomes more and more critical. Indeed in the elderly where blood sugar even in the non-diabetic tends to be higher and to fluctuate more than normal there may also be a contribution to senile cataract by the osmotic damage of accumulating sorbitol. In these and less well defined cortical cataracts where osmotic damage is a prime cause of the opacity there is a loss of protein from the lens, a leakage in of sodium and a loss of potassium. Accompanying this hydration is a loss in the activity of the glycolytic enzymes of the cortex and thus presumably a lowering in the availability of ATP.

Nuclear Cataracts

By contrast with the type of cataracts just discussed where protein loss is a major factor there are those which are found in the nucleus of the lens and where the protein is retained but becomes insoluble and coloured. Table 11.1 shows the changes in the nuclear cataracts as they become progressively more coloured and hence opaque. In human by contrast with most animals the lens colour is a pale yellow which steadily deepens with age to become, with the development of a nuclear cataract, brown of a steadily deepening shade and eventually virtually black. It is possible that metabolites of tryptophan including the glucoside of 3-hydroxy kynurenine (found only in the primate lens) are responsible for the tanning of the lens proteins. During the formation of the insoluble cross-linked proteins it is the γ-crystallin which is first involved followed by the β and then the α compounds. During the tanning process the reduced glutathione which is already reduced in amount by the ageing process becomes very much less and the sulphydryl groups of the proteins become oxidized to the disulphide which cross-links and insolubilizes the protein. The high concentration of SH groups within the γ-crystallin probably explains the reason why this protein is the first to become insoluble. Of special interest is the oxidation of the methionine side chains of the lens protein to methionine sulphoxide. This oxidation follows that of the SH groups and reflects the increase in the cross-linking within the lens protein. Glyceraldehyde 3 phosphate dehydrogenase which requires SH groups for its function (see p. 111) decreases in activity but this decrease has not been related to the cataract formation.

The Retina

The importance of the retina as the main sensing device of the body can hardly be overestimated; more than a third (about a million) of all the nerve fibres entering or leaving the central nervous system do so by way of the optic nerve. Even with this enormous number of nerve fibres, the number of receptor cells in the retina is much greater, so that each nerve fibre receives the impulses from about 100 photosensitive cells.

The Biochemistry of the Visual Process

A good deal is known about the biochemistry of the visual process in the rods, which are concerned with vision at low light intensities, whilst very little is known about the biochemistry of the cones, which are responsible for vision at high light intensities and colour vision. However, evidence has been obtained for the existence of three pigments having their absorption maxima in the red, blue and green regions of the spectrum respectively and which are located in separate cones.

Vitamin A (retinol) plays a fundamental role in the visual processes of the rods; in the retina its *all trans* isomer can undergo either of two reactions. It can be oxidized by NAD^+ in the presence of alcohol dehydrogenase to the corresponding aldehyde called *all trans*-retinal and be further converted by retinal isomerase to Δ^{11}-*cis*-retinal. Alternatively, *all trans*-vitamin A may isomerize to a form with the *cis* configuration at the Δ^{11} position which can also be oxidized by NAD^+ and alcohol dehydrogenase to give Δ^{11}-*cis*-retinal. In the dark Δ^{11}-*cis*-retinal combines with a protein called opsin to form the purple visual pigment rhodopsin. In the light, rhodopsin is bleached and dissociates into opsin and *all trans*-retinal; i.e. light alters the configuration of the retinal from Δ^{11}-*cis* to *all trans* and the complex subsequently dissociates. The bleaching of rhodopsin is accompanied by an increase in the number of free —SH groups, whilst the resynthesis of rhodopsin is associated with a decrease in the number. These findings suggest that Δ^{11}-*cis*-

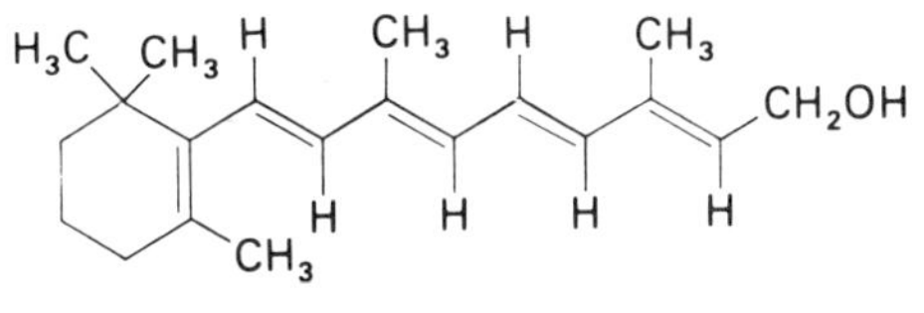

vitamin A_1 (*all-trans*) 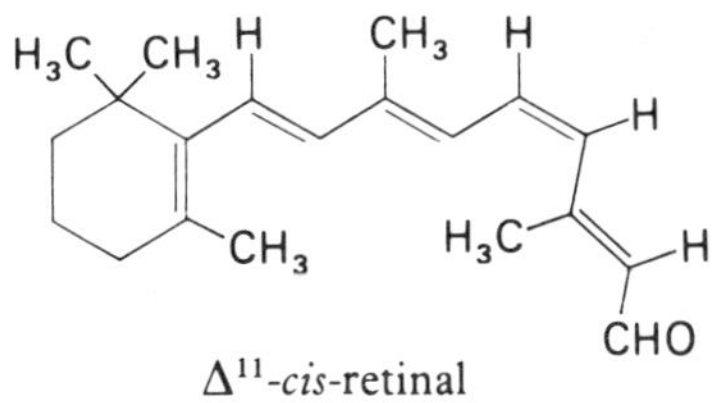 Δ^{11}-*cis*-retinal

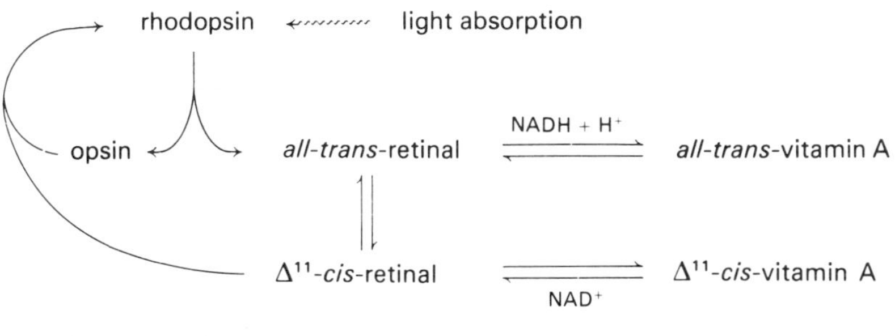

Figure 11.1 Reactions involved in the photodissociation of rhodopsin

retinal combines with an —SH group in opsin to form rhodopsin. The *all trans*-retinal liberated during the bleaching of rhodopsin can be isomerized by retinal isomerase to give Δ^{11}-*cis*-retinal which is able to reenter the visual cycle. This outline of photochemical events in rod vision leaves unanswered a whole range of important questions. How, for example, are the rearrangement and dissociation of rhodopsin within the lipoprotein lamellae of the outer segment of the rods connected with the electrical changes which eventually give rise to an impulse in the optic nerve? At present little is known about these dark-reactions. In the dark Na^+ ions enter the outer segment of the rods and are expelled by sodium pumps located in the membrane of the adjacent segment which are fuelled with ATP from the numerous mitochondria nearby. On illumination current flow is reversed, the current intensity being determined by the number of photons received. It may be that Ca^{2+} ions released from the lamellae as rhodopsin undergoes a conformational change are responsible for altering ionic permeabilities and reversing current flow.

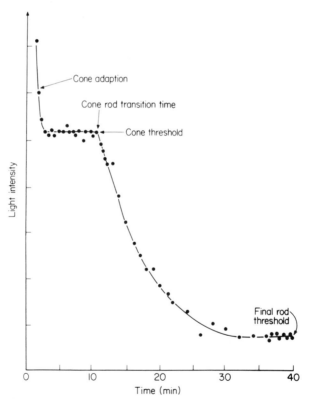

Figure 11.2 Normal curve of dark adaptation showing cone threshold, cone–rod transition time and final rod threshold. From *Vitamin A Requirements of Human Adults*, MRC Special Report Series, H.M. Stationery Office, 1949

Night Blindness

It is clear that a deficiency of vitamin A will lead to defective rod vision. This can be demonstrated by the following experiment. After a preliminary exposure to intense light (to bleach the rhodopsin) the subject is placed in the dark and the lowest light intensity that he can detect is measured from time to time during a period of about an hour. The results obtained in this way are shown in Figure 11.2. First cone adaptation occurs until a plateau is reached which measures the lowest intensity of light to which the cones are sensitive (cone threshold). The cone–rod transition time is a measure of the time taken for sufficient of the opsin to recombine with Δ^{11}-*cis*-retinal for the visual process of rods to operate after bleaching. The final rod threshold represents the minimum intensity of light that the fully dark-adapted eye can detect. In vitamin A deficiency both the cone–rod transition time and the rod threshold are increased. The rate of recombination of opsin and Δ^{11}-*cis*-retinal depends upon the product of the concentration of the two components and, as the concentration of Δ^{11}-*cis*-retinal is reduced in vitamin A deficiency, it follows that the cone–rod transition time will be lengthened in vitamin A deficient subjects. Similarly, in the dark, the sensitivity of the response to test illuminations depends upon the amount of rhodopsin available to be decomposed by the light. In the fully dark-adapted eye of the vitamin A deficient subject, the amount of rhodopsin present is less than in the fully dark-adapted eye of a normal subject; hence, the former has a higher final rod threshold than the latter. It has also been found that the peripheral rods lose their rhodopsin more rapidly than the central rods in vitamin A deficiency.

General Biochemistry of the Retina

The retina has the highest rate of respiration of any mammalian tissue, and also a higher rate of glycolysis than most other tissues; even under highly aerobic conditions it still accumulates lactic acid. As in other nerve tissues, the store of glycogen is very small so that continued metabolism is almost entirely dependent upon the supply of glucose in the blood. Interrupting the blood supply to the retina for even a few minutes causes irreversible damage.

The effect of alcohols on the metabolism of the retina is of particular interest. Since alcohol dehydrogenase is a necessary part of the rhodopsin cycle and is a relatively unspecific enzyme, it might be expected that alcohols would interfere with the metabolism of the retina by competing for this enzyme. It has been shown that drinking ethanol causes an impairment of night vision, even when taken in small quantities. However, from the clinical point of view the effect of methanol on the retina is much more important. Methanol is lethal to man if drunk in sufficient quantity, the fatal dose being about 65 gm, and quantities as small as one teaspoonful can cause blindness. It is generally believed that the toxicity of methanol is due not to methanol itself but to formaldehyde which is produced from it metabolically. This is another instance of a lethal synthesis (see p. 144). Formaldehyde is believed to be the toxic agent because it, but not methanol, is an extremely potent inhibitor of respiration and glycolysis in the

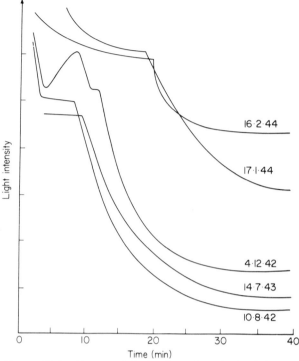

Figure 11.3 Curves of dark adaptation obtained from a subject maintained on a diet deficient in vitamin A from 10 August 1942 until 16 February 1944. From *Vitamin A Requirements of Human Adults*, MRC Special Report Series, H.M. Stationery Office, 1949

isolated retina. With isolated retinal mitochondria a concentration of formaldehyde of 1 mM or upwards almost completely abolishes oxidative phosphorylation without affecting respiration markedly. Electron microscopy shows that the rods are packed with mitochondria, accounting for the very vigorous respiration of the retina, and suggesting that much of its energy is supplied by oxidative phosphorylation. If oxidative phosphorylation is inhibited by formaldehyde, the supply of ATP will fail and the coupling of the photochemical process with neural transmission will be abolished. Furthermore, it is possible to explain how the administration of ethanol immediately following the ingestion of methanol can prevent the toxic effects of the latter. Ethanol competes successfully with methanol for the available alcohol dehydrogenase, yielding acetaldehyde instead of formaldehyde and acetaldehyde does not uncouple oxidative phosphorylation.

References

The Eye. Davson, H. (ed.). Vols. 1–4. Academic Press, New York, 2nd edn., 1969.
Physiology of the Eye. Davson, H. Churchill-Livingstone, Edinburgh, 3rd Edn., 1972.

SECTION 12

Reproduction and Development

CHAPTER 36

The Biochemistry of Sex

The reproduction of mammals requires the cooperation of a number of specialized tissues which are responsible for (1) the fertilization of the mature ovum, (2) providing the appropriate environment for foetal development and (3) supplying carbohydrate, fat and protein for the nutrition of the young animal after birth. The organs most directly concerned are the gonads, the uterus and the mammary glands; in this section, certain aspects of their biochemistry will be considered.

In essence, reproduction consists of the transmission of the characteristic potentialities of a single cell to its progeny: this is true even for the multicellular organism, in which those properties potential in the fertilized egg are expressed in the many-celled complexes of the fully differentiated organism. In biochemical terms, the transmission implies that the parent cell is able to produce daughter cells capable of manufacturing those patterns of enzymes appropriate to their spatial and temporal position in the life history of the organism, since the enzyme complement will determine both the structural development and the metabolic activities of the cells and organs.

In man as in other higher animals, the fertilized egg is produced by the fusion of a sperm cell from the male and an ovum from the female. We will begin our discussion of reproduction and development by considering some characteristic aspects of the biochemistry of males and females.

The Determination of Sex

Sexual determination itself is a consequence of the interplay between genetic and hormonal factors. Normal human females have 22 autosomes and 2 X chromosomes, normal human males have 22 autosomes, an X chromosome and a Y chromosome; but the extent of differentiation of sexual behaviour ('maleness' or 'femaleness') manifested during development may also be influenced by early exposure to sex hormones. Experiments with animals have suggested that development of masculine 'mating behaviour' depends on an exposure of the brain at an early critical postnatal stage to androgen (normally produced by the testes of the new-born male); without this exposure, a female behavioural pattern will emerge. Currently, attempts are being made to detect biochemical changes in the brain in response to the presence of the androgen, principally by searching for new species of RNA or enhanced protein synthesis in particular parts of the brain.

Even in the adult, sex-related differences in enzymic compositions can be detected which are distinct from those concerned directly with reproductive functions. It is probable that these differences also are an expression of both genetic and hormonal influences. Examples are:

(1) Kidney transaminidase which is twice as active in mature male rats as in females, but the female has twice the activity of ornithine transaminase.

(2) Hexobarbital oxidase (a microsomal detoxifying enzyme) which is almost five times more active in the male rat.

(3) Glutathione peroxidase activity which is about 80 per cent greater in mature female rats than in males. The rates of GSH oxidation and lipid peroxidation are also greater; the latter process uses unsaturated fatty acids of the liver that are more concentrated in the female.

There is, therefore, no doubt that there is a considerable specificity of enzymic composition distinguishing males and females.

The Metabolism of Spermatozoa

During mammalian fertilization the male's genetic message is carried to the ovum by the spermatozoon. Mature spermatozoa neither grow nor divide and have a stable metabolic pattern. During the maturation, however, there are marked changes in their biochemical composition. The immature sperm contains RNA, which disappears almost completely during maturation because there is then no further requirement for protein synthesis. A group of enzymes concerned with the degradation of RNA (acid phosphatase, alkaline phosphatase and $5'$-nucleotidase) also disappear. Glycogen is removed during maturation so that the activity of the mature sperm is largely dependent on external carbohydrate for its energy provision.

Structurally sperm are divided into three parts. The head contains the DNA (about $2 \cdot 5$ to $3 \cdot 0 \times 10^{-9}$ mg/cell) and a hydrolytic enzyme that enables the sperm to penetrate the ovum. The middle section may be thought of as the sperm 'motor' and is most active metabolically. It consists of a central core of fibres surrounded by mitochondria (Figure 12.1), has a high respiratory rate and provides ATP for locomotion. This ATP is used by the third section of the sperm, the tail, which possesses a series of fibrils responsible for producing an ATP-dependent, whip-like movement. The fibrils have an ATP-ase activity, but its detailed properties have not yet been defined.

Seminal Fluid

In vivo the sperms live in the seminal plasma, a fluid produced by the combined secretions of several accessory glands. It contains high concentrations of substances found only in low concentration elsewhere in the body. Characteristic components of the seminal plasma are:

(a) Citric acid—derived from the secretions of the prostate gland and present in the semen of man at concentrations of 5–25 mg/ml.

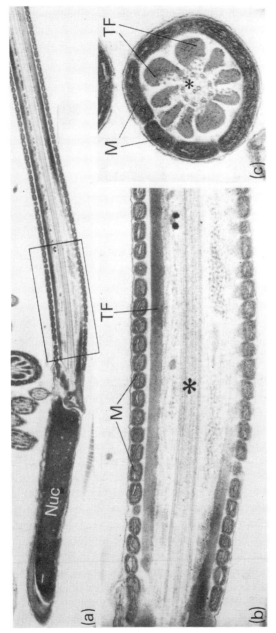

Figure 12.1 Electron micrograph of rat sperm (a) The head piece and part of the middle piece. The nucleus contains DNA for transmission to progeny. (b) This shows the area enclosed by the rectangle in (a) at higher magnification. Mitochondria (M) form a sheath around the filaments forming the central core of the mid-piece. (c) A transverse section of the mid-piece showing the mitochondrial sheath (M), the array of 9 thick filaments (TF) and the central array of thin filaments in a 9 + 2 arrangement (*). By kind permission of Dr D. W. Gregory

(b) Acid phosphatase. The prostatic fluid is one of the richest sources of acid phosphatase in the body, the amount of the enzyme in the gland increasing about five-hundred-fold when the male matures. Normally the enzyme does not pass into the blood stream but it does so in cancer of the prostate or when metastases from a prostatic carcinoma develop in bone. Therefore, measurements of the plasma acid phosphatase are useful as a diagnostic aid during the clinical treatment of prostatic carcinoma. Injections of androgens also increase the blood level of this enzyme whereas castration or oestrogens cause large decreases.

(c) Prostaglandins. These are a group of complex acid substances which are potent smooth-muscle contractile agents. They are derived from polyunsaturated fatty acids such as arachidonic; a typical structure (prostaglandin E_1) is

$$CH_3-(CH_2)_4-\underset{OH}{\overset{H}{\underset{|}{C}}}H-CH=CH-\overset{H}{\underset{HO-CH}{\underset{|}{C}}}\underline{\quad\quad}\overset{H}{\underset{\underset{\underset{H}{|}}{\overset{|}{C}}}{\underset{|}{C}}}-(CH_2)_6-C\overset{O}{\underset{OH}{\diagdown}}$$

Prostaglandins are found in many types of animal tissue (for example, in high concentration in seminal plasma and platelets, but also in menstrual fluid, kidney, thymus, brain and lung).

The high concentration in semen points to a function in stimulating smooth muscle contraction in the female reproductive tract after coitus, perhaps aiding the movement of sperm to the oviduct. Prostaglandins also stimulate the contraction of the uterus at parturition, as they are concentrated in the amniotic fluid released at this time.

In addition to these influences on the reproductive processes, prostaglandins have a great variety of other biochemical effects. Certain prostaglandins appear to be responsible for the increase of vascular permeability which occurs during the inflammatory response; this reaction is inhibited by aspirin which blocks the formation of prostaglandin. It is also known that prostaglandins can either decrease (in adipose tissue and bladder) or increase (in most other tissues) the level of cyclic AMP by altering the activity of adenyl cyclase. As a consequence of the lowering of cyclic AMP levels in adipose tissue, following the inhibition of adenyl cyclase, lipolysis in that tissue is also inhibited. On the other hand, some prostaglandins stimulate the adenyl cyclase of the adrenal gland, testes and corpus luteum, thereby increasing steroidogenesis in these tissues.

(d) Choline, phosphoryl choline, glyceryl phosphoryl choline, presumably derived from the metabolism of the phospholipids of the sperm or the accessory organs.

(e) Spermine, secreted by the prostate gland, is a nitrogenous base present at concentrations of 3–4 mg/ml of semen. Its precise function is unknown, but it has been shown to stabilize preparations of cells, mitochondria, ribosomes, bacteriophage and some enzymes *in vitro*. Some of these effects may be due to its ability, as a polyamine, to complex with nucleic acids.

$$H_2N-(CH_2)_3-NH-(CH_2)_4-NH-(CH_2)_3-NH_2$$
spermine

(f) *Fructose.* Seminal fructose comes from the vesicular and epididymal secretions and reaches concentrations of about 2–4 mg/ml of semen. It is one of the most important constituents of the semen since it provides the chief metabolizable substrate available for locomotory activity. It is formed by the accessory glands, principally the seminal vesicles. The concentration of seminal fructose reflects the concentration of blood glucose and, as a consequence, is affected by insulin. The conversion of glucose to fructose in the vesicles is dependent on the availability of testosterone and may occur in two ways. Glucose may be phosphorylated to glucose-6-phosphate which is isomerized to fructose-6-phosphate from which fructose is released by a phosphatase. Alternatively, glucose may be reduced by $NADPH + H^+$ and sorbitol dehydrogenase to sorbitol, a sugar alcohol; this can be reoxidized to fructose by an NAD^+ dependent enzyme (aldose reductase). It appears likely that testosterone increases the amount of both the sorbitol dehydrogenase and the aldose reductase in the vesicles.

The fructose of the seminal plasma provides the energy necessary for motility under anaerobic conditions, as, for example, after deposition in the female reproductive tract. Fructose is converted to fructose-6-phosphate by hexokinase then utilized in the glycolytic pathway with the accumulation of lactic acid. Under aerobic conditions the lactate formed may be oxidized in the citric acid cycle of the mitochondria in the sperm; this ability to respire (using not only lactate but perhaps also the seminal citrate and fatty acids derived from the intracellular phospholipids) will be essential when the supply of fructose deposited with the sperm is exhausted (probably about 20 minutes after ejaculation). In this time, many of the sperm will have moved into the uterus, where oxygen may be somewhat more readily available. Thus the significance of the initial fructolysis may be to enable the sperm to move away from the original concentrated deposit in the genital tract where oxygen tension is relatively low.

Fertilization

The motility of the sperm depends on factors other than the availability of fructose. Firstly, the pH optimum is about 7; if the medium is more acid motility decreases, if more alkaline it increases but only for a short time. Secondly, sperms in concentrated suspension are inactive, and will become much more active on moderate dilution. Further dilution inactivates them. Sperms are viable for long periods of time if frozen to $-79°$ in 10 per cent glycerol.

Apart from the motility of the sperms (which will affect the number that reach the ova) successful fertilization depends on the ability of the sperm to penetrate the ovum.

The sperm head appears to contain two hydrolytic activities (a hyaluronidase and probably a proteinase) which are important in assisting penetration into the ovum. Before this penetration can commence, however, the sperm must spend some hours in the uterus, for completion of the process called 'capacitation'. In this, the sperm heads lose a protective coating (probably of mucopolysaccharides) which is provided by the epididymis and secretory glands. The coating is removed by a β-amylase present in the uterine secretions, so that the uterus itself has a role in determining whether penetration of the ovum by sperms will occur. The glucosamine liberated from the degraded carbohydrate (and also from uterine mucopolysaccharides) can be used by sperm as an energy source, which will be especially important when seminal fructose and citrate are exhausted.

Hormonal Control of Spermatogenesis (Figure 12.2)

The male sex hormones (the androgens) are formed primarily in the testis, though the adrenals also synthesize small amounts of these substances. The principal hormone testosterone, is required for the maintenance of spermatogenesis, in which successive generations of sperm are produced in a highly coordinated fashion. The release of the hormone in the testis is controlled by gonadotrophic hormones—the luteinizing hormone (LH) and the follicle-stimulating hormone (FSH) derived from the pituitary. LH is a protein with two dissimilar polypeptide chains while FSH contains both protein and a carbohydrate component. The action of LH on the interstitial cells of the testis is mediated by cyclic AMP; the hormone stimulates the adenyl cyclase of this tissue. The production of LH and

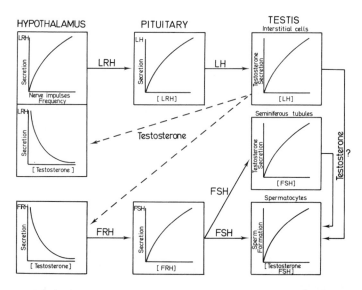

Figure 12.2 Interactions of hormones in spermatogenesis. LRH—Luteinizing hormone releasing hormone. LH—Luteinizing hormone. FRH—Follicle stimulating hormone releasing hormone. FSH—Follicle stimulating hormone

FSH is in turn controlled by the formation of the gonadotrophic-releasing hormones (LRH and FRH) in the hypothalamus.

The testosterone level in the blood is believed to regulate the level of both LH and FSH by a feedback control of the neurosecretory cells which secrete these gonadotrophic releasing hormones (LRH and FRH). Testosterone is necessary for the growth of male gonads, the continuing function of the accessory glands in the reproductive tract and, as shown in the diagram, the production of sperm. In general terms, the hormone promotes active protein synthesis by stimulating the synthesis of RNA (ribosomal and messenger). In the testis (and also in muscle, kidney and liver) this effect appears to depend on an interaction of the hormone with specific extranuclear receptor proteins in the cytosol of the target tissues. Subsequently, the protein steroid complex moves to the nucleus and activates RNA synthesis, presumably by stimulating RNA polymerase.

Testosterone and other androgens are synthesized from cholesterol in the testis. The androgens are oxidized and conjugated with glucuronic acid for excretion in the urine. Adult males excrete about 20 mg of 17-ketosteroids daily.

Ovarian Development

The hormonal control of ovarian development is illustrated in Figure 12.3. Maturation of the oocyte and follicle is controlled by the follicle-stimulating hormone (FSH) and the luteinizing hormone (LH), which are produced by the pituitary in response to the secretion of the 'follicle-stimulating hormone releasing hormone' (FRH) and the 'luteinizing hormone releasing hormone' (LRH) by the hypothalamus. Growth and maturation of the follicle produces increasing secretion of oestrogen, whilst the corpus luteum produces the steroid hormone progesterone. LH stimulates the adenyl cyclase of the interstitial cells and corpus luteum and thereby increases their content of c-AMP. Oestrogens and progesterone exercise a feedback control on the formation of LRH and FRH. These effects are concentration dependent but become inhibitory when the concentrations of LRH and FRH exceed critical values. Nerve stimulation from the central nervous system maintains a constant low level of secretion of both FRH and LRH. These interactions produce the following changes in plasma–hormone levels (Figure 12.4; refer also to Figure 12.3).

(a) Initial elevated levels of FSH and LH (days -12 to -7), which produce

(b) an increase and then a decline in the oestrogen level (days -3 to $+1$).

As a consequence

(c) LH and FSH levels also pass through peak values (days -2 to $+2$)

(d) at the peak of LH production, progesterone levels begin to increase and ovulation occurs. Subsequently, in the unfertilized female, progesterone levels decrease and the cycle is repeated.

If fertilization occurs, the placental development following implantation of the fertilized ovum permits an additional production of the hormone progesterone because the placenta produces the peptide hormone chorionic gonadotrophin.

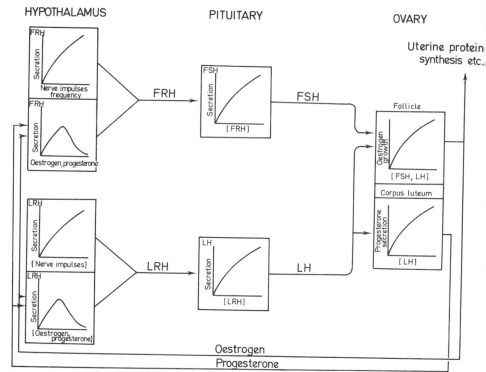

Figure 12.3 Interactions between hormones during the female reproductive cycle. FRH—Follicle stimulating hormone releasing hormone. FSH—Follicle stimulating hormone. LRH—Luteinizing hormone releasing hormone. LH—Luteinizing hormone

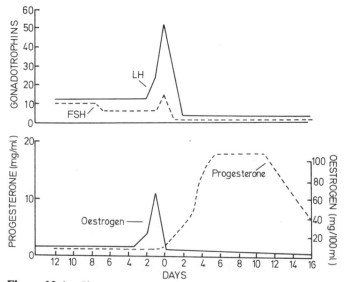

Figure 12.4 Changes in plasma concentrations of hormones during the menstrual cycle

This acts on the corpus luteum to maintain the secretion of oestrogen and progesterone. Subsequently, the placenta itself synthesizes both progesterone and oestrogen which, during pregnancy, are responsible for maintaining the processes of gestation and, as is apparent from Figure 12.3, also prevent the initiation of new cycles of ovarian development.

The Action of Oestrogen and Progesterone

There is now general agreement that oestrogen is bound by specific receptor proteins in the cytoplasm of target cells, for example in the uterus. The steroid–protein complex is then transported to the nucleus and the production of new types of RNA begins. New ribosomal and messenger RNA appear in the cytoplasm, forming new polyribosomes, and a phase of active protein synthesis qualitatively different to that occurring before hormone activation is commenced. It has been suggested that it is the protein element of the complex which plays the key role in activating transcription of nuclear DNA, the binding of the steroid being required to transform the protein into a suitable 'active configuration'. A similar basic pattern of operation appears to hold for progesterone.

Oestrogen produces growth of the adult ovary and, in the uterus, stimulation of protein synthesis as indicated above. Moreover, secretion of LRH and FRH is influenced by the levels of this hormone. Progesterone is required for many different processes, notably movement of the ovum in the fallopian tubes, preparation of the uterine wall for implantation of the fertilized ovum and inhibition of uterine contractions to permit the continued development of the foetus. As a consequence of the hormonal actions, the uterus undergoes striking changes during pregnancy when there is a considerable increase in its weight. One of the most interesting features of this change is that it involves a relatively rapid synthesis of the fibrous proteins collagen and elastin (see pp. 264–268). Collagen forms about 6 per cent of the weight of the non-gravid uterus. During pregnancy the weight of the uterus increases about ten-fold, but that of the collagen only six-fold, so that at term, the uterus has 3–4 per cent collagen. After parturition, the weight of the uterus and its collagen content decline rapidly in parallel, falling in some four weeks to about the values present in the non-gravid uterus and continuing to decrease for the next 6–8 weeks. The pattern of changes in the amount of elastin in the uterus is similar; this protein accounts for only about 1 per cent of the weight.

Variations in the amount of uterine cathepsin, which increases about six-fold during pregnancy to a value maintained for the first few weeks after birth, suggest that this enzyme may be concerned with the removal of uterine proteins during involution. Its inactivity before birth points to a localization in lysosomes (p. 418), from which it may be liberated by hormonal changes or by the partial anoxia which follows the collapse of the uterine blood vessels after birth.

Development of the Oocyte

The development of the oocyte is characterized by the storage of protein and RNA for later use. It is likely that much of the protein ('yolk protein') is derived ultimately from the liver, though some endogenous synthesis certainly occurs during maturation. On the other hand ribosomal, transfer and messenger RNA are formed by the maturing cells and conserved. At the end of the growth phase, metabolic activity declines to a very low level at which it remains until the ovum is fertilized in the oviduct. The maturation of the mouse oocyte *in vitro* can be supported by a medium containing pyruvate (or oxaloacetate) but not glucose, phosphoenol pyruvate, lactate or TCA cycle intermediates; in the human too, pyruvate is a key nutrient. It is not clear why energy metabolism is restricted in this way, though it is known that maturing oocytes contain morphologically atypical mitochondria, without a well-developed system of transverse cristae. It has been suggested that the cell membranes and mitochondrial membranes of the oocyte may be impermeable to all but a very restricted range of substrates.

CHAPTER 37

Foetal and Perinatal Biochemistry

Changes after Fertilization

Development of the embryo can be supported by a much wider range of substrates than that of the oocyte (lactate, pyruvate, TCA cycle intermediates can all serve in this way) and mitochondrial structure also changes with the appearance of transverse cristae. It is clear that oxidative metabolism via the citric acid cycle is important throughout the preimplantation period. Some enzyme activities increase (e.g. hexokinase, aldolase, malic dehydrogenase) while others decline (phosphofructokinase, lactic dehydrogenase). There are also qualitative changes in enzymes, e.g. the lactic dehydrogenase present before implantation is of the H type and synthesis of the M type does not begin until implantation. Glycogen reserves are also accumulated (in the mouse embryo) between fertilization and implantation, while the amount of protein declines.

There are also profound changes in the metabolism of nucleic acids. Experiments with amphibian eggs have made it possible to follow these processes in some detail. In this technique, a recipient egg is enucleated and another nucleus, withdrawn from a donor cell, is inserted with a glass pipette. Subsequently, it is possible to define the results of the mutual interactions between the egg cytoplasm and the nuclei taken from cells at various stages of development. Young oocytes synthesize RNA but not DNA whilst mature ova synthesize DNA but neither RNA nor protein. Embryonic nuclei return to the synthesis of all types of RNA and protein synthesis becomes increasingly important. If an embryonic nucleus is transplanted into egg cytoplasm, it ceases to synthesize RNA and, after about 20 minutes, begins DNA synthesis. Subsequently (after a few hours) RNA synthesis is resumed in a sequence characteristic of normal development. The nucleus, in effect, can be made to repeat a normal developmental programme by exposure to a younger cytoplasm. These experiments provide strong evidence for an important role of cytoplasmic components in influencing nuclear activity and, by extension, imply that after fertilization the activity of the sperm nucleus will also be regulated by the egg cytoplasm, especially as both egg and sperm nuclei concentrate cytoplasmic proteins in the early stages of development. It is also believed that cytoplasmic components are distributed unevenly throughout the egg, so that after cleavage embryonic nuclei will be exposed to different cytoplasmic environments. These differences, in that they could elicit different responses from the nucleus, may be the basis for the subsequent differentiation of the organism.

Foetal Energy Metabolism

Various changes in energy metabolism have been followed in foetal development up till parturition (Figure 12.5). In the liver, there is a steady increase in enzymic complexity, perhaps reflecting an exposure to new substrates. Although the glucose levels of the foetal plasma are relatively low (about half that of the

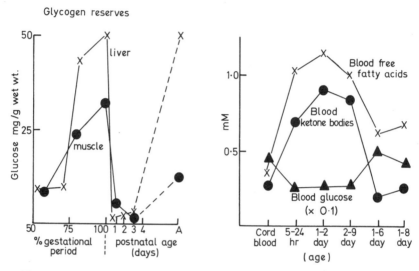

Figure 12.5 Foetal and perinatal changes in tissue glycogen and plasma substrates

mother) it is believed that carbohydrate is the main energy source during foetal development. Glucose provided by the mother is used as an energy source and is also stored as glycogen. In the liver, the concentration of glycogen at birth is similar to the adult level, whilst in muscle and heart the levels reached are considerably greater (five to ten times) than those in the adults. Enzymes involved in glycogen synthesis, including phosphoglucomutase, UDPG-pyrophosphorylase and glucosyl transferase, also increase in activity during this period of hepatic glycogen accumulation, whereas some other enzymes of carbohydrate metabolism such as glucose-6-phosphatase and dehydrogenase remain low. Plasma free fatty acids are lower than those of the mother (about 0.3 mM rather than 1 mM) and it seems likely that they do not make a large contribution to foetal energy metabolism. During foetal development there is a considerable increase in triglyceride which may comprise up to 16 per cent of the fresh weight of the animal at birth; foetal liver can convert acetate or citrate to fatty acids from the age of about 15 weeks. Thus, at birth, the metabolic fuel reserves of the human are as follows: total lipid, 161 g/kg body weight; total carbohydrate, about 11 g/kg body weight (one-third in the liver and the rest in muscle).

Changes in Energy Metabolism at Birth

Immediately after birth, as the animal becomes fully dependent on its internal reserves, blood glucose levels fall as do the glycogen reserves which are mobilized to provide glucose for the blood (see Figure 12.5). Glucose-6-phosphatase activity also increases sharply at this time and glycogenolysis occurs under the influence of adrenalin or glucagon. After several days, when maternal feeding is established, the blood glucose concentration rises steadily to adult levels.

Conversely, there are rapid increases in plasma-free fatty acids at birth (see Figure 12.5) following mobilization of the triglyceride reserves, with a concomitant release of glycerol and an increase in the level of circulating ketone bodies. The increased availability of ketone bodies may contribute to the pronounced ability of the brains of new-born infants to tolerate very low blood glucose levels for some time (see p. 326). Weaning brings about another series of changes in food intake and biochemical development. In general terms, the relatively high-fat diet of the milk is replaced by one in which carbohydrate contributes most to the energy supply. In man, the change is somewhat less dramatic than in the rat, as the ratios of calories provided in the milk of these two species by fat and lactose are 27 (rat) and 1·5 (man).

Enzyme Changes During Early Development

All the changes in the environment occurring during late foetal life, birth, perinatal development and weaning are mirrored in changes in the enzymic composition of the tissues. Detailed information is available for such changes in the liver of the rat. They may be summarized as follows (see Figure 12.6).

Just before and at birth, there are pronounced increases in

(i) A form of aldolase which acts readily on fructose-1-phosphate, and, therefore, enables the animal to metabolize fructose after its phosphorylation (see p. 204).

(ii) Glucose-6-phosphatase, required for utilization of glycogen reserves (see p. 455).

(iii) Fructose-1,6-diphosphatase. This increase is somewhat slower, and does not reach its maximum level until about 2 weeks after birth. It is a key enzyme in the development of gluconeogenic capacity.

(iv) Galactokinase which reaches a peak shortly after birth, and is essential for the metabolism of the galactose derived from milk lactose.

In the same period, there are pronounced decreases in

(v) Pyruvate kinase, reflecting the decreased availability of carbohydrate (endogenous reserves are rapidly depleted and, in the rat, only about 5 per cent of the calories are derived from the milk as carbohydrate).

(vi) Citrate cleavage enzyme, which is relatively unimportant on a high-fat diet (milk) with limited supplies of carbohydrate.

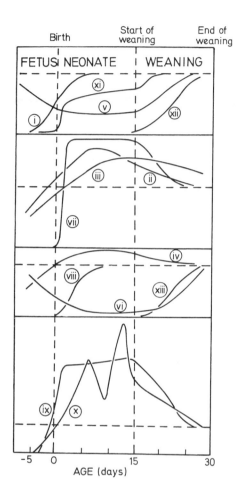

Figure 12.6 Diagram of developmental changes in enzymes in rat liver. The horizontal dashed lines through each section represent the normal adult activities. (i) 'Aldolase'. (ii) Glucose-6-phosphatase. (iii) Fructose-1,6-diphosphatase. (iv) Galactokinase. (v) Pyruvic kinase. (vi) Citrate cleavage enzyme. (vii) Phosphopyruvate carboxykinase. (viii) Glycerol kinase. (ix) β-hydroxyacyl-CoA dehydrogenase. (x) HMG-CoA synthetase. (xi) Argininosuccinate synthetase. (xii) Glucokinase. (xiii) Malic enzyme

Just after birth, there are pronounced increases in

(vii) Phosphopyruvate carboxykinase, which appears for the first time at this stage, to initiate gluconeogenesis.

(viii) Glycerol kinase, required for gluconeogenesis from glycerol freed by the utilization of triglyceride reserves.

(ix) and (x) β-Hydroxyacyl-CoA dehydrogenase and HMG-CoA synthetase, required for the oxidation of fatty acids and formation of ketone bodies (see pp. 230 and 233).

(xi) Argininosuccinate synthetase, the rate-limiting enzyme in the urea cycle.

Finally, at weaning, there are pronounced increases in

(xii) Glucokinase which opens up all the metabolic pathways for glucose, subsequent to its phosphorylation in a reaction which will be regulated largely by the concentration of glucose available (see p. 199)

(xiii) $NADP^+$-malate dehydrogenase (the malic enzyme) and citrate-cleavage enzyme (vi) both of which are needed for lipogenesis (see Figure 5.3).

(v) Pyruvate kinase, involved in the increased glycolytic flux which follows the greater availability of carbohydrate.

CHAPTER 38

The Control of Fertility

As Figure 12.2 shows, maintenance of spermatogenesis depends on the continuing presence of testosterone elaborated by the seminiferous tubules under the influence of FSH. Thus, interference with either the production of FSH or testosterone will render the male infertile. As yet, no way has been found to inhibit either of these activities without undesirable side-effects. However, attention is being given to the use of steroids at concentrations which lower the level of FSH and LH but do not interfere with the secondary sex characteristics. Derivatives of testosterone are effective in this way, but they impair liver function as well. Small doses of oestrogen can also inhibit FSH and LH production, and male 'pills' containing oestrogen and androgen (to maintain secondary sexual characteristics) are being tested.

Other investigations centre on the development of anti-spermatogenic agents, i.e. organic compounds which interfere directly with spermatogenesis. Many such compounds are known, but without exception those developed to date produce extensive genetic damage.

The Female

The use of steroid hormone contraceptives has been strikingly successful in controlling fertility since its introduction in 1956; by 1968 it was estimated that almost 15,000,000 women were using them, with a failure rate of 0·1 pregnancies per 100 woman-years exposure. Commonly, they are either derivatives of 19-nortestosterone or 17-α-hydroxyprogesterone acetate.

The oestrogen component acts by inhibiting ovulation either by suppressing the peak of LH secretion or by preventing the necessary synergism between the action of FSH and LH in promoting follicle growth (see Figures 12.3 and 12.4). The action of the progesterone derivatives is less well understood.

As the oestrogen components produce some undesirable effects, namely, liver damage, thromboembolism and raised levels of plasma glucose and insulin, continuing interest is being shown in devising contraceptive pills in which these effects are minimized and in the development of alternatives such as post-coital contraceptives. These are compounds which prevent the implantation and development of the fertilized ovum. No such compounds satisfactory for human use have yet been devised.

CHAPTER 39

The Metabolism of the Mammary Gland

The metabolic changes which occur in the mammary gland at the onset of lactation are probably larger than those in any other organ during normal development. For each volume of milk secreted, about 400 to 500 volumes of blood are circulated through the gland. When no milk is being secreted metabolism is very low but at the height of lactation each cell is producing in each second 4,000,000 to 6,000,000 molecules of carbohydrate, fat and protein. At the same time there are considerable increases in the activity of many enzymes. The enlargement of existing mitochondria and the formation of new ones at parturition is reflected in a five-fold increase in the activity of succinic dehydrogenase and cytochrome oxidase. The amounts of the enzymes of the pentose phosphate pathway, glutamic dehydrogenase, transaminases and arginase all increase to a maximum during lactation, falling rapidly again at involution. The maximum increase in protein synthesis is accompanied by a large increase (about ten-fold) in the RNA content of the gland.

There is an almost exactly similar increase in the activity of aspartate transcarbamylase, a key enzyme in pyrimidine synthesis (see p. 185) suggesting that an important regulatory step in RNA synthesis is the increased availability of pyrimidines resulting from increased synthetic activity. The formation of this enzyme is under the control of oestradiol, progesterone and growth hormones.

The Biosynthesis of Milk Protein

Casein, β-lactoglobulin and α-lactalbumin (B protein) make up 90 per cent of the milk protein and, as they do not occur in the blood, must be formed in the gland itself. 70–90 per cent of the milk proteins are formed from the amino acids of the blood; the remaining non-essential amino acids are synthesized in the gland itself.

The Biosynthesis of Milk Fat

Milk fat which has a characteristic composition for each species of mammal comes from two quite separate sources. Part is synthesized in the gland itself, part is transferred directly from the blood. The quantities derived from each source are somewhat variable and depend on the state of nutrition of the animal. Isotope studies have shown that acetate can be used to synthesize fatty acids in the gland; and the substantial arterio-venous differences detected for β-hydroxybutyrate make it probable that C_2 units may also be derived from this compound.

Dietary lipids also make their appearance in the milk. Measurement of arterio-venous differences across the gland show that the chylomicron fraction and the very low density lipoproteins contribute to the milk lipids; free fatty acids of the blood are not involved. During lactation there is a very large increase in the activity of clearing factor lipase, and there is little doubt the triglycerides presented to the gland are first hydrolysed and then resynthesized before entering the milk. Although isotope studies showed that the glycerol of the milk fat came mainly from blood glucose, there is no increase in plasma glycerol in the blood leaving the gland. Therefore the glycerol liberated by the clearing factor lipase is probably oxidized within the gland.

Glucose Degradation

Glucose utilization by the pentose phosphate pathway is particularly important in the metabolism of the mammary gland. As much as 60 per cent of the total glucose degraded may pass through this pathway, probably to form the NADPH + H^+ required for fat synthesis. Glucose is also glycolysed in the gland and any lactate formed is oxidized to CO_2 and water.

The Biosynthesis of Lactose

Lactose is a disaccharide in which galactose is joined by a β-linkage to carbon 4 of glucose; it is a reducing sugar as carbon 1 of the glucose is unsubstituted. It is almost confined to the mammary gland and milk, though small amounts may appear in the urine during lactation. Glucose required for lactose synthesis is derived mainly from the blood, though it is probable that other carbon sources are utilized for gluconeogenesis in the gland itself.

The pathway for lactose synthesis is shown in Figure 12.7. The enzyme lactose synthetase that catalyses lactose synthesis has been extensively studied recently and has been resolved into two proteins called A and B, neither of which have any activity in the synthesis of lactose from UDP galactose and glucose but when the two were mixed together they formed a complex which was active. The A-protein was the larger of the two and the smaller B-protein was identified as α-lactalbumin which is a major constituent of milk. However, it turned out that protein A although devoid of lactose synthetase activity did have an activity in its own right carrying out the following reaction:

$$\text{UDP-galactose} + N\text{-acetylglucosamine} = \text{UDP} + N\text{-acetyllactosamine}$$

With increasing concentrations of α-lactalbumin the formation of lactose starts and increases but the synthesis of acetyllactosamine declines until virtually lactose alone is the product. Thus the presence of α-lactalbumin changes the acceptor for the galactose of UDP galactose from acetylglucosamine to glucose. Thus an enzyme normally involved in the synthesis of glycoproteins can be changed to

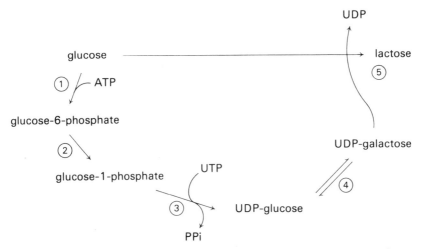

Figure 12.7 The intermediates and enzymes of the lactose biosynthetic pathway in the lactating mammary gland. The enzymes are denoted as follows: (1) hexokinase; (2) phosphoglucomutase; (3) UDP-glucose pyrophosphorylase; (4) UDP-galactose-4-epimerase; (5) lactose synthetase

an enzyme with a specialized function for lactose synthesis. α-Lactalbumin has therefore been called a specifier protein since it has no catalytic activity by itself but changes the substrate specificity of the A-protein. The A-protein of lactose synthetase is found in many tissues other than the mammary gland, for example in the liver and in the brain of the chick. What is interesting is the fact that the addition of α-lactalbumin to any of these N-acetyllactosamine synthetases makes them capable of synthesizing lactose although of course this normally never happens.

Thus the specification of lactose synthetase activity by α-lactalbumin in the mammary gland is unique to that gland because the synthesis of α-lactalbumin is unique to that gland. Although A-protein increases somewhat during pregnancy, there is no abrupt change after parturition. By contrast, the α-lactalbumin level remains very low during pregnancy but shows a very rapid rise after parturition.

Hormonal Control of Lactation

The hormonal control of lactation is outlined in Figure 12.8. The development of mammary gland function depends on interactions between a number of different hormones particularly cortisol and insulin, which direct the formation of new cells capable of responding to a third hormone, prolactin. Mammary gland in tissue culture can be shown to synthesize A-protein and α-lactalbumin in response to a mixture of insulin, hydrocortisone and prolactin, but if progesterone is added as well A-protein continues to be synthesized but the synthesis of α-lactalbumin is suppressed. The abrupt decrease of progesterone *post partum* triggers the onset of lactation and lactose synthesis. Here then is the new type of control of a metabolic process where the synthesis of a protein necessary

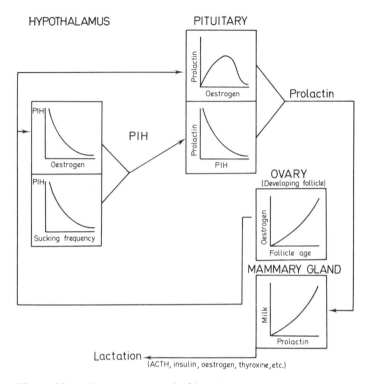

Figure 12.8 Hormonal control of lactation. PIH—prolactin release inhibiting hormone

for one purpose (nutrition of the young) is switched on hormonally at the appropriate time and this protein in turn switches on another enzymic activity required for the nutrition of the offspring.

Figure 12.8 also emphasizes the important roles of hypothalamus and pituitary glands in regulating the activities of lactation. Note that as suckling becomes less intense, the prolactin-release inhibiting hormone, will be secreted in larger amounts and the hormonal stimulus for continued milk production will diminish.

CHAPTER 40

Genetic Defects

The transmission of information to progeny is not always achieved with complete precision. If the offspring in which faulty transmission has occurred are viable, they often exhibit 'genetic defects'. Such defects influencing the metabolism of carbohydrates have already been discussed (p. 206). Another particularly instructive group of such defects relates to the metabolism of amino acids.

Phenylketonuria

This is a condition in which phenylpyruvic acid is excreted in the urine. The condition is inherited through a single recessive gene and large amounts of phenylalanine accumulate in the tissues. It has been established that phenylalanine hydroxylase (catalysing the conversion of phenylalanine to tyrosine, see Figure 12.9) which normally develops in the liver a few weeks after birth, fails to appear; for these patients therefore tyrosine is an essential amino acid. Both sexes are equally susceptible to the disease, which may be detected several weeks after birth, when the plasma phenylalanine concentration increases some thirty-fold over the normal and phenylpyruvic acid is excreted in the urine, which in consequence gives a green colour when tested with ferric chloride. After about 6 months mental symptoms appear and the majority of the patients are idiots or imbeciles; about 1 per cent of the mentally defective population are phenylketonurics. Besides their mental deficiency the patients are usually undersized, underpigmented, extremely nervous and have a series of persistent limb movements which may continue for hours at a time. In younger children, some of the nerves lack normal myelin. All these symptoms are the consequence of the inability to convert phenylalanine to tyrosine.

The urine of the phenylketonurics does not contain any abnormal metabolites but rather greatly increased concentrations of the normal metabolites, phenylalanine, phenyllactic, phenylpyruvic and phenylacetic acids (the last often conjugated with glutamine). Whether suspected persons are heterozygote carriers of the disease can be tested by measuring the plasma tolerance to a dose of phenylalanine. The discovery of the biochemical defect in phenylketonuria has made possible a treatment based on the limitation of dietary phenylalanine; this is a difficult procedure as the need to diminish the intake of this amino acid must be balanced against the requirement for phenylalanine for protein synthesis. Only if the controlled diet is begun at a very early age are the clinical symptoms

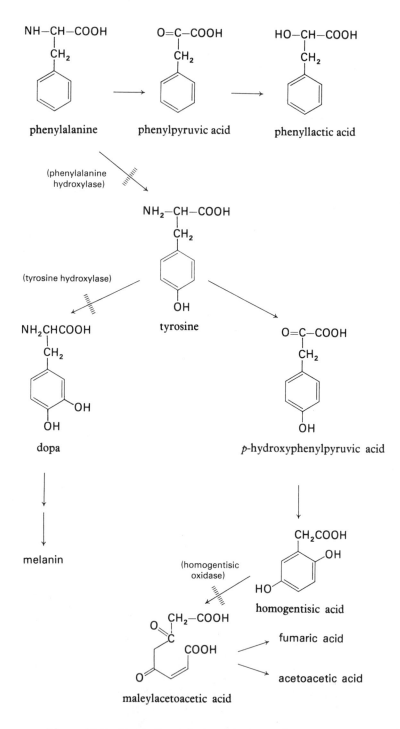

Figure 12.9 Metabolism of phenylalanine and tyrosine

amenable to treatment; pigmentation becomes normal, nervous restlessness diminishes and there is a great improvement in mental capacity. In patients diagnosed and put on a low phenylalanine diet soon after birth, development is apparently completely normal. In older patients there is no improvement in the mental condition following restriction of dietary phenylalanine.

Alcaptonuria

This is a rare, hereditary, metabolic disease in which homogentisic acid (a normal metabolite occurring during the metabolism of phenylalanine and tyrosine, Figure 12.9) accumulates and is excreted in the urine. The urinary homogentisic acid is often oxidized to a dark, melanin-like substance and thus the urine itself becomes darkly coloured; however, if the urine is acid or contains much ascorbic acid, this darkening is greatly delayed and may not be observed at all. Under these conditions, homogentisic acid may be detected by its ability to reduce copper and silver salts and molybdates. Besides appearing in the urine, the accumulated homogentisic acid may cause pigmentation of the cartilage and other connective tissues and, in later years, arthritis usually develops in the affected areas.

Alcaptonuria is caused by the failure of the enzyme homogentisic acid oxidase (Figure 12.9) to develop in the liver after birth. There is no known treatment for the condition as, even in starvation, homogentisic acid is still formed from the tyrosine and phenylalanine derived from the breakdown of endogenous protein.

Albinism

In this condition the melanocytes fail to synthesize the pigment melanin which is responsible for skin colour; some or all of the melanocytes may be involved. The genetic defect is such that the enzyme tyrosine hydroxylase (tyrosinase) which forms 3,4-dihydroxyphenylalanine (DOPA) from tyrosine is absent (Figure 12.9). As adrenaline synthesis is unaffected, it seems that only the melanocytes are involved directly.

Cystinuria

Cystine is an extremely insoluble amino acid and sometimes gives rise to renal stones. In the inherited metabolic disease cystinuria the urine contains more cystine than normal; it precipitates in the urine and may also be deposited in the kidney to form stones. In this condition there is a defect of reabsorption by the kidney tubules. Lysine, ornithine and arginine are also excreted in these cases, but as these compounds are soluble they normally escape notice.

Sickle-cell Anaemia

In this condition the erythrocytes become sickle-shaped; this may be a permanent configuration, in which case the patient is also severely anaemic, or it may develop at a relatively low oxygen tension. The sickle-cell effect is not very marked in the young, but becomes fully apparent only when foetal haemoglobin (haemoglobin F) is replaced by the normal adult haemoglobin A after about 6 months of post-natal development. Electrophoresis of the haemoglobin of the sickle cells from non-anaemic patients revealed two different pigments; haemoglobin A and an abnormal form of haemoglobin (haemoglobin S). This discovery led to the suggestion that the difference between the two types of sickling was that one occurred in individuals homozygous for haemoglobin S, and the other in heterozygotes. Analysis of the haemoglobin pattern of two non-anaemic 'sickle-cell' parents and of their two children confirmed this suggestion. Neither parent was anaemic but their red cells responded to lowered oxygen tension by forming sickles. Hence both parents were considered heterozygous for the sickle-cell trait. The elder child showed no sickling under any circumstances and had normal haemoglobin. The younger had only the sickle-cell haemoglobin and was anaemic.

The sickling of the cells is due to a physical property of the haemoglobin S. The oxygen capacities of haemoglobin S and haemoglobin A are about the same, as are the solubilities of the oxidized forms; however, on reduction, the solubility of haemoglobin A is decreased by about one-half, while that of haemoglobin S is diminished fifty-fold so that it comes out of solution in a gel-like mass. The greater the amount of haemoglobin S, the greater the distortion of the cell, so that the tendency for sickling is greater in the homozygote than in the heterozygous carrier. Most of the ill effects of the condition result from the increase in blood viscosity which follows the distortion of the erythrocytes (after gelling of the haemoglobin S) and the destruction of these distorted cells. As the blood viscosity increases, the rate of blood flow diminishes and the extent of deoxygenation increases, so that the sickling becomes more pronounced. If the sickle cells block small blood vessels the resulting stasis will cause a further lowering of the oxygen tension and increase sickling. The blocking of such vessels will prevent blood flow to areas of the tissue which may become necrotic. Finally, the abnormally shaped sickle cells tend to be engulfed by the reticuloendothelial cells, thus shortening the life span of the affected erythrocytes. One of the oddest consequences of the sickle-cell condition is that in some African countries it appears to offer protection against malignant malaria! This is probably because the erythrocytes infected with the malarial organism tend to stick to the walls of the blood vessels and therefore to become deoxygenated. If the host is a sickle-cell heterozygote, these deoxygenated cells will sickle and because of their changed shape will be phagocytosed. Thus there is a selective removal of the malarial organism in the heterozygotes.

The biochemical differences between haemoglobin A and haemoglobin S have been worked out. As it was known that the haem group was the same in each, the difference had to be sought in the structure of the globin itself. By treatment with trypsin (see Figure 1.27), the molecules were broken into comparatively large peptides whose amino acid composition could be more readily investigated. About 28 polypeptides were obtained, which were spread out in a characteristic pattern on paper by electrophoresis and chromatography. The pattern is called the

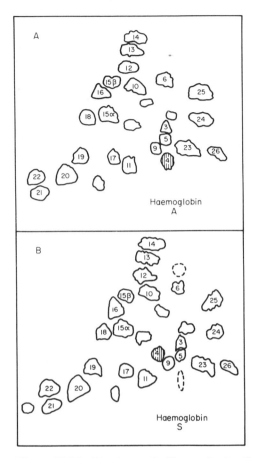

Figure 12.10 Tracings of 'fingerprints' of haemoglobins A and S. The enclosed areas represent peptide fragments of the proteins released by trypsin digestion. The only difference between the two proteins is in the peptide represented by the shaded area. From Baglioni, *Biochim. Biophys. Acta.*, 1961

$$\text{Hb A} \quad NH_3^+\text{—Val—His—Leu—Thr—Pro—Glu—Glu—Lys} \ldots$$
$$\uparrow$$

$$\text{Hb S} \quad NH_3^+\text{—Val—His—Leu—Thr—Pro—Val—Glu—Lys} \ldots$$
$$\uparrow$$

Figure 12.11 Amino acid sequence at the N-terminal end of the polypeptide chain of haemoglobin. Proteolysis with trypsin splits the chain at the arrow

'fingerprint' of the haemoglobin. Figure 12.10 shows the 'fingerprints' of haemoglobin A and haemoglobin S—the shaded areas indicate where they differ. The peptides corresponding to these shaded areas were eluted and their detailed amino acid composition studied. These analyses showed that the only difference between the two haemoglobins is the substitution of a valine for a glutamic acid unit normally found in the haemoglobin A (Figure 12.11). Thus, changing one amino acid in 280 is responsible for the different properties of the two haemoglobins and for all the clinical consequences which arise therefrom. Presumably this difference in turn is brought about by one alteration in the base sequence of the DNA.

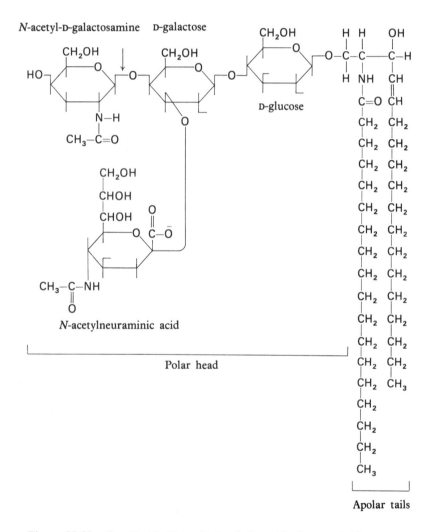

Figure 12.12 Ganglioside GM$_2$; the bond cleaved by hexosaminidase A is indicated by the arrow

Tay–Sachs Disease

The symptoms of this condition appear after about six months, and untreated sufferers die in two to five years. There is a rapid mental and motor deterioration, leading to blindness and general paralysis. Histological examination of nerve tissue reveals a distortion of nerve structure and an accumulation of lipid-rich granules which, as they contain acid phosphatase, are believed to be derived from lysosomes (see p. 419). The granules contain large quantities of the complex lipid ganglioside 'GM2' (Figure 12.12). Cerebral levels of GM2 can be up to 300 times greater than normal. Its abnormal accumulation in brain lysosomes is due to a deficiency of the lysosomal enzyme hexosaminidase A which cleaves a beta-linked *N*-acetyl galactosamine residue from GM2. This cleavage marks an early stage in the normal catabolic removal of the ganglioside.

The Tay–Sachs disease is transmitted as a recessive unit, more prevalent in the Jewish community (heterozygote frequency 0·03) than in the non-Jewish community (heterozygote frequency 0·003). Antenatal diagnosis can be performed, as assays for hexosaminidase A can be carried out on amniotic cells.

CHAPTER 41

Terminal Differentiation

The development of the human being to maturity proceeds, like that of other higher organisms, with the appearance of a set of individual tissues each with its own definable characteristics of composition, structure, function and metabolic capacity. The cells of the early embryo are totipotent, but from them appear groups of cells with a more restricted potential for development, destined to form tissues and organs. The final appearance of the cells of tissues and organs, with their limited and specific range of capacities, is called 'terminal differentiation'.

In our discussion of basic metabolic processes, some stress has been laid on the evidence for widespread similarities between the enzymic and structural composition of the cells of different tissues, though it has also been noted that there are characteristic patterns of enzymes in different tissues. Thus, terminally differentiated cells contain proteins (enzymes) that are ubiquitous in the organism, others of wide but not universal distribution and frequently a unique protein or proteins found only in the particular type of cell. It should be noted, however, that the development of a particular enzyme in different tissues follows its own organ specific pattern (see Figure 12.13). A number of studies of the appearance of cell-specific proteins during differentiation have been reported; these include the formation of myosin (in muscle); thyroglobulin (in thyroid glands); haemoglobin (in erythroid cells); casein, lactoglobulin and lactalbumin (in mammary gland).

Most commonly, the appearance of a new organ-specific enzyme or protein occurs in a sudden burst of synthetic activity, so that the adult level is reached abruptly, in hours or days. There is often also a pattern of sequential increases in groups of enzymes occurring at definite developmental stages, for example in the developing rat liver, just before and just after birth and at the end of the suckling period (see Figure 12.6). More complex functions too may often develop rapidly at critical periods. For example, the swimming ability of young rats emerges rapidly in the second week of post-natal life, reminiscent of the enzyme changes at the end of suckling. It has indeed been suggested that this change is related to the electrical excitability of the brain and as such is a reflection of the concomitant increase in the amount of the membrane-bound sodium–potassium ATP-ase in the brain tissues. In some differentiating systems, there is evidence for an initial low level of synthesis of the characteristic protein, followed by a dramatic upsurge under the influence of specific stimulators as differentiation proceeds. For example, the milk proteins are produced in minute amounts in the mammary glands of virgin mice. Culture of the cells with insulin, hydrocortisone and

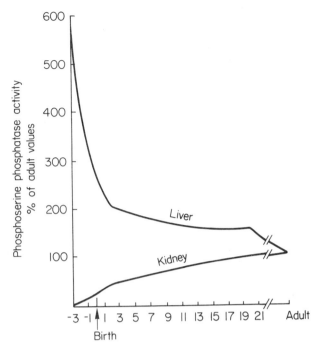

Figure 12.13 Developmental changes in the activity of phosphoserine phosphatase in rat and liver kidney. Redrawn from Greengard, *Essays in Biochemistry*, 1971

prolactin produces a capacity for greatly increased synthesis of all three proteins, at different times after exposure to the hormones. Actinomycin D (see p. 181) prevents the accumulation of the proteins, but only if it is present early in the developmental cycle; this observation may be interpreted as indicating a requirement for the initial formation of a relatively long-lived messenger RNA, necessary for the particular syntheses. In order to provide the capacity for greatly accelerated protein synthesis in differentiation, the genome often contains many copies of genes coding for ribosomal RNA. There is however no evidence of a similar multiplication of the genes for specific messenger RNAs, which may of course be transcribed a number of times.

In the mammary gland, it seems clear that the cells that are stimulated by the hormone mixture are already differentiated in the sense that the characteristic synthetic capacities are already active, albeit at a low level. The tissue is sometimes said to be 'protodifferentiated', i.e. capable of final differentiation in response to an external stimulus. Clearly, however, there must be a preceding phase in which an undifferentiated cell line gives rise to such protodifferentiated cells. It is not possible to provide any detailed insight into the molecular basis of this transformation. As for the final phase of differentiation, in which there is a relatively massive synthesis of specific proteins under hormonal stimulus, it is probable that the increase in the number of specific hormone receptor sites on the cell surface is responsible for the greatly increased sensitivity to the hormone. It is also known that differentiating cells can increase the range of external stimuli to which they will respond during development. For example, tyrosine transaminase can be formed in the liver of foetal rats at 18 days of gestation in response to cyclic AMP but not to glucagon. Two days later, glucagon stimulates the synthesis, i.e. the cell has developed a responsiveness to the hormone. After a further two days, hypoglycemia is a sufficient stimulus (acting through glucagon release). Finally, the ability of the tissue to respond to cyclic AMP in this way disappears in the adult; instead, increased synthesis can now be triggered by an ACTH-cortisone release or a high-protein diet.

These considerations make it clear that the appearance of an enzyme during development is governed by a multiplicity of interacting factors which must operate in an appropriate stepwise fashion before the protein is produced, in some cases even though the necessary messenger RNA is already present in the cell. Thus, induction of some liver proteins (such as ferritin) is not prevented by actinomycin D, and therefore must result from the translation of a long-lived messenger RNA. In short, 'an active gene is a necessary, but not sufficient condition for protein synthesis'; other factors can regulate the timing and extent of the translation of the genes message, and these factors may be particularly important in the developmental sequences of differentiation.

Cell Surfaces and Differentiation

The formation of tissues frequently demands that collections of like cells should aggregate specifically. Attempts to describe this process have centred on interaction of cell surfaces. It appears that glycoproteins (see p. 302) at the cell surface may be particularly important in such interaction. The major glycoproteins of the erythrocyte membrane have a hydrophobic C-terminal half, while the carbohydrate residues are linked to the N-terminal half. It is suggested that the hydrophobic section is buried in the lipid membrane structure, leaving the glycopeptide region exposed on the surface. It is also known that there is a high concentration of glycosyl transferases in the cell surface, and the intriguing suggestion has been made that specific cell interactions result from the binding of the exposed glycopeptide on one surface by the 'binding site' of the glycosyl transferase on another.

CHAPTER 42

Senescence

In man, as in all higher animals, the process of development after maturity culminates in senescence and death. The probability that a human being will die increases with his age, in a relationship which can be defined by a 'survival curve', relating the percentage of the population surviving at any age to that age. This curve also shows (a) an initial sharp decline, reflecting the relatively high rate of mortality of infants (b) a long period of almost constant low probability of death during youth and maturity and (c) finally a phase of rapidly increasing probability of death during senescence. It is also true that the *potential life-span* of human beings has remained strikingly constant, at a characteristic value for man, despite variations in diet and other external circumstances. Increases in the average length of life have been achieved by delaying the onset of the phase of rapidly increasing mortality in the community rather than by extending the life-span itself. In general terms, in the animal kingdom there appears to be an inverse correlation between life-span and the metabolic rate (i.e. the number of calories expended in metabolism per unit body weight) and a direct correlation between life-span and the size of the brain in relation to that of the body. This second factor has probably been particularly important in determining the relatively long life-span of man.

All these observations raise questions about the process of senescence. Is it due to accidental changes in the structure of living tissues, which could in principle be avoided? Or is it an inescapable consequence of the way in which the developmental programme of the organism is encoded and realized? The question cannot yet be answered decisively, but it is worthwhile to review briefly some of the attempts being made to explain senescence at a molecular level.

There appears to be little support for the notion that senescence in man is genetically programmed, in the sense that the life-span is limited as the result of particular changes designed to terminate life and switched on under genetic control at a specific stage of development. Instead, we must think of a process which involves the accumulation of 'errors' in the genome itself and in the molecular mechanisms by which it is expressed (i.e. in the repair and replication of DNA, in the transcription of RNA in the synthesis of proteins; see Section 3) to such an extent that the range of adaptive mechanisms available to cells is unable to cope with the consequent attenuation of metabolic activity. This explanation implies that there should be detectable changes in the properties of DNA in the cells of senescing organisms. There is evidence that the DNA of the chromosomes is, as a whole, less readily transcribed in senescent cells. This change is due to a diminished availability of the DNA within the chromosomes (where it is normally encased in a protein coat, largely of histones) which limits transcription, and also to intrinsic changes in the structure of the DNA itself. These changes are believed to reflect alterations in the base sequence of the DNA molecules (i.e. mutations) brought about by such influences as disordering due to thermal agitation, mechanical damage (e.g. during the mechanical movements involved in the unwinding of DNA molecules during replication), metabolic processes (e.g. the peroxidation reactions and the production of free radicals) and errors in DNA synthesis caused by the accumulation of errors in the enzymes and RNA species involved in this process (see below). As a consequence of these changes in the structure of DNA molecules, the information encoded by the DNA is gradually distorted and finally may be lost, especially as cells have a limited capacity for DNA repair. There should be changes in senescing cells in the properties of the RNA molecules and enzymes required for preparing amino acids for protein synthesis (i.e. in t-RNA and in the t-RNA-aminoacylating enzymes). There is a growing body of evidence that this is so. Age-related changes in t-RNA have been reported for a variety of cell types. In fact, variations in the t-RNA types of cells may be involved in all developmental processes, not only in senescence. Such variations have been shown to occur in regenerating rat liver, in hypophysectomized rats treated with growth hormone and in metamorphosing tadpoles treated with thyroxine. On the basis of such results a general theory for the control of development has been postulated based on the genetically determined appearance and disappearance at specific stages of development of particular types of t-RNA and their acylating enzymes, which together will determine whether or not particular types of messenger RNA can be translated. In essence, this hypothesis supposes that the appearance of the set of proteins characteristic of a particular developmental stage (including senescence) is regulated by the availability of appropriate types of t-RNA. As it has been shown experimentally that the availability of particular t-RNA molecules can determine the relative rates of synthesis of the α- and β-chains of haemoglobin by a rabbit reticulocyte preparation, the proposal appears to be a feasible one. However, the relationship between the age-dependent variation in t-RNA species and t-RNA acylating enzymes and the processes of senescence itself remains to be clarified. Finally, it is predicted that senescing cells should have an unusually high proportion of defective protein molecules, i.e. ones in which the amino acid sequence is abnormal. This concept is basic to a theory of ageing termed the 'error catastrophe' theory. This proposes that, even if no changes occur in the DNA of the cell, the known frequencies of error in transcribing the DNA into RNA and of translating the message of m-RNA into protein, will produce a proportion of protein molecules with abnormal amino acid sequences. If these defective proteins are involved in general metabolism or structure, they can be eliminated ultimately without great harm to the organism, but if they are involved in the transcription or translation processes themselves, then their defective functioning (a consequence of their abnormal shape) will cause an accelerating accumulation of errors, leading finally to a situation of 'error catastrophe' in which there is a rapid loss of function (in senescence) and death. The experimentally testable proposal is that senescing cells should contain an unusually high proportion of abnormal protein molecules. A number of studies with different types of cell preparations has suggested that this is so. For example, when lung fibroblasts are grown in culture through a number of cycles (i.e. are allowed to senesce in culture), it was shown that the heat lability of the enzymes glucose-6-phosphate dehydrogenase and 6-phosphogluconate dehydrogenase increased with the age of the cells. Increased heat lability was equated with an increased proportion of abnormal protein molecules in the extracts. The heat-labile enzyme also had an altered substrate specificity. It is claimed that this is evidence for the increased rate of formation of deformed proteins during ageing; misincorporation frequencies of about 10^{-3} would be necessary to produce these results in the oldest cells. Similar studies with the ageing fungus *Neurospora* showed that defective enzyme (in this case glutamic dehydrogenase) which accumulates is sufficiently similar in shape to the normal enzyme to react with the same anti-serum, though it is much less active in catalysis. Thus all these results appear to provide evidence for the accumulation of defective proteins during senescence.

The ideas which have been developed in this discussion can be summarized briefly in the scheme shown in Figure 12.14. On the right is described the invariable accumulation of erroneous proteins during the normal cycle of protein synthesis; this becomes important only if an altered DNA polymerase is produced as this in turn can cause an accelerating rate of error by producing mutations in DNA which may or may not be lethal. The same effect can be produced by mutation in the normal DNA of the cell.

This discussion has concentrated on age-related changes involving DNA, RNA and protein formation. However, the importance of other biochemical processes in senescence is also being considered. For example, changes in the structure of the connective tissues have been examined. It seems that there is a considerable

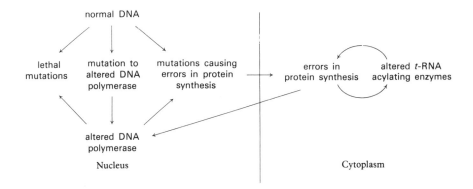

Figure 12.14 Possible factors leading to senescence

increase in the amount of cross-linking of collagen and elastin molecules with age and, in the ground substance of the connective tissue itself, a reduction in water content, changes in the chemical composition which decrease the water holding capacity and decreased rates of fluid transfer. These observations suggest that the flow of nutrients to and excretory products from the organs of the body may be impaired in older animals. What significance, if any, this has for the process of senescence is not yet clear. Finally, attention has been paid to the accumulation of pigment granules (lipo-fuschin or 'age-pigment' granules) in ageing tissues. These granules contain lipid and protein and are highly insoluble. Again, the details of their formation and the connection between their accumulation and senescence is not clear, though their formation appears to require cross-linking (by products of the peroxidation of unsaturated fatty acids) between a variety of cell components. It has been suggested that these age pigment granules cause cell damage and therefore senescence.

References

Metabolic and Endocrine Physiology an Introductory Text. Tepperman, J. Year Book Medical Publishers Chicago, 3rd edn., 1973.

Enzymic Differentiation in Mammalian Tissues. Greengard, O. *Essays in Biochemistry*, 7, 159–205, 1971.

Toward Molecular Mechanisms of Developmental Processes. Rutter, W. J., Pictet, R. L. and Morris, P. W. *Annual Review of Biochemistry*, 42, 601–646, 1973.

Nuclear Transplantation and Regulation of Cell Processes. Gurdon, J. B. *British Medical Bulletin*, 29, 259–262, 1973.

Prenatal Diagnosis of Genetic Disease. Friedmann, T. *Scientific American*, 1971, November, p. 34.

SECTION 13

Hormones and Endocrine Tissues

CHAPTER 43

General Biochemistry of Hormones

Hormones are substances released into the circulation in response to a change in the environment of the secretory cells of the particular endocrine gland. Once released, hormones are carried around the body by the circulation and affect those tissues whose cells contain receptors able to interact with them. As a result of hormone/receptor interactions, the relative rates of various metabolic processes are altered and a new level of physiological activity is instituted.

It must not be thought that chemical communication occurs only between endocrine glands and their target tissues. Neurotransmitters perform the same role between neurones and their post-synaptic cells whilst other substances, such as prostaglandins, released from cells that would not normally be classified as endocrine, modify the activities of neighbouring cells. Hormones, neurotransmitters and similar effectors are usually able to affect their target cells at very low concentrations. However, it must not be forgotten that some cells release massive amounts of metabolites which can cause profound changes in the metabolism of other tissues. For example, the release of fatty acids by fat cells can greatly alter the balance of metabolism in the liver. Interactions on such a massive scale are not considered to be hormonal; such designations are clearly somewhat arbitrary and it must be recognized that 'hormonal' control is merely a specialized facet of the responses evoked in cells by materials released from other cells.

Although much is known about the synthesis, storage and secretion of many hormones, their actions on target cells are much less well understood. The reasons for this lie in the highly integrated nature of cell metabolism: all the components of a living cell are in a state of dynamic equilibrium with one another so that a change in the activity or concentration of one component will result in alterations of differing magnitudes in the activities or concentrations of other components. Therefore, although hormones alter the metabolism of their target cells by reacting with a limited number of their components, it is clear that other components will be affected indirectly and to varying extents. A consequence of this dynamic balance of cell metabolism is that it is difficult to distinguish the primary change induced by the hormone from the secondary, tertiary and quaternary changes that follow from it. However, some progress has been made. Many hormones appear, as their primary action, to activate adenyl cyclase attached to the cell membrane and thereby to alter the intracellular concentration of c-AMP. Since c-AMP is an allosteric effector for many enzymes, profound metabolic changes ensue when its concentration is altered. Steroid hormones bind to a receptor protein in the cytoplasm which subsequently enters the nucleus, but how this is related to steroid induced changes in protein synthesis is unknown.

Adenyl cyclase, the membrane-bound enzyme responsible for forming c-AMP from ATP

$$ATP + H_2O \rightleftharpoons 3',5'\text{-cyclic AMP} + PP$$

has its activity modulated by a variety of hormones (Figure 13.1). It is not clear whether the hormones bind directly to specific sites on the catalytically active enzyme or whether they attach to receptor sites on regulatory subunits. The general consensus of opinion seems to be that, in many instances, there are several regulatory subunits in functional contact with a catalytic centre and that each subunit is specific for a particular hormone. On this basis the responsiveness of a cell to a given hormone would depend upon whether or not it possessed the appropriate receptor unit. Some hormones activate adenyl cyclase whilst others are inhibitory. The response of the enzyme to a hormone depends upon the nature of the receptor it possesses. Adrenaline, for example, activates some adenyl cyclases and inhibits others. c-AMP modifies the activity of a variety of enzymes, in particular a group of protein kinases responsible for phosphorylating enzymes and proteins and thereby modifying their activity (see p. 437). On ac-

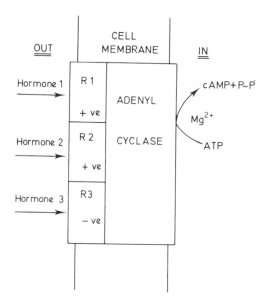

Figure 13.1 Possible relationship between adenyl cyclase and hormone receptors. c-AMP synthesis occurs only at the internal face of the membrane whereas the hormone receptors (R_1 etc.) face outwards to the extracellular fluid. In this figure hormones 1 and 2 activate adenyl cyclase whilst hormone 3 is inhibitory

count of its intermediary role in hormonal control, c-AMP is often referred to as 'the second messenger'.

c-AMP is degraded to AMP by phosphodiesterase, an enzyme inhibited by methyl xanthines such as caffeine, theophylline and theobromine. Thus the cytoplasmic level of c-AMP is determined by the relative activities of adenyl cyclase and phosphodiesterase.

Many hormones produce an easily recognizable physiological response very rapidly after their arrival at the target cells. For example, noradrenaline and vasopressin both cause an increase in blood pressure almost as soon as they are injected intravenously, whilst oxytocin released in the suckling reflex causes an immediate flow of milk from the mammary glands. Similarly, the concentration of blood glucose begins to fall very shortly after insulin is administered, and adrenaline and glucagon cause the rapid mobilization of glucose from glycogen. Some hormones induce changes that take much longer to become obvious. Responses of this type are brought about by the sex hormones which control the onset of puberty and changes accompanying the oestrus cycle.

Rapid physiological responses induced by hormones follow from the activation of existing cellular enzymes and do not require the synthesis of new proteins. Slow responses, in contrast, usually involve the initiation of a new pattern of protein synthesis leading to a more or less profound reorganization of the enzymic machinery of the cell. Some hormones induce both rapid and slow responses; thus insulin is not only responsible for quickly increasing the ability of muscles to take up glucose, it also alters the enzyme pattern of liver. The interplay between hormones is such that slow-acting hormones can modify the response of tissues to fast-acting ones. Thyroxine, for example, increases the amount of adenyl cyclase in adipose tissue and thereby increases the ability of catecholamines and glucagon to effect a rapid mobilization of free fatty acids.

Hormonal regulation of metabolism presupposes that hormone action can be terminated as well as initiated; thus metabolic pathways exist that can convert hormones into compounds devoid of, or having much reduced, physiological activities.

CHAPTER 44

The Adrenal Cortex

The adrenal cortex secretes a number of steroid hormones which, for convenience, may be divided into three groups; (i) the glucocorticoids, cortisol and corticosterone, which influence carbohydrate, lipid and protein metabolism and suppress some inflammatory processes; (ii) the mineralocorticoids, aldosterone and deoxycorticosterone, whose activity is described on p. 408, and (iii) the sex steroids comprising oestrogens, androgens and progestins. It must not be thought that the terms gluco- and mineralocorticoid are mutually exclusive, steroids that are predominantly glucocorticoids may have weak effects on sodium retention whilst potent mineralocorticoids may have some ability to stimulate glucose and glycogen synthesis.

The corticosteroids are all synthesized from cholesterol, most of which is derived from the circulation, although some is formed in the gland from acetate by the pathway outlined on p. 239. Within the cells, cholesterol is stored as its esters in lipid droplets that decrease in number when hormone production is stimulated; the hormones are not stored to any significant extent, human adrenal cortex, for example, contains only sufficient cortisol to sustain normal secretion for $1\frac{1}{2}$ minutes.

The reactions leading from cholesterol to the various corticosteroids are many and complex and will not be discussed in detail. Suffice it to say that many of them are hydroxylations requiring oxygen, NADPH + H$^+$, a flavoprotein, non-haem iron and cytochrome P$_{450}$. These reactions are illustrated in Figure 13.2.

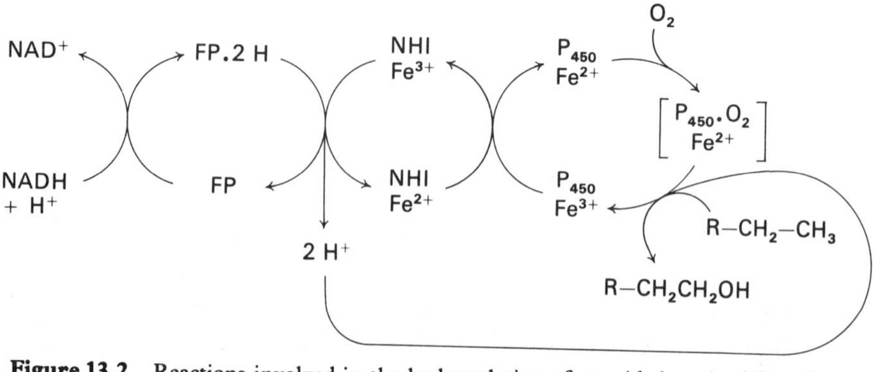

Figure 13.2 Reactions involved in the hydroxylation of steroids by mixed function oxidases. FP = flavoprotein; NHI = non-haem iron; P$_{450}$ = cytochrome P$_{450}$

The NADPH + H$^+$ used in these reactions is supplied mainly by the two dehydrogenations of the pentose phosphate pathway (see p. 208) which is particularly active in the adrenal cortex and other tissues synthesizing steroid hormones. Whilst many of the hydroxylations occur in the endoplasmic reticulum, some take place within the mitochondria but still use reducing equivalents generated in the cytoplasm. This transfer of reducing power from cell sap to mitochondria seems to be made possible through the agency of cytoplasmic and mitochondrial malic enzymes (Figure 13.3). The first reactions in the

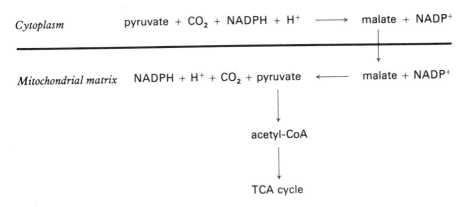

Figure 13.3 Reduction of mitochondrial NADP$^+$ by cytoplasmic NADPH + H$^+$

biosynthesis of steroid hormones from cholesterol take place in the mitochondria and are concerned with the removal of a six carbon fragment from the side arm. Biosynthetic relationships between cholesterol and some of the corticosteroids are shown in Figure 13.4.

The rates of synthesis of glucocorticoids in the zona reticularis and zona fasiculata of the adrenal cortex are controlled by the adrenocorticotrophic hormone (ACTH) secreted by the anterior pituitary gland. It seems probable that the immediate effect of the trophic hormone is to stimulate the formation of c-AMP by adenyl cyclase and that this is rapidly followed by an increased production of steroid hormones. However, the relationship between c-AMP and hormone formation is still uncertain. ACTH also enhances RNA and protein synthesis in the adrenal cortex but these changes are slow to appear compared with the rapid increase in hormone output. Aldosterone production in the zona glomerulosa of the adrenal cortex is controlled by the renin/angiotensin system described on p. 320.

Physiological Function of the Corticosteroids

Loss of corticosteroid secretion as a result of the destruction of the adrenal cortices by disease invariably leads to death unless some form of replacement therapy is undertaken. Despite this dramatic demonstration of their importance, it cannot be said that there is any process that has an absolute requirement for

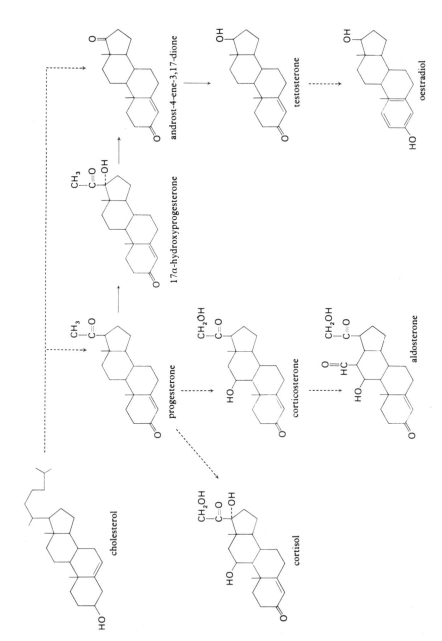

Figure 13.4 Biosynthetic relationships between cholesterol and steroid hormones

corticosteroids. It seems, rather as though they exert a 'permissive' effect on already existing processes. Thus sodium reabsorption by the kidneys is not totally defective in the absence of mineralocorticoids; there is merely a failure to reabsorb some 2 per cent of the sodium filtered at the glomerulus. Nevertheless, the cumulative effect is marked sodium depletion, water loss and haemoconcentration. Loss of corticosteroid production can, to a considerable extent, be overcome by supplementing the diet with NaCl or, better still, by administering a mineralocorticoid. However, full compensation requires the additional provision of a glucocorticoid. For example, fasting adrenalectomized animals are unable to maintain their stores of liver glycogen as well as intact animals unless glucocorticoids are supplied.

Glucocorticoids

It has to be confessed that, despite much work, the role of the glucocorticoids in human metabolism is still far from clear. There seems to be no doubt that they enhance the ability of hormones such as the catecholamines and growth hormone to stimulate lipolysis and fatty acid release from the adipose tissue, although the mechanism by which they act is not understood. Glucocorticoids also oppose the facilitating action of insulin on glucose entry into fat cells. By these actions, the glucocorticoids contribute to the mobilization of the major energy reserve of the body. In diabetics the absence or ineffectiveness of insulin leads to an imbalance between esterification and lipolysis in the adipose tissue and consequently to massive mobilization of fatty acids and ketoacidaemia. Adrenalectomy, by reducing the effectiveness of lipolytic hormones, greatly ameliorates these aspects of the diabetic state.

Injection of glucocorticoids into intact rats causes involution of muscle, connective tissue and lymphoid tissue thereby making amino acids from peripheral proteins available for gluconeogenesis in the liver. In addition to providing the liver with substrates drawn from involuting tissues, treatment with glucocorticoids leads to changes in the enzyme pattern of the liver. Thus, after giving cortisol, the rates of RNA and protein synthesis begin to rise and after about 3 hours it is possible to begin to detect increases in the activities of individual enzymes. Amongst the earliest enzymes to increase in amount are tyrosine aminotransferase and tryptophan oxygenase; the oxygenase may be important in diverting amino acids from protein synthesis towards energy metabolism by destroying the essential amino acid tryptophan. Other enzymes increasing in quantity over varying period of time after cortisol administration are pyruvate carboxylase, phosphoenolpyruvate carboxykinase, fructose-1,6-diphosphatase, glucose-6-phosphatase and alanine-amino transferase. All of these enzymes are concerned in the conversion of glucogenic compounds into glucose. The five enzymes of the urea cycle also increase in amount and thereby enhance the ability of the liver to dispose of unwanted α-amino groups. The increased enzyme activities seen after glucocorticoid administration are all the result of increased protein synthesis *de novo* and do not occur in the presence of actinomycin D which prevents DNA-dependent synthesis of RNA. At present we can only speculate about the relationship between these various changes. The early increases in enzyme activities may result from a direct action of the glucocorticoids on RNA or protein synthesis whilst the later changes may be caused by alterations in metabolism, such as increases in substrate levels, that are consequent upon the early changes. In any event the glucocorticoids are hormones that modify the enzymic composition of their target tissues.

The effects of administering glucocorticoids to normal men are less dramatic than is the case with rats; there is an increase in the steady level of blood glucose but no significant change in nitrogen loss. This suggests that gluconeogenesis from protein may not be increased by glucocorticoids in normal men but rather that increased fatty acid mobilization increases the steady state level of blood glucose by cutting down the glucose consumption of muscles by antagonizing the effect of insulin on glucose entry to the muscle fibres. Furthermore, an increase in blood glucose would tend to stimulate insulin release and so counteract the tendency of glucocorticoids to cause amino acid release from peripheral tissues. This sequence of events is of significance in starvation, as glucocorticoid release in association with an intact insulin system will delay the onset of increased protein degradation in muscle and lymphoid tissue. If this were uncontrolled it would severely impair the ability to survive in a hostile environment. It must also be remembered that under these conditions blood glucose, and hence insulin, levels are maintained, in part, by the brain switching from glucose to ketone bodies as its major fuel. The fact that, even during prolonged starvation, the loss of protein nitrogen does not increase significantly above the basal level of about 4.5 g per day until the main energy reserve of the body (triglyceride) is severely depleted, is a reflection of the part played by glucocorticoids in organizing the orderly mobilization of energy stores.

Aldosterone

Aldosterone (Figure 13.4) is secreted by the zona glomerulosa of the adrenal cortex and causes the cells bordering the distal tubules of the kidney to reabsorb sodium at a faster rate from the modified ultrafiltrate of plasma present in the lumen. This results in the retention of sodium which is accompanied by a retention of water, especially in the extracellular compartments of the body. Hence aldosterone has antidiuretic properties whilst aldosterone antagonists such as

aldactone or spironolactone

aldactone are useful as diuretic agents. Besides stimulating sodium reabsorption, aldosterone also stimulates potassium excretion although this effect is not as marked as that on sodium reabsorption.

Experiments on sodium transport by toad bladder indicate that there is a lag period of about 90 minutes following the administration of aldosterone before the rate of sodium transport begins to rise and that the rise can be inhibited by actinomycin D. Hence aldosterone is another slow-acting hormone that exerts its effects by stimulating protein synthesis. This may be contrasted with the rapid rise in sodium reabsorption that follows the administration of vasopressin to the same preparation.

The rate of aldosterone synthesis and of its release is controlled by angiotensin II which is formed when the blood pressure in the renal artery falls (see p. 320). By increasing the amount of extracellular water, aldosterone causes the blood volume to rise with a concomitant improvement in the blood pressure.

Inactivation of Corticosteroids

Corticosteroids are inactivated in the liver by a wide variety of processes including reduction and conjugation, especially with glucuronic acid (see p. 229) and to some extent with sulphate (see p. 228). The conjugated steroids, in contrast to the hormones, are not well bound by plasma proteins and, in humans, pass into the urine with ease. Conjugated steroids also enter the bile and are lost with the faeces or reabsorbed.

CHAPTER 45

The Adrenal Medulla

The hormones released from the adrenal medulla are the catecholamines, noradrenaline and adrenaline, which are synthesized from tyrosine by the sequence of reactions shown in Figure 13.5. Adrenaline is formed in those cells exposed to blood draining from the adrenal cortex through a portal system of capillaries. The relatively high concentration of corticosteroids in that blood induces the synthesis of phenylethanolamine N-methyl transferase which is responsible for converting noradrenaline to adrenaline. Noradrenaline is the end product of catecholamine biosynthesis in those cells receiving a direct arterial blood supply and which are, therefore, not exposed to high levels of corticosteroids. Within the chromaffin cells constituting the adrenal medulla, adrenaline and noradrenaline are concentrated in subcellular organelles termed chromaffin granules which consist of a lipoprotein membrane bounding an internal matrix composed of catecholamines, ATP and protein. Dopamine-β-hydroxylase, the enzyme which converts dopamine to noradrenaline, is located in both the granule membrane and in the matrix. The other matrix proteins, the chromogranins, are not enzymatically active and their role is uncertain although they have often been thought to be concerned, together with ATP, in maintaining the high catecholamine content of the matrix by forming some kind of complex. Catecholamine entry to the matrix phase of the granules is mediated by a Mg^{2+}-ATP-dependent pump located within the membrane. However, once they are incorporated into the matrix, the stored amines can be maintained in a concentrated state without the continuous expenditure of metabolic energy.

Dopamine-β-hydroxylase is the only enzyme involved in catecholamine biosynthesis to be located in the storage granules; the other enzymes are all soluble components of the cytoplasm. This distribution implies that noradrenaline formed in the granules from dopamine must migrate to the cytoplasm to be methylated to adrenaline and then re-enter the granules for storage.

The store of catecholamines within the adrenal medulla is very large, equivalent to about a month's normal requirements. Despite this, the rate of catecholamine synthesis increases when the gland is stimulated to secrete. This may be related to the fact that tyrosine hydroxylase, the first and rate-limiting enzyme in catecholamine biosynthesis, is inhibited by dopamine, noradrenaline and adrenaline. Acetylcholine released from the endings of splanchnic nerve fibres induces catecholamine release from chromaffin cells by increasing the influx of Ca^{2+} ions. These ions initiate a process in which the membranes of chromaffin granules

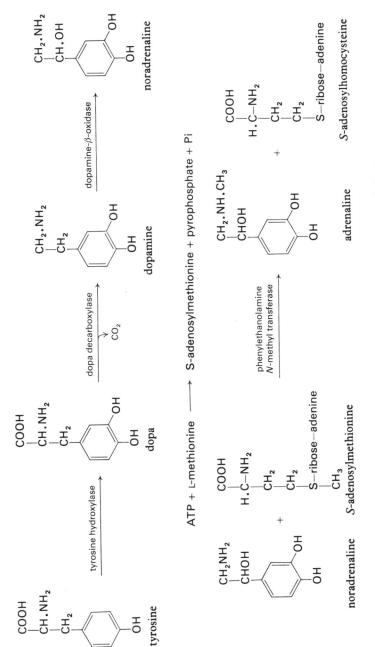

Figure 13.5 Reactions involved in the biosynthesis of catecholamines

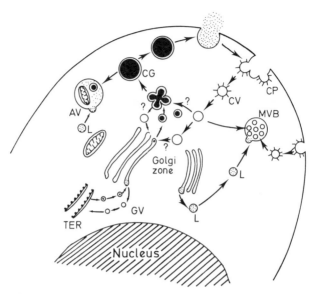

Figure 13.6 Possible events in the life-cycle of a chromaffin granule. Macromolecular components of the chromaffin granule (CG) are synthesized in the endoplasmic reticulum, transferred from the transitional endoplasmic reticulum (TER) to cisternae of the Golgi apparatus in Golgi vesicles (GV). Immature granules bud off from Golgi cisternae, and may fuse together to give the larger, mature chromaffin granule. Mature secretory granules, if not used for secretion, are digested by lysosomes (L) in autophagic vacuoles. Secretion from the chromaffin granule occurs by exocytosis, which leaves the granule membrane attached to the plasma membrane. The granule membrane is converted into one or more coated pits (CP) which give rise to coated vesicles (CV) in the cytoplasm. It is not known whether the membrane so retrieved can be reused, in which case it may be filled with secretory products in the Golgi region; or whether it is immediately degraded by lysosomal enzymes, perhaps in multivesicular bodies (MVB). Modified from A. D. Smith, *The Scientific Basis of Medicine Annual Reviews*, 1972

close to the cell membranes fuse with it to form open sacs from which the entire contents of the granule matrix can escape into the extracellular space (Figure 13.6). Thus the secretory products of the adrenal medulla are not only catecholamines but include ATP and its catabolites and proteins. This type of secretory process, in which the membrane of a storage granule remains within a secretory cell whilst the matrix contents are expelled, is termed exocytosis. Following the secretion of its contents, the membrane of the chromaffin granule has to be retrieved from its association with the cell membrane. The mechanism whereby this is done is unknown, but some suggestions are made in Figure 13.6. Depending upon circumstances, the relative proportions of adrenergic and noradrenergic chromaffin cells activated can be varied neurogenically; thus

Table 13.1 Some effects of catecholamines on a variety of tissues mediated via changes in the activity of adenyl cyclase

Liver	Increased glycogenolysis
	Decreased glycogen synthesis
	Increased gluconeogenesis
	Decreased triglyceride secretion
	Increased protein catabolism
	Increased synthesis of phosphoenolpyruvate carboxykinase
White adipose tissue	Increased lipolysis
	Increased glycogenolysis
	Activation of phosphofructokinase
Heart muscle	Increased glycogenolysis
	Increased lipolysis
	Increased Ca^{2+} uptake resulting in an increased contractile force (positive inotropic effect)
Smooth muscle	Increased glycogenolysis
	Contraction or relaxation depending on the muscle, e.g. relaxation of tracheal smooth muscle; contraction of vas deferens smooth muscle

adrenaline is released predominantly when the level of blood glucose falls whilst noradrenaline is released preferentially during exposure to stressful external situations.

Metabolic Actions of the Catecholamines

The catecholamines induce a wide variety of metabolic changes in many tissues (see Table 13.1). These changes all follow from a primary action of the catecholamines on adenyl cyclase located in the cell membrane of the receptor cell. In most instances the catecholamines enhance its activity and thereby increase the concentration of c-AMP in the target cells. In some instances, however, they have an inhibitory effect on the enzyme and diminish c-AMP levels. Some metabolic consequences of increases in c-AMP concentrations induced by catecholamines are discussed on p. 439 and their involvement in ganglionic transmission is described on p. 341.

Catecholamines and Insulin Secretion

Catecholamines exert a powerful inhibitory action on the release of insulin which normally occurs in response to increases in the level of blood glucose and ketone bodies. This inhibition is a consequence of catecholamines inhibiting adenyl cyclase and thereby reducing c-AMP concentrations in the β-cells of the pancreas. Inhibition of insulin secretion maintains glucose and fatty acids and ketone bodies (mobilized by catecholamines) at higher levels than if insulin secretions had proceeded normally.

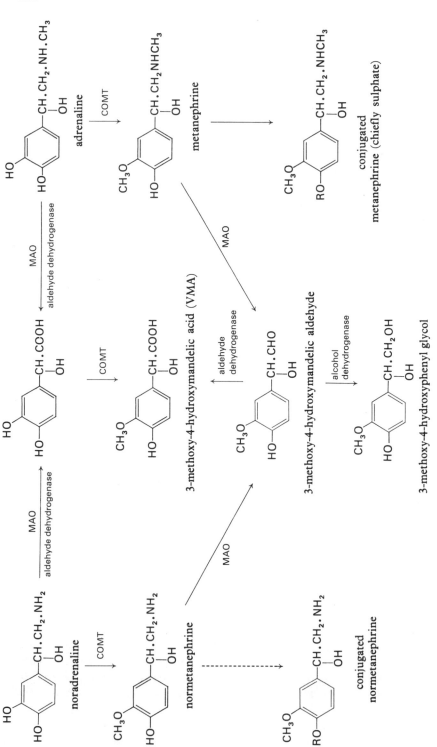

Figure 13.7 Metabolism of catecholamines by monoamine oxidase (MAO), catechol-*O*-methyl transferase (COMT) and conjugation

Inactivation of Catecholamines

Catecholamines are inactivated by the combined actions of catechol-*O*-methyl transferase (COMT) and monoamine oxidase (MAO). Aldehydes formed by MAO are either oxidized to the corresponding acids by aldehyde dehydrogenase or reduced to their alcohols by alcohol dehydrogenase. Some of the metabolites are conjugated with sulphate or glucuronic acid in the liver, this is especially so for metanephrine (see Figure 13.7).

Much of the noradrenaline released from noradrenergic neurones is retrieved by the nerve endings and used again; this stratagem both conserves transmitter and terminates its physiological action.

Catecholamines are continuously oxidized, in the cells where they are formed, by MAO so that the size of the amine store depends upon the balance between synthesis, secretion and endogenous breakdown. Inhibiting MAO with drugs such as iproniazid can greatly increase the concentration of biogenic amines in both the brain and peripheral tissues.

CHAPTER 46

The Thyroid

The thyroid hormones, chiefly thyroxine (T_4) and triiodothyronine (T_3) are formed as modified amino acid residues in thyroglobulin, an iodinated glycoprotein composed of two subunits and having a molecular weight of 680,000. The protein subunits are formed in the rough endoplasmic reticulum where mannose is also incorporated, they subsequently pass to the Golgi region where galactose is added and the glycosylated polypeptides are packed into vesicles which migrate to the apical pole of the cell. Somewhere in the region of the apical membrane about 15 of the 120 tyrosine residues in the polypeptides are iodinated (Figure 13.8) and thyroglobulin is released directly from the vesicles into the follicular lumen (Figure 13.9).

Iodide is concentrated by active transport from the blood into the follicular cells where it is converted, by iodide peroxidase, into an active species of iodine (possibly the iodinium ion I^+) able, with the aid of iodinase, to iodinate appropriate tyrosine residues in the polypeptide precursors of thyroglobulin. The

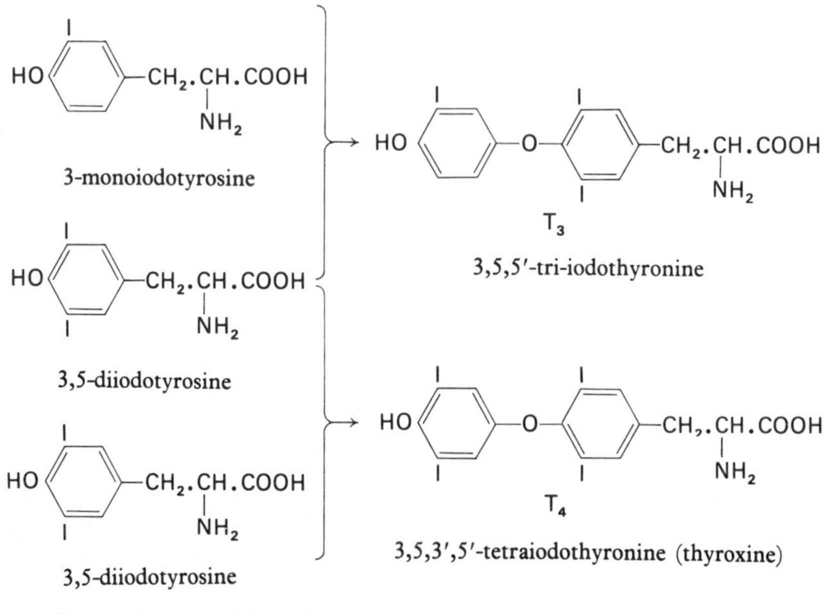

Figure 13.8 Possible pathways for the biosynthesis of thyroid hormones

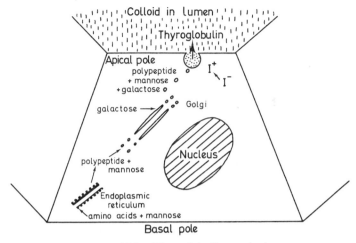

Figure 13.9 Thyroglobulin synthesis

daily uptake of iodine amounts to approximately 50 μg. Iodination can be inhibited by blocking iodide uptake with thiocyanate and nitrate and by inhibiting iodide peroxidase with thiocarbamides such as thiourea and thiouracil. These last two drugs have found clinical application in the treatment of hyperthyroidism.

thiourea

thiouracil

Hormone secretion begins with the follicular cells taking up colloid droplets containing thyroglobulin by a process of endocytosis (see Figure 13.10) which is greatly stimulated by the thyroid stimulating hormone (TSH) released by the anterior pituitary gland. TSH is a hormone which acts via c-AMP whose synthesis it increases. Iodide uptake is also enhanced by TSH. Within the cells, the colloid laden vesicles fuse with organelles called lysosomes which contain a range of hydrolytic enzymes with pH optima around pH 5. Thyroglobulin is hydrolyzed by the lysosomal proteases and the peptide bonds incorporating T_3 and T_4 into the protein are split. On reaching the basal pole of the cell, the contents of the vacuoles are released; the hormones enter the blood and any undegraded thyroglobulin passes into the lymph. Secretion probably occurs by exocytosis (see Figure 13.10), as only that mechanism appears to account for the release of a molecule as large as thyroglobulin.

Thyroxine accounts for most of the organic iodine released by the thyroid with T_3 making up the remainder, possibly up to 20 per cent. In the plasma thyroxine combines with two types of carrier protein, one an α-globulin and the other, of less quantitative importance, an albumin. The greater biological potency of T_3 as

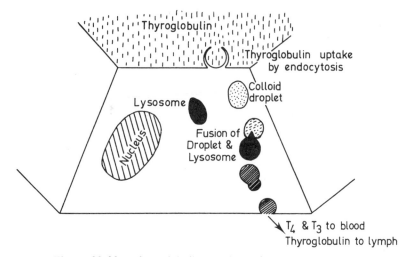

Figure 13.10 Thyroglobulin uptake and hormone release

compared with thyroxine may be a consequence of the former not being bound by plasma protein.

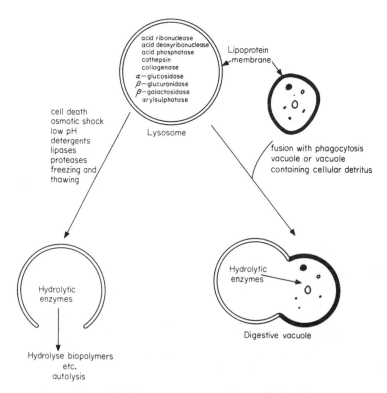

Figure 13.11 Lysosome, structure and function

Lysosomes

Lysosomes are found in most cell types: they contain a spectrum of hydrolases that enable them to break down a wide variety of biopolymers (see Figure 13.11). They seem to be responsible for degrading material taken up by endocytosis and pinocytosis as well as for the breakdown of intracellular organelles such as mitochondria and storage vesicles. After death, the release of the hydrolytic enzymes, caused by a fall in pH damaging the lysosomal membranes, accelerates the rate of autolysis. During life, of course, rates of breakdown and resynthesis remain essentially in balance.

Metabolic Effects

Hyperthyroid individuals are characterized by a metabolic rate that may be 50 per cent higher than normal; this in turn causes a marked loss of weight. Other

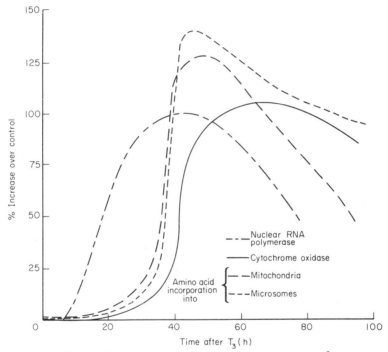

Figure 13.12 Idealized version of the time-course response of some activities in nuclei, mitochondria and microsomes from livers of thyroidectomized rats after a single injection of 20 μg triiodothyronine (T_3)/100 g body weight. The main features are that mitochondrial respiration (here illustrated only by cytochrome oxidase activity) reached a peak value after amino acid incorporation into protein by microsomes and mitochondria; that the protein synthetic capacity of both mitochondria and microsomes increased simultaneously and after a long lag period; that the DNA dependent RNA polymerase in nuclei showed a response several hours before a stimulation was observed in the cytoplasmic protein synthesis and was well within the latent period of BMR stimulation. From J. R. Tata in *Actions of Hormones on Molecular Processes*, Wiley, New York, 1964

symptoms of hyperthyroidism are tachycardia, high blood pressure, hyperglycaemia and a negative nitrogen balance. In hypothyroidism the reverse obtains; the metabolic rate is depressed and body weight increases. Hypothyroidism at birth develops into cretinism unless thyroxine replacement therapy is given.

Thyroid hormones have been found to affect many biochemical parameters both *in vivo* and *in vitro*. It would appear that the first change that can be observed following the injection of physiological doses of thyroid hormones (20 μg T_3/100 g) to thyroidectomized animals is an increase in the activity of DNA-dependent RNA polymerase in the nucleus. This is followed by an increased incorporation of amino acids into mitochondrial and microsomal proteins (Figure 13.12). Subsequently the basal metabolic rate (BMR) rises in parallel with the activity of cytochrome oxidase. The rise in BMR is not accompanied by an uncoupling of oxidative phosphorylation so that the P:O ratio remains around 2·5. If high doses of thyroid hormones are given (1700 μg T_3/100 g), the BMR increases after a lag period of only 2 hours compared with 30 hours under physiological conditions and the P:O ratio is reduced to 0·7. This agrees with the uncoupling of oxidative phosphorylation that is observed when mitochondria are incubated *in vitro* with relatively high concentrations of thyroxine. Following physiological doses of T_3, the mitochondria of skeletal muscle are found to be enlarged, more numerous and to contain a more dense array of cristae than normal. This is probably the morphological expression of the increased incorporation of amino acids into mitochondria and of the increased BMR. Evidence has been obtained that thyroxine and some other slow-acting hormones stimulate the formation in the nucleus of both messenger RNA and ribosomal RNA (DNA acts as a template in the synthesis of both these species of RNA) and this leads to an increase in the number of polysomes present in the cytoplasm. The new polysomes are tightly bound to the microsomal membranes and are active in protein synthesis. The increase in membrane-bound polysomes is accompanied by an increase in phospholipid incorporation into the microsomal fraction of the cell and to a proliferation of the endoplasmic reticulum.

Inactivation

The thyroid hormones are removed rapidly from the circulation by the liver which converts them to their less active pyruvic acid derivatives by transamination. The hormones and their pyruvic acid analogues can also be deiodinated by a dehalogenase present in liver and skeletal muscle.

CHAPTER 47

Pancreatic Hormones

Insulin

Insulin consists of two polypeptide chains (A and B) linked together by two disulphide bridges (see Figure 1.26); it is formed in the β-cells of the pancreatic islets from a single polypeptide precursor, proinsulin. The A-chain of insulin is derived from the last 21 amino acids at the carboxyl terminus of proinsulin whilst the B-chain consists of the first 30 amino acids from the amino terminus (Figure 13.13). Thus the formation of insulin involves the removal of about 30 amino acids from the central region of the proinsulin chain. At each end of this connecting segment are a pair of basic amino acids (Lys-Arg and Arg-Arg) which are the sites of attack of the proteolytic enzyme(s) responsible for forming insulin. The formation of polypeptide hormones from larger precursors is not uncommon; glucagon appears to be formed in this way and possibly also oxytocin and vasopressin. Proinsulin unfolds when reduced in 8 M urea but returns to its native configuration in high yield when allowed to reoxidize. Reoxidation of reduced insulin, on the other hand, gives low recoveries of active hormone. It would appear, therefore, that the formation of proinsulin followed by peptide ex-

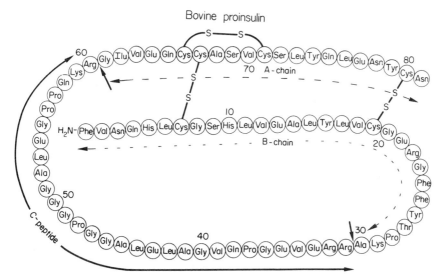

Figure 13.13 Primary structure of bovine proinsulin

cision allows correct S—S bond formation to occur more readily than if insulin were to be formed by the interaction of two polypeptide precursors.

Proinsulin, formed by the ribosomes of the rough endoplasmic reticulum, passes into the lumen of the reticulum and thence to the Golgi region for packaging into secretory granules. Cleavage of proinsulin to insulin seems to occur at about the time of granule formation. Secretion of insulin and the connecting peptide occurs by Ca^{2+}-dependent exocytosis similar to that described for the adrenal medulla (see p. 410). There is evidence that the induction of insulin secretion by some agents may involve the activation of adenyl cyclase whilst the inhibitory effect of

Table 13.2 Substances affecting insulin secretion

Substances inducing insulin secretion	Substances inhibiting insulin secretion
Glucose	Adrenaline
Fructose	Noradrenaline
Amino acids (Arg; Lys; Leu; Phe; Ala; Val)	
Butyrate	
Octanoate	
Ketone Bodies	
Glucagon	
Secretin	
Pancreozymin	
Tolbutamide[a]	

[a] Tolbutamide used clinically to treat diabetics with some residual β-cell function

$$H_3C-\bigcirc-\underset{\underset{O}{\|}}{\overset{\overset{O}{\|}}{S}}-N-\overset{\overset{O}{\|}}{\underset{H}{C}}-\overset{}{\underset{H}{N}}-(CH_2)_3-CH_3$$

tolbutamide

adrenaline seems to be mediated through an inhibition of that enzyme. Table 13.2 lists agents that induce and inhibit insulin secretion.

Metabolic Effects of Insulin

Insulin is released when there is an adequate supply of fuels in the circulation for the provision of energy. These fuels may be newly arrived from the diet or, like ketone bodies, be derived from endogenous sources. The major actions of insulin are to encourage the laying down of triglyceride and glycogen as energy stores and to maintain protein synthesis at a level that at least balances protein catabolism. However, on occasions it may be necessary to prevent increases in circulating fuelstuffs inducing insulin release. Such an instance would be when the levels of blood glucose and free fatty acids have been increased by catecholamines released in the 'flight or fight' reaction. It would clearly be pointless to mobilize these reserves only to lay them down again. Inhibition of insulin release by catecholamines prevents such a futile manoeuvre. Sometimes the sphere of insulin action has to be limited in order to conserve the resources. Thus during prolonged starvation insulin secretion continues and is responsible for preventing excessive protein catabolism but at the same time it has to be prevented, by glucocorticoids, from inhibiting lipolysis and favouring lipogenesis in the adipose tissue.

Effect of Insulin on Glucose Uptake

The entry of glucose into muscles, adipose tissue and lactating mammary gland is stimulated by insulin. In contrast, the uptake of glucose by liver, brain and non-lactating mammary gland is not affected by insulin.

The glucose concentration within cells other than those of the liver, skin and possibly kidney is low. Thus the movement of glucose into muscle, adipose tissue and brain cells is a spontaneous, downhill process which is restricted only by the poor permeability of the cell membrane; it does not take place against a concentration gradient, hence it does not require an input of energy. It is now generally believed that one of the primary actions of insulin is to increase the permeability of cell membranes so that glucose can move down its concentration gradient more rapidly. Liver cell membranes are freely permeable to glucose with the result that the concentration of glucose inside the parenchyma cells approximates to that in the blood; thus in this instance there is no force causing a net movement of glucose across the membrane in either direction.

It has been suggested that glucose enters cells by combining with some component in the cell membrane which then diffuses across the membrane and liberates the glucose at the internal surface, i.e. glucose enters the cell by carrier-mediated diffusion. Evidence supporting this view includes the demonstration that the rate of uptake of glucose is independent of the external glucose concentration provided that the latter exceeds a certain value; i.e. the uptake mechanism can be saturated in the same way that an enzyme can be saturated by its substrate. Furthermore, glucose uptake can be inhibited competitively by some other hexoses and pentoses and finally the mechanism for sugar transport exhibits some degree of specificity; for example, glucose but not sorbitol can enter muscle cells. As all these properties are similar to those possessed by enzymes, it is argued that glucose entry must be mediated by an enzyme-like carrier.

Insulin can be shown to alter membrane permeability by using sugars such as 3-methyl glucose and arabinose which are not metabolized and which therefore accumulate within the cells when their entry is facilitated by insulin.

Insulin Coma

The amount of glucose used by tissues is roughly proportional to their weight. Thus, since the total weight of muscle in a 70 kg man is about 25 kg, whilst the brain and the liver weigh 1·5 kg and 1·7 kg respectively, it follows that muscle will use about 90 per cent of the glucose available to those three tissues. The total quantity of glucose in the blood is 30 mmoles and the consumption of glucose by resting muscle in the absence of insulin is about 15 mmole/kg/h. Therefore the 25 kg of muscle will consume the blood's supply of glucose in 5 minutes. Since a steady state exists, the rate of glucose utilization by the muscle must be equal to the rate at which glucose enters the blood from other sources, notably the liver. The three- to four-fold increase in the permeability of muscle cell membranes towards glucose that is brought about by insulin means that the muscles can use the blood's supply of glucose in about 1 minute. If the production of glucose is limited, then the blood glucose level must fall until the concentrations of glucose inside and outside the muscle reach values that limit glucose transport into the muscle to the amount that can be produced and passed into the blood. That is, the blood glucose level must fall as the permeability increases. This lowering of the level of blood glucose by the muscle will bring about a fall in the rate of brain metabolism that could culminate in insulin coma and death.

Effect of Insulin on Enzymes and Metabolic Pathways

In addition to its effects on glucose uptake, insulin stimulates amino acid uptake and protein synthesis in many tissues, including liver, adipose tissue, skeletal and cardiac muscle and mammary gland.

Administration of insulin to rats made diabetic with alloxan leads to an increased synthesis in the liver of glucokinase, glycogen synthetase, phosphofructokinase, pyruvic kinase, glucose-6-phosphate dehydrogenase and citrate cleavage enzyme. Glucokinase catalyses the same reaction as hexokinase but has a much higher K_m for glucose (about 15 mM) and is not inhibited by glucose-6-phosphate. Consequently it is of importance in removing glucose from the circulation when concentration of the latter rises above resting levels, for example, after meals rich in carbohydrate. Glucose-6-phosphate formed by glucokinase can be both laid down as glycogen and converted to fat. Fat formed under the influence of insulin is exported from the liver as very-low-density lipoproteins (VLDL).

In addition to increasing the rate at which glycogen synthetase is formed *de novo*, insulin also stimulates the conversion of the relatively inactive, phosphorylated, form of the enzyme into the more active, non-phosphorylated, form. Besides inducing key enzymes involved in the utilization of glucose, insulin reduces the activity of glucose-6-phosphatase, fructose-1,6-diphosphatase, pyruvic carboxylase and phosphoenol pyruvate carboxykinase. These are the four rate-limiting enzymes involved in gluconeogenesis.

In adipose tissue, insulin increases glucose uptake and lipogenesis and decreases the rate of lipolysis. The overall effect is to encourage fat deposition. Lipogenesis is clearly favoured by the increased availability of glucose for fatty acid and L-α-glycerophosphate formation. Insulin also indirectly causes pyruvic dehydrogenase to change from an inactive, phosphorylated form into an active dephosphorylated form. A consequence of this activation is an increased rate of supply of acetyl-CoA for fatty acid synthesis. Insulin also lowers the rate of lipolysis by reducing the concentration or effectiveness of c-AMP formed in response to lipolytic hormones such as the catecholamines and glucagon. Since c-

AMP activates a protein kinase responsible for activating the 'hormone' sensitive lipase, it is clear that any reduction in its concentration or effectiveness will decrease the rate of lipolysis. The activity of clearing factor lipase (lipoprotein lipase) is increased by insulin and this facilitates triglyceride storage by enhancing fatty acid uptake from circulating triglycerides.

Consequences of Insulin Deficiency

The importance of insulin to the overall well-being of the body is illustrated dramatically by disturbances seen in patients suffering from uncontrolled diabetes mellitus. Insulin lack induces hyperglycaemia on account of decreased glucose use and increased gluconeogenesis. Hyperglycaemia leads to the characteristic diuresis, electrolyte loss and dehydration associated with the disease as there is a limit to the ability of the kidneys to reabsorb glucose and concentrate urine. Dehydration leads to haemoconcentration, reduced blood volume, hypotension, diminished renal blood flow and anuria. Adipose tissue, lacking insulin control, releases fatty acids in excessive amounts with the result that ketone body production rises to unphysiological levels and causes ketoacidaemia, ketonuria and hence sodium loss. Protein catabolism exceeds synthesis in the absence of insulin with the result that glucogenic amino acids become available for gluconeogenesis and hence exacerbate the hyperglycaemia. Furthermore, the loss of cellular proteins with their fixed anionic charges leads to a loss of cellular, and hence body, potassium.

It is hardly surprising that uncontrolled diabetes is invariably fatal, leading as it does to metabolic acidosis, loss of electrolytes, dehydration, circulatory failure and anuria.

Inactivation

Insulin is inactivated by the enzyme glutathione insulin transhydrogenase which uses reduced glutathione to reduce the disulphide bonds in insulin and so splits the protein into its component α- and β-polypeptide chains. The site of destruction is primarily in the liver and the half-life of insulin is about 30 minutes under normal circumstances. There is also some evidence that adipose tissue has enzyme systems able to degrade insulin.

Glucagon

Glucagon is a polypeptide hormone released from the α-cells of the pancreatic islets when the fuel status of the circulation is poor. In many ways its actions may be regarded as the opposite of those of insulin. It mobilizes energy stores by accelerating lipolysis in adipose tissue and increasing glycogenolysis in liver. Its ability to mobilize liver glycogen at physiological concentrations is greater than that of the catecholamines and it does not have their effects on the circulatory system. Glucagon is also a powerful activator of gluconeogenesis and may play an important role in maintaining blood glucose during periods of starvation. Glucagon also induces insulin release: although at first sight this seems to be counterproductive, it may be that insulin and glucagon always work against each other to establish the level of the various fuels in the circulation and to prevent violent departures from normal values. This view is encouraged by the recent observation that insulin stimulates glucagon release.

References

Chemical Mediation of Hormone Action. Bitensky, M. W. and Gorman, R. E. *Annual Review of Medicine*, **23**, 263–284, 1972.

Membrane Receptors. Cuatrecosas, P. *Annual Review of Biochemistry*, **43**, 169–214, 1973.

Metabolic and Endocrine Physiology—An Introductory Text. J. Tepperman. Year Book Medical Publishers, Chicago, 3rd edn., 1973.

SECTION 14

The Control of Metabolic Processes in Cells

CHAPTER 48

The Control of Metabolic Processes in Cells

The control of the processes whereby cells take up material from their immediate environment, metabolize it, and then dispose of the unwanted material obviously comprises a large field and many of the topics will only be mentioned in passing. In the higher organism comparatively few of the cells of the body make any direct contact with the outside world. Except for the cells of the intestinal mucosa, the individual cells of the multicellular organism have a more or less stabilized environment which is the cooperative product of the rest of the cells of the body. Even in this comparatively stable internal environment, however, there may be changes of sufficient magnitude to cause a large response by the cells. An example of the sort of changed environment that cells may experience is in the increase from about 1 mM in the amount of lactate in the plasma during exercise, which rises to about four times the value in the resting condition. Another example is the doubling of the glucose content of the plasma (5 mM) after a carbohydrate meal, whilst during starvation, or in response to adrenaline, the free fatty acid content (0·5 mM) of the plasma can increase about three-fold. Also the saline constituents of the plasma can change, reflecting the amount of water loss or of salt ingestion. These variations in internal environment usually last only a matter of hours but the point to be made is that, chemically, the internal environment is variable. In contrast, the chemical composition of the external environment is relatively constant, as illustrated by the constancy of atmospheric carbon dioxide and oxygen, the gaseous substrates consumed by most organisms. Similarly, the salt concentration in the environment of the fish, or of other marine organisms, is relatively stable compared with the changes in salt concentration that can occur in the body fluids of higher organisms.

In the intact animal, therefore, there must be constant adaptation to the changing internal environment which itself is a result of the complex interactions of hormonal and nervous stimuli as well as the gross intake of material from outside the body.

The Passage of Materials into Cells

Irrespective of the nature of the organism the first thing that must happen when a cell reacts to its environment is for a piece of that environment to be taken up into the cell and it is at the cell membrane that the first opportunity exists for the cell to exercise choice and control. We must, therefore, consider the passage of materials through cell membranes.

Water, which is, by far, at the highest concentration of any bodily constituent (about 55 M), interacts by hydrogen or hydrophobic bonding with other molecules in solution and with the 'soluble' groups of 'insoluble' molecules. Taking the solute osmolarity of the body as about 300 m osmolar, it follows that there are more than 100 molecules of water for every other molecule of solute and therefore small changes in the behaviour of the water molecules can have potentially large effects on the other molecules (see p. 52). Water itself moves passively in and out of cells in response to the changes in the osmolarity of the intra- and extra-cellular fluid but the rate at which it does so will be influenced by the interactions with the groups bordering the channels through which the water penetrates, the size of the channels and the interaction of the water molecules with each other. The greater the total interaction the slower the rate of movement. Similar considerations apply to the movement of water within the cell, which will be limited by the permeability of the membranes which surround organelles, or which, like the endoplasmic reticulum, partition off parts of the cell. Moreover, water-soluble materials (such as ATP and ADP) cannot pass freely around cells. Many experiments have shown that the intracellular membranes are effective barriers to free movement. Matters are complicated by the fact that the intracellular membranes are composed not of static molecules but of enzymically active proteins intermingled with phospholipids whose conformation and charge are dependent on the metabolic state of the cell. Also, organelles and membranes within a cell do not have fixed positions but are constantly moving. Consequently the preferred path of diffusion of a molecule is constantly subjected to spatial and temporal modification. Any detailed analysis of this complex situation has so far proved impossible.

The Regulation of Energy Production

To maintain its life the cell must synthesize its energy currency (ATP) at a rate sufficient for (a) the synthesis of cellular components, (b) the maintenance of cellular organization and (c) the maintenance of the intra- and extracellular solute differentials. For efficiency, the synthesis and demand for ATP must be balanced; only a small reserve is maintained to satisfy urgent requirements during the short period before the pathways of ATP synthesis (glycolysis and oxidative phosphorylation) adjust themselves to the altered situation. The requirement for ATP arising under (a) is roughly equivalent to the rate of synthesis of new macromolecules and this probably makes the most variable demands on the energy supply. The requirement under (b) is not at all clear but the maintenance of cell structure against cellular disorganization will require an input of energy to prevent the natural increase in entropy. Under (c) the requirement for ATP to maintain ionic gradients is probably constant for the body as a whole, although the ionic shifts accompanying the passage of an impulse in excitable tissues will result in local high demands. In isolated tissues it can be shown that the maintenance of ionic concentrations uses ATP and results in the subsequent stimulation of glycolysis or oxidative phosphorylation to produce more ATP.

Control by Carrier Molecules

All metabolic flow is restricted by the availability of carrier molecules such as ADP, NAD^+, CoA, etc. These carrier molecules are present at comparatively low concentrations and ultimately set limits on the extent of metabolic flow.

The average ATP concentration is only about 5 mM and this determines the rate at which energy can be used. The consumption and resynthesis of ATP is continuous even when resting. Under these conditions ADP is maintained at a steady-state concentration of about 1 mM. The rates of substrate level or oxidative phosphorylation are determined by the concentration of ADP available to receive the high-energy phosphate. Because of the reciprocity of the concentrations of ATP and ADP, the rate of provision of chemical energy in the form of ATP will always tend to increase to meet its rate of utilization. Thus in all cases the rate of provision of ATP by glycolysis is controlled by the concentration of ADP as is the provision of respiratory ATP (except in the uncoupled state—see p. 141). This principle of controlling the metabolic flow by the availability of a limited amount of a carrier molecule is widespread in living systems. For example, the amount of NAD^+ available also controls the rate of glycolysis since glyceraldehyde dehydrogenase cannot proceed without it. $NADH + H^+$ formed by the dehydrogenation step reduces the concentration of NAD^+ and automatically slows the reaction. Thus we have two successive steps in glycolysis tightly controlled by the limitation of a carrier molecule, NAD^+ at the dehydrogenation step and ADP at the phosphorylation step. Similar controls are found with all hydrogen carriers $NADP^+$, FAD, etc. and with radical carriers CoA, pyridoxal phosphate, etc. In all cases the concentration of the unladen carrier controls the rate of the donating pathway and in all cases the concentration of the laden carrier determines the rate of the utilization pathway and, of course, the concentrations of the laden and unladen carriers are reciprocally related. In general, all the carrier molecules are at concentrations of less than 5 mM in the cell. Often several pathways are donor pathways and compete for the available empty carrier and also several pathways are users competing for the loaded carrier. However much the donor and user pathways change in activity they are ultimately restricted to rates determined by the carrier concentrations.

The Control of Substrate Flow and Enzyme Activity by Small Molecules

In any metabolic pathway, for example glycolysis, there are enzymes of different activities and it follows that without any other constraints the flux of substrates through the pathway will be controlled by the least-active enzyme. In living systems the flux through metabolic pathways is variable for two reasons. Firstly, the concentration of substrate is usually insufficient to saturate the enzyme, therefore the activity will vary with substrate concentration. Secondly, many enzymes alter their activity in response to changes in the concentration of molecules other than their substrate. In some cases, such a molecule attaches itself to a specific allosteric binding site (see p. 87) and causes a complete change in the shape of the curve relating substrate concentration to enzyme activity. It frequently happens that the effector molecules are also carrier molecules which may also be substrates of the particular enzyme. Enzymes subject to this intense modulation have been called regulator enzymes and frequently they are also enzymes of inherently low activity, even when they are fully activated. The flux through such an enzyme will depend on the concentration of the substrate, the modulated activity of the enzyme and the availability of any required carrier molecule. Starting from glucose-6-phosphate phosphofructokinase is the enzyme of lowest activity in the glycolytic sequence, even when it is fully activated.

$$\text{fructose-6-phosphate} + \text{ATP} \longrightarrow \text{fructose-1,6-diphosphate} + \text{ADP}$$

The Control of Glycolysis in Muscle

Above a very low concentration of ATP, phosphofructokinase is progressively inhibited by the nucleotide and at the concentration of ATP found in the normal resting cell (about 5 mM) the enzyme is about 80 per cent inhibited. The inhibition is relieved (or the enzyme is activated) by ADP, fructose-6-phosphate, orthophosphate and in particular AMP. On the other hand, citrate strengthens the inhibitory effect of ATP and it would appear that, in real life, it is the balance of ATP and citrate on the one hand and AMP on the other that determines the activity of phosphofructokinase.

As a controlling substance ATP is inherently unsound since its concentration cannot alter very much without altering the yield of energy that can be obtained when it is transformed into ADP. Thus a change in the ratio of ATP/ADP from roughly 100/1 (4·95/0·05 mM) to roughly 4·0/1 (transforming 1 mM ATP to an equivalent amount of ADP) will result in 13 per cent less energy becoming available from the hydrolysis of the ATP. Changes in ATP are therefore kept to a minimum and the minor changes are signalled by a chemical amplification system resulting in changes in AMP, the activator of phosphofructokinase. The amplification works through the extremely active enzyme myokinase (adenylate kinase) which carries out the reaction

$$2\,\text{ADP} \rightleftharpoons \text{AMP} + \text{ATP}$$

Since the total amount of nucleotides is fixed, the change in any one of the three must be compensated by appropriate changes in the concentrations of the other two. The equilibrium constant of the reaction is such that, with the physiological concentration of the ATP at about 5 mM, ADP will be maintained round 0·1 mM and AMP around 0·001 mM. A doubling of the ADP concentration by the breakdown of a very small amount of ATP (about 2 per cent of total) results in a four-fold increase in the concentration of AMP. Thus the very small change in the concentration of the phosphofructokinase inhibitor (ATP) results in a very large increase in the phosphofructokinase activator (AMP).

Under conditions where fat is being oxidized, there is less need to use the glycolytic mechanism to form ATP and therefore the inhibitory role of the ATP is reinforced by another chemical signal mechanism. This signal is the change in the concentration of citrate, a three-fold increase in which halves the already inhibited glycolytic rate. The increase in citrate occurs because of the extra load thrown on citrate synthetase as a result of the increase in the concentration of acetyl-CoA formed by fat oxidation. The citrate formed within the mitochondria diffuses out to interact with the cytosolic phosphofructokinase. Where measurements have been made of the concentrations of citrate *in vivo* in relation to the rates of glycolysis, it has been found that the two parameters move inversely.

In muscle the next stage of carbohydrate metabolism where small molecules have a key effect on enzyme activity is in disposal of pyruvate. Pyruvate produced by the pyruvate kinase reaction has two routes open to it. It may stay in the cytoplasm where it may be reduced to lactate at rates dependent on the availability of the loaded carrier molecule NADH + H^+. Alternatively the pyruvate may pass into the mitochondria where once again the availability of carrier molecules will determine its fate. Pyruvate may be converted to acetyl-CoA by the enzyme pyruvate dehydrogenase (see p. 123). Two carrier molecules, NAD^+ and CoA, are required for this complex reaction and since the concentration of the NAD^+ + NADH + H^+ is at least ten times higher than that of Acyl-CoA + CoA, the rate of utilization of the acetyl-CoA will determine the rate at which the pyruvate can be used. Both acetyl-CoA and CoA directly affect the activity of the pyruvate dehydrogenase. Acetyl-CoA inhibits and CoA stimulates. The oxidation of fatty acids will increase the amount of acetyl-CoA and thus reduce the amount of the carrier (CoA) available to the pyruvate dehydrogenase. Acetyl-CoA will also inhibit the activity of the enzyme. This reinforces the inhibition of glycolysis already brought about by the inhibiting effect of citrate on phosphofructokinase. Measurements of acetyl-CoA and CoA *in vivo* support the concept that the concentration of acetyl-CoA is itself controlling the rate at which it is being formed from pyruvate.

The Control of Glycolysis in Liver

In liver the control of glycolysis is similar to that in muscle as far as phosphofructokinase is concerned but pyruvic kinase is also subject to control by small molecules. The liver has most of its pyruvate kinase in a form that is relatively inactive except in the presence of small amounts of fructose-1,6-diphosphate. The enzyme is inhibited by physiological concentrations of ATP and some amino acids such as alanine. *In vivo* it would appear that the activator (fructose-1,6-diphosphate) is always at a concentration high enough to ensure full activation but the concentration of ATP is always at a level to ensure some inhibition. Despite this, it appears unlikely that the rate of glycolysis *in vivo* will be limited by the activity of pyruvate kinase since its intrinsic activity is very much greater than that of phosphofructokinase.

The pyruvate formed by the pyruvate kinase reaction has three routes open to it instead of the two in muscle. These are (i) the formation of lactate, (ii) conversion to acetyl-CoA and (iii) carboxylation to oxaloacetate. Since lactate is constantly being formed by muscle at rates proportional to the body's overall muscular activity, there will be a tendencey for lactate entering the liver from the muscles to drive the reaction

$$\text{lactate} + \text{NAD}^+ \longrightarrow \text{pyruvate} + \text{NADH} + \text{H}^+$$

to the right and prevent the transformation of liver pyruvate to lactate.

The route of pyruvate through pyruvate dehydrogenase is controlled by carrier molecule concentrations in the same way as in muscle.

The third route of metabolism open to liver pyruvate is carboxylation to oxaloacetate catalysed by the mitochondrial enzyme pyruvate carboxylase.

$$\text{pyruvate} + \text{ATP} + \text{CO}_2 \xrightarrow{\text{acetyl-CoA}} \text{oxaloacetate} + \text{ADP} + \text{Pi}$$

ADP is an inhibitor (competitive with ATP) and acetyl-CoA is absolutely required for the reaction. This reaction, then, depends on the concentrations of ATP and acetyl-CoA, both of which must be present together. The ATP varies because of its usage in the reaction but the acetyl-CoA varies largely because of the changes in the main source of carbon being oxidized. If fatty acids are the main source of oxidizable material then the concentration of acetyl-CoA will increase. Under these circumstances there is a relative lack of carbohydrate and therefore the pathway of gluconeogenesis must be encouraged and the formation of oxaloacetate is the first step in this direction. The decision about the further flow of oxaloacetate towards gluconeogenesis or towards the citric acid cycle is again made by the availability of carrier molecules. ATP is an inhibitor of citrate synthase and GTP is the phosphate carrier molecule required for the formation of phosphopyruvate by the enzyme phosphopyruvatecarboxykinase.

$$\text{oxaloacetate} + \text{GTP} \longrightarrow \text{phosphopyruvate} + \text{GDP} + \text{CO}_2$$

The concentration of GTP is defined by the concentration of ATP through the reaction

$$\text{ATP} + \text{GDP} \rightleftharpoons \text{ADP} + \text{GTP}$$

for which the equilibrium constant is approximately unity. ATP is once again a participant in a reaction sequence and a switching device directing the carbon flow in the correct pathway. The phosphopyruvate, in some way not discovered, is protected against the activity of pyruvic kinase and then flows back up the glycolytic pathway to the level of fructose-1,6-diphosphate. At this stage the phosphate is removed by fructose-1,6-diphosphatase to give fructose-6-phosphate. Fructose-1,6-diphosphatase is an enzyme inhibited by AMP and therefore controlled in a reciprocal way to phosphofructokinase. The reciprocal control of these two enzymes is necessary to prevent the continuous synthesis and breakdown of fructose-1,6-diphosphate and consequent breakdown of ATP that would ensue if both enzymes had unrestricted access to the substrate. The fructose-6-phosphate that is made is rapidly isomerized to glucose-6-phosphate apparently in an uncontrolled fashion. Glucose-6-phosphate is formed in the liver by three routes, from blood glucose by hexokinase or glucokinase (see p. 107), from glucose-1-phosphate (formed by the action of phosphorylase on glycogen) and from fructose-6-phosphate. All other tissues have these routes open to them except that the special phosphorylating enzyme, glucokinase, is usually not available. Also, in most tissues the net production of fructose-6-phosphate does not occur because it cannot be formed from pyruvate as the pathway of gluconeogenesis is incomplete. In the liver, glucose-6-phosphate has four competing pathways open to it (i) dephosphorylation to form glucose catalysed by glucose-6-phosphatase (present virtually only in the liver and kidney), (ii) dehydrogenation to give 6-phosphogluconic acid (absolutely controlled by the availability of the carrier molecule $NADP^+$) catalysed by glucose-6-phosphate dehydrogenase, (iii) isomerization to glucose-1-phosphate catalysed by phosphoglucomutase and finally (iv) isomerization back to fructose-6-phosphate catalysed by phosphohexoseisomerase.

Control by Phosphorylation and Dephosphorylation of Enzymes

Apart from inhibiting or accelerating enzyme activity, small molecules (in particular the adenosine nucleotides) can react with enzymes to form stable molecules which may have very different enzyme activities (and different responses to metabolite concentrations) from the original enzyme. Phosphorylation or dephosphorylation of the enzyme is brought about by a set of protein kinases on the one hand and a set of phosphatases on the other. In many cases the protein kinase is dependent on the presence of cyclic AMP for its activity. The protein kinases and the phosphatases are enzymes whose specific substrates are also enzymes. This is not unusual; for example, the proteinases may have enzymes as substrates which they may degrade and inactivate irreversibly or from

Table 14.1 Interconvertible enzymes

Muscle phosphorylase	*b* activity repressed by endogenous modulators
	b form assayable in presence of AMP
Liver phosphorylase	*b* form inactive, activity not elicited by AMP
Phosphorylase *b* kinase	*b* form active *in vivo* at high Ca^{2+}, low ATP
	b form assayable at high pH
Phosphorylase *a* phosphatase	*a* and *b* forms appear to exist
	Activity stimulated by glucose
Muscle glycogen synthetase	*b* form activated *in vivo* by elevated G-6-P
Liver glycogen synthetase	*b* form inactive at tissue ligand concentrations, assayable in presence of G-6-P
Hormone-sensitive lipase	*b* to *a* reaction catalysed *in vitro* by protein kinase
	Activated *in vivo* by lipolytic hormones
Pyruvate dehydrogenase	Phosphatase reaction (*b* to *a*) promoted by cyclic AMP
	Fatty acids promote *a* to *b* reaction *in vivo*

a is the more active form of the pair of enzymes.

which they may remove an amino acid sequence (pepsinogen, trypsinogen) to bring about irreversible activation. In the cases we shall consider, the process is reversible and responsive to the needs of the organism as indicated by hormonal signals or signals from changing substrate levels. Table 14.1 summarizes some of the enzymes that exist as interconvertible pairs.

Glycogen Synthetase

The flow of the carbon from glucose-1-phosphate towards glycogen synthesis is controlled by the availability of UDP-glucose and the activities of the two forms of glycogen synthetase (see Figure 14.1). The more active form of the enzyme pair (*a* form) is non-phosphorylated and its activity is little affected by glucose-6-phosphate. The less active form (*b*) has an activity in the absence of glucose-6-phosphate about one-tenth that of the *a* form but with saturating concentrations of glucose-6-phosphate its activity can become some 80 per cent of

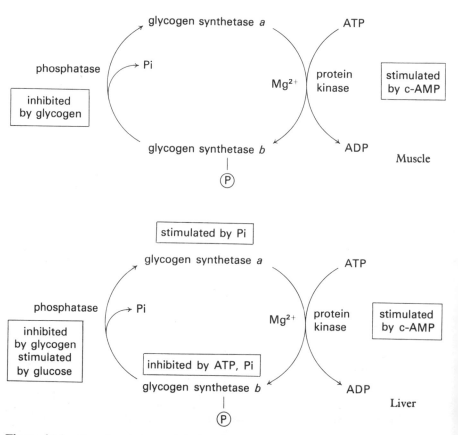

Figure 14.1 Phosphorylation and dephosphorylation of muscle and liver glycogen synthetases. *a* is the more active form of the pair of enzymes

the *a* form. The situation in real life, however, is somewhat different. In both muscle and liver the *b* form of the enzyme is virtually inactive since it is strongly inhibited by the *in vivo* concentration of adenosine phosphates and orthophosphate. There appears to be some difference in the *a* forms of the synthetase in muscle and liver in that some inhibitory effect is still exercised by phosphates on the muscle enzyme but in the liver the *a* form is activated by orthophosphate. In both tissues it is the amount of the *a* enzyme that determines the rate of glycogen synthesis. The actual activity of the enzyme will depend on the balance of the metabolites within the tissue. For example, the concentration of UDPG in the muscle (0·03 mM) is much lower than that in the liver (0·5 mM). In muscle the K_m for UDPG is about 0·3 (*a* enzyme) and in liver the K_m may be as low as 0·07 mM (*a* enzyme). The standing concentrations of the *a* form of the enzyme (dephosphorylated form) and the *b* form (phosphorylated form) of the enzyme depend on the relative activities of the glycogen synthetase phosphatase (which forms the *a* form by removing a phosphate from the *b* form) and the protein kinase which uses ATP to phosphorylate the *a* enzyme back to the *b* enzyme. It is thus important to know how the activities of the phosphatase and the kinase are controlled. The phosphatase in both liver and muscle is inhibited by glycogen in a concentration-dependent manner and thus as the glycogen reserves are replenished the amount of *a* enzyme decreases. In the liver (see Figure 14.1) the amount of glycogen required to inhibit the phosphatase is much larger than in muscle and moreover the presence of glucose apparently activates the phosphatase. These observations probably account for the ability of the liver to accumulate a much higher concentration of glycogen than the muscle. The protein kinase (that inactivates glycogen synthetase by phosphorylating it in both muscle and liver) is activated by cyclic AMP whose concentration can increase under the influence of the two hormones glucagon and adrenaline, which are the hormones that start the process of stimulating phosphorylase activity. The role of the antagonistic hormone insulin that promotes glycogen deposition in both liver and muscle is not clear although in muscle there are suggestions that its effects involve the protein kinase. In the physiological situation it would appear that the quantity of glycogen present in the muscle controls the rate of glycogen formation whereas in the liver the glucose availability is probably more important. The main hormonal control via adrenaline and glucagon operates to switch off glycogen synthesis by increasing the level of cyclic AMP and consequently activating the protein kinase.

Phosphorylase

Phosphorylase, the enzyme responsible for the phosphorolytic breakdown of glycogen in muscle (Figure 14.2) is controlled in a similar fashion to glycogen synthetase. There are two forms of the enzyme, a phosphorylated one (*a* form) and a non-phosphorylated form (*b* form). In contrast with glycogen synthetase, it is the *a* form (active form) that is phosphorylated. Thus there is a reciprocal relationship between the activity of glycogen synthetase and phosphorylase

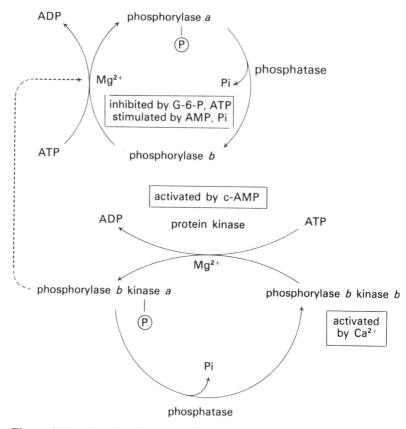

Figure 14.2 Phosphorylation and dephosphorylation of muscle phosphorylase. *a* is the more active form of the enzyme pairs

which is related to the degree of phosphorylation of the enzyme pairs. As with glycogen synthetase the inactive form is susceptible to changes in metabolite concentrations. In this case the *b* form is inhibited by glucose-6-phosphate and ATP and stimulated by AMP and orthophosphate. In resting muscle the metabolite concentrations are such that phosphorylase *b* is virtually inactive. The phosphorylation of the *b* to the *a* form results in an enzyme uncontrolled by metabolites and therefore an increase in phosphorylase activity occurs. The phosphorylation of the *b* enzyme is brought about by an enzyme phosphorylase *b* kinase which itself exists in two forms, an active phosphorylated form (*a* form) and an inactive non-phosphorylated form (*b* form). The phosphorylation of the phosphorylase *b* kinase (to make it active) is brought about by the same protein kinase that phosphorylates glycogen synthetase to inactivate it. The inactive phosphorylase *b* kinase may also be activated by an increase in the concentration of Ca^{2+} without any phosphorylation of the enzyme occurring. Once again then the *b* form of the enzyme is susceptible to activation by a metabolite whereas the active *a* form is not. The control of the protein kinase is precisely the same as it was when it was concerned with the inactivation of glycogen synthetase. A hormonal stimulation of adenyl cyclase (by adrenaline) increases the formation of cyclic AMP from ATP and the increased concentration of the nucleotide activates the protein kinase. The active protein kinase transfers the high-energy phosphate of ATP to (presumably) a serine residue on the phosphorylase *b* kinase. This active phosphorylase kinase in turn uses another molecule of ATP to phosphorylate a serine on phosphorylase *b* to give the fully active enzyme.

The dephosphorylation cycle starts with an undefined muscle phosphatase which removes the phosphate group from phosphorylase *b* kinase and thus inactivates it. The removal of the phosphate group from phosphorylase *a* is brought

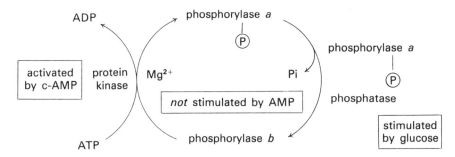

Figure 14.3 Phosphorylation and dephosphorylation of liver phosphorylase. *a* is the active form of the enzyme pairs

about by a specific phosphatase that has not yet been purified but which is thought to exist in active and inactive forms with activation brought about by the presence of ATP and magnesium. The control of these dephosphorylation processes are at present uncertain. The understanding of the control of phosphorylase in liver (Figure 14.3) is by no means as complete as that which has been described for muscle. The *a* and *b* forms of the enzyme, interconvertible by phosphorylation and dephosphorylation, occur but the *b* form remains virtually inactive even in the presence of AMP and it is probable that the only way of activating liver phosphorylase *in vivo* is via the hormonal mechanism. Since no clear evidence is available for the presence of a liver phosphorylase *b* kinase it may be that protein kinase phosphorylates phosphorylase *b* directly. There is evidence that glucose inhibits liver phosphorylase activity (compare the stimulation of glycogen synthetase activity) and it is believed that this effect is again by the stimulation of the appropriate phosphatase.

Phosphorylase *a* and phosphorylase *b* can exist as tetramers or dimers. In the phosphorylation/dephosphorylation sequence dimers only are concerned; an *ab* dimer is the intermediate form.

Pyruvate Dehydrogenase

The control of this enzyme by the concentrations of acetyl-CoA and CoA have been described (see p. 435) but superimposed on this is a phosphorylation/dephosphorylation cycle (Figure 14.4) in which the phosphorylated enzyme is the inactive *b* form as in glycogen synthetase. Of the total pyruvate dehydrogenase complex (comprising pyruvate dehydrogenase, transacetylase and lipoyl dehydrogenase) it is the pyruvate dehydrogenase that is phosphorylated and dephosphorylated. The total complex contains the protein kinase that does the phosphorylation as a firmly bound constituent but the phosphatase is much more readily dissociated during the isolation of the complex. Pyruvate and ADP inhibit the kinase-catalysed inactivation of pyruvate dehydrogenase and in starvation and diabetes the *a* form is greatly reduced. Insulin leads to the conversion of the *b* form to the *a* form in fatty tissue and in conditions of high blood fatty acids conversion of the *a* to the *b* form is favoured.

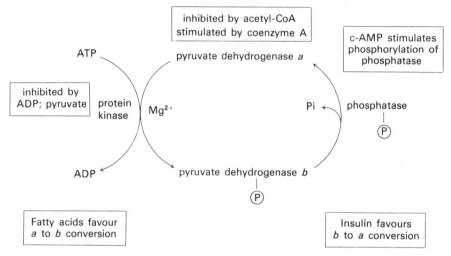

Figure 14.4 Phosphorylation and dephosphorylation of pyruvate dehydrogenase. *a* is the more active form of the enzyme pair

Hormone Sensitive Lipase

The activity of this enzyme in fat cells is increased by many hormones including adrenaline and noradrenaline and decreased by insulin and prostaglandin. The change in activity is roughly correlated with the concomitant changes in the levels of cyclic AMP in the tissue. In adipose tissue of low lipase activity there is an increase on incubation with cyclic AMP and ATP and if the endogenous protein kinase is inhibited then the activation is dependent on added protein kinase. The unphosphorylated *b* form of the lipase still has substantial activity; conversion to the *a* form causes a doubling of the activity only.

Control of the Amount of Enzyme in Tissues

It has been pointed out that the activity of an enzyme can vary according to the composition of the fluid in which it finds itself and also in response to minor chemical modifications such as phosphorylation. The enzymic activity may fall to very low levels in response to chemical modification and there is no reason why it should not fall to zero. Under such circumstances the enzyme has apparently vanished since enzymes can, by definition, be measured only by their activity. However, the protein molecule is still in existence and can be measured by immunological means. With some enzymes that have easily dissociated prosthetic groups, there may always be, in a tissue, a proportion of enzymically inactive (but potentially active) apoenzyme. Then the measurement of the enzyme activity *in vitro* may well obscure the situation *in vivo* since it is likely that excess prosthetic group will be present in the assay medium and all the previously inactive protein will become enzymically active. Usually there is a pretty strict correlation between the enzyme activity measured and the amount of the enzymic protein, so that an increase in the measured enzyme activity of the tissue may be taken as an increased level of the enzymic protein in the tissue. However, for certainty the increase in enzymic activity must be shown by independent tests to be accompanied by an increase in the amount of the relevant protein. In very few cases have these critical measurements been made and what follows must be read bearing this in mind.

Vitamins and Enzyme Activity

Normally the enzyme activities of the body reflect the need to deal with the current metabolic situation. Sometimes however, the body's response may be limited because of the shortage of an essential chemical that cannot be manufactured by the body and which must be eaten in the diet (essential amino acids, essential fatty acids and vitamins, Table 14.2). This is particularly well illustrated in lack of the vitamins of the B group which play specific roles as prosthetic groups in many enzymes.

In the further consideration of enzyme quantities we shall be considering the normal healthy mammal not limited in any way by chronic dietary insufficiencies.

Protein Turnover

Apart from the structural proteins such as collagen, proteins in the body are being constantly synthesized and broken down. Both the rate of synthesis and the rate of breakdown may vary independently. The amount of enzyme measured in a tissue at any time is the result of these two processes. By the use of isotopes that can be incorporated into proteins an estimate can be made of the average length of time that an individual molecule will remain as an active molecule. This is usually expressed as the half-life of the enzyme and is usually of the order of a few days. Certainly by the end of a week much of an individuals enzymic protein has been renewed. The means and the control of the process of protein degradation is still obscure. All tissues contain digestive enzymes; however, these enzymes, which have been equated with the lysosomal enzymes (see Figure

Table 14.2 Functions of non-synthesizable 'essential' nutrients

Essential amino acids	Biochemical role
Valine	Protein synthesis
Methionine	Protein synthesis, methyl donor S-adenosyl methionine (see p. 239)
Threonine	Protein synthesis
Leucine	Protein synthesis
Isoleucine	Protein synthesis
Phenylalanine	Protein synthesis, thyroxine synthesis (see Figure 13.8), adrenaline synthesis (see Figure 13.5)
Tryptophan	Protein synthesis, 5 H.T synthesis (see p. 347) nicotinic acid synthesis
Lysine	Protein synthesis
Arginine [a]	Protein synthesis, component of urea cycle (see p. 220)
Histidine [a]	Protein synthesis, histamine synthesis (see p. 269)
Essential fatty acids	
Linoleic acid	Membrane components, prostaglandin synthesis (see p. 341)
Vitamins	
A	Visual pigments (see p. 356)
B	
Thiamine	Synthesis of thiamine pyrophosphate, part of pyruvate dehydrogenase complex, etc. (see p. 124)
Lipoic acid	Part of pyruvate dehydrogenase complex (see p. 122)
Riboflavin	For flavin enzymes as H carrier (see p. 128)
Nicotinic acid	For NAD^+ and $NADP^+$ synthesis (see p. 109)
Pyridoxal	For transaminases, amino acid decarboxylases etc. (see p. 214)
Pantothenic acid	For CoA synthesis (Figure 2.19)
Biotin	For many CO_2 fixation enzymes (see p. 200)
Folic acid	In C_1 metabolism (see p. 189)
Cobalamine	In methylmalonyl-CoA metabolism (see p. 203)
C	
Ascorbic acid	Collagen synthesis (see p. 267)
D	
Calciferol	Calcium absorption (see p. 166)
K	Blood clotting (see p. 293)

[a] Essential in infants.

13.11), are unspecific and experiments have shown that enzymes are degraded specifically. Recently some proteinases have been discovered that selectively attack enzyme classes; for example, pyridoxal-requiring enzymes or flavine-requiring enzymes. These enzymes attack the apoenzyme more readily than the holoenzyme and their activity is conditioned by the composition of the diet. There is also evidence that an enzyme in combination with its substrate is less susceptible to proteolysis than in the absence of substrate. Such a finding, if general, would mean that heavily used enzymes are automatically preserved whilst those less heavily used would decrease.

Factors Causing Changes in Enzyme Activity

As might be expected, a change in the dietary intake is matched by a change in the enzyme make-up of the appropriate tissues to deal with the altered circumstances. Although all tissues are capable of altering their enzyme pattern there are three tissues that bear the brunt of coping with dietary vagaries. The first is the intestinal tract which alters the balance of its digestive enzymes to meet the peculiarities of the diet. For example, the change in lactase in response to milk feeding. The second tissue is the liver which via the portal system receives the mixed hydrolysate from the intestine (with the exception of the chylomicra). The major adjustments to the hydrolysate are made in the liver and the mixture passing into the blood is arranged to meet the needs of the rest of the body as nearly as possible. In times of starvation the liver still has to make adjustments to the metabolite mixture leaving by the hepatic vein but the supply route of materials is via the hepatic artery instead of the portal system and the composition is determined by the contributions made by the various organs in their venous outflow. To illustrate this point, the liver of a well-fed human will be receiving lactate from the muscles via the hepatic artery, after prior passage through the lungs and the heart. In starvation the liver will also receive this lactate but it will be accompanied by fatty acids derived from the adipose tissues. The oxidation of the fatty acids in the liver will increase the supply of acetyl-CoA which will switch off pyruvate dehydrogenase, stimulate pyruvate carboxylase and ensure an efficient utilization of the lactate for gluconeogenesis. The venous outflow of the liver then contains glucose, acetoacetate and triglyceride. The heart and lungs are unique in that they sample the outpourings of other tissues prior to adjustment by the liver, and are the first tissues to receive the products of liver metabolism. In spite of this there is very little change in the enzyme pattern of the heart and lungs in response to the changes in metabolite flow; all the major changes are made in the liver.

In many cases the changes in enzyme quantity are mediated through changes in the quantities of hormones secreted which in turn change because of the changed metabolite pattern of the blood. The functions of the hormones have been dealt with in detail in Section 13. Here we shall be concerned with the more general aspects of hormonal control and in particular with the antagonistic roles of glucocorticoids and insulin. In general glucocorticoids stimulate the synthesis of the key gluconeogenic enzymes whereas insulin reduces the amounts of those enzymes and reciprocally stimulates the synthesis of the glycolytic enzymes and those of the pentose phosphate pathway. The effects of insulin outweigh those of the glucocorticoids. In many cases, dietary changes produce enzyme changes which can be logically explained as a result of hormone secretion. For example, an increase in the carbohydrate intake (particularly coupled with a calorie excess) causes an increase in insulin secretion with an increase in the enzymes described above. This will result in a flow of carbon in the direction of fat deposition and replenishment of glycogen stores. Similarly, starvation results in a lowering of insulin, a secretion of corticosteroids and thus an increase in the enzymes of gluconeogenesis directing carbon flow into glucose synthesis and allowing fatty acids to be released from the fat depots for immediate oxidation and energy production. However, this simple picture of corticosteroid or insulin dominance does not explain all the enzyme changes that are brought about by changes in metabolic pressure. For example, some changes occur even when the adrenals are removed. Also in the normal animal, by feeding a substance such as glycerol as the main article of the diet, it is possible to increase simultaneously, the enzymes of glycolysis and of gluconeogenesis. There appears to be no doubt that the ingestion of a large amount of a particular foodstuff will produce enzymic changes appropriate to deal with the material independent of any hormonal involvement although normally both mechanisms work in harmony. In general, enzymic changes in response to diet can be predicted on common-sense principles. For example, a high carbohydrate diet results in an increased activity of phosphofructokinase and pyruvate kinase in the liver, an increase in acetyl-CoA carboxylase, the enzymes of the pentose phosphate pathway, malic enzyme and the citrate cleavage enzyme. All of these are concerned with increased throughput of the carbohydrate into fatty acid synthesis and subsequent transport of the fatty acids to the fat depots for storage as triglycerides. A diet high in protein will cause increases in liver transaminases, in glutamate dehydrogenase and in the enzymes of the urea cycle. All of these changes allow access to the carbon skeletons of the amino acids for purposes other than protein synthesis and permit disposal of the excess ammonia. With high-protein diets there is also an increase in the gluconeogenic enzymes of the liver (pyruvate carboxylase, PEP carboxykinase, fructose-1,6-diphosphatase and glucose-6-phosphatase) since suitable carbon skeletons from the amino acids must be used to replenish the deficiency of carbohydrate. A diet high in fat also results in an increase in gluconeogenic enzymes which must, in this case, make use of the glycerol of the triglycerides. As might be expected, there is a fall in the activity of the enzymes of fatty acid synthesis and there is also an increase in the enzymes of ketone body formation. The changes in the enzymes of the liver during starvation are somewhat similar to those occurring with a high-protein or a high-fat diet. The enzymes of gluconeogenesis increase, as do transaminases, and lipid is broken down thereby increasing the levels of ketone bodies. However, the difference is that in starvation the breakdown of endogenous proteins and lipids and the use of the products is coordinated in the most economical way for the body. The evidence of the enzymic changes is derived largely from experimental animals, mainly the rat, and there is very little information about the enzymic changes in the liver of the normal man. However, the changes in blood metabolites during starvation (see p. 467) suggest that man does not respond to this condition by an increase in the utilization of body protein for gluconeogenesis. Rather the use of fat and fat products is greatly increased and gluconeogenesis from protein drops markedly. The protein catabolized falls to the minimum amount to provide ammonia for the neutralization of excreted organic acids. The carbon skeletons of these deaminated amino acids and the glycerol of the fat are converted to glucose. The brain consumes the minimal amount of glucose and greatly increases its usage of ketone bodies.

In experimental animals clear differences can be demonstrated between the way in which the body handles the non-essential amino acids and the way in which it handles the essential amino acids. In general the enzymes degrading the non-essential amino acids are present in the liver at high concentrations that change parallel with the dietary intake. The transaminases metabolizing the essential amino acids are at low concentrations in the liver and gut but at high activity in the kidney and skeletal muscle. If these activities were high in the liver and the gut, the blood leaving the liver would be devoid of essential amino acids and there would be nothing left for the other organs. The arrangement described allows an even distribution to the non-hepatic tissues which can oxidize those surplus to the requirements for protein synthesis. There is no evidence that the enzymes catabolizing the essential amino acids increase in response to a dietary excess, but on the other hand they do respond by a major decrease when essential amino acids are in short supply.

Enzyme Changes During Development

Many of the enzyme changes which accompany the transition from the foetus to the free living infant are given on p. 375. In a few cases the control mechanism achieving the changes has been established. For example, the process of hepatic gluconeogenesis is switched on by the synthesis of phosphopyruvate carboxykinase at the time of birth. Premature delivery of the young rat by Caesarean section switches on the same rate of synthesis as the natural delivery. The enzyme is detectable by about one hour after delivery. Within the uterus it is possible to start the synthesis of the enzyme by injection of adrenaline, glucagon or cyclic AMP. Injection of insulin or glucose on the other hand prevents the synthesis of the enzyme. The complete sequence of events is as follows: the burst of muscle activity accompanying birth coupled with the deprivation of the maternal glucose supply produces a hypoglycaemia which in turn causes a reduction in the level of insulin and an increase in the supply of adrenaline and glucagon. The adrenaline and glucagon together potentiate the formation of cyclic AMP which activates phosphorylase and allows access to the hitherto untapped stores of glycogen in the liver. Also the cyclic AMP derepresses the synthesis of phosphopyruvate carboxykinase which can then complete the chain of gluconeogenic enzymes. This enables the lactate produced by the burst of muscle activity to be resynthesized into glucose. Subsequent to these hormonally initiated changes, dietary factors in the shape of the milk containing high levels of fat set the pattern of the liver enzymes to one of low fat synthesis and high gluconeogenesis. At the time of weaning, the diet of the rat changes to a predominantly carbohydrate one and there is a consequent large increase in the enzyme activities associated with fatty acid synthesis (citrate cleavage, acetyl-CoA carboxylase, pyruvic kinase, glucose-6-phosphate dehydrogenase and malic enzyme) which all show a peak about ten days after weaning but in their subsequent decline remain substantially above the pre-weaning level. Of the enzyme group, malic enzyme responds in a different way from the others, appearing immediately in response to early weaning (whereas the others are delayed somewhat) and rising at the weaning time even when the diet given is one high in fat (the rest of the group of enzymes remain low with a continued high-fat diet). The inducer for the malic enzyme appears to be the hormone thyroxine which becomes effective about three days after birth and increases in potency up to the normal weaning time. The reason for the special control of this enzyme in the rat is not clear. Other hormones also affect the levels of the fat-synthesizing enzymes. For example, with the onset of sexual maturity, glucose-6-phosphate dehydrogenase and the malic enzyme become almost twice as high in the female as in the male. Castrating the males increases the level of the enzymes of fat synthesis and spaying the female decreases them. Implantation of pellets of the appropriate sex hormone can reverse these changes. Such enzyme changes offer an explanation for the increase in the adipose tissue of eunuchs. As might be expected the enzymes of fat utilization and ketone-body formation in the liver move during the weaning period in a reciprocal pattern to the enzymes of fat synthesis.

Enzyme Changes in Organs Other Than the Liver

The kidney shows changes in its pattern of gluconeogenic enzymes similar to those of the liver when the dietary carbohydrate is restricted. However, most other organs show little response to dietary changes. Perhaps this is not surprising since the liver smooths out dietary anomalies and presents other tissues with a relatively constant blood as far as major nutrients are concerned. However, there are changes in the enzyme make-up of tissues as the animal ages. For example, in the muscle of the foetus there is a progressive increase in the relative amount of the M-type isoenzyme of lactate dehydrogenase following innervation. In the brain the ability to oxidize fatty acid decreases with age so that the adult rat has only one-tenth of the activity of the five-day-old animal. The utilization of ketone bodies in the brain increases after birth but declines after weaning so that the adult and the new-born animal have roughly the same enzymic capacity to deal with ketone bodies. By contrast, the utilization of ketone bodies by the heart remains comparatively low until weaning and then increases some three-fold.

Diurnal Variations in Enzymes

Many enzymes have now been found to show diurnal rhythms in the liver. Part of the rhythm is associated with the feeding schedule and part with the day length. Glucokinase and tyrosine transaminase are two enzymes showing swings of up to 100 per cent and therefore measurements of these and similar enzymes have to be made under standardized conditions if comparative studies are to be made. Similar diurnal swings are found in the glycogen content of the liver but these changes are reciprocals of the changes in glucokinase activity.

References

Intermediary Metabolism and Its Regulation, Larner, J. Prentice-Hall Inc., New Jersey, 1971.

Regulation in Metabolism, Newsholme E. A. and Start, C. John Wiley and Sons, London, 1973.

Interrelationship of Mammalian Hormones and Enzyme Levels *in vivo*. Pitot, H. C. and Yatvin, M. B. *Physiological Reviews*, **53**, 228–325, 1973.

Cellular Responses to Cyclic AMP. Bitensky, M. W. and Gorman, R. E. *Progress in Biophysics and Molecular Biology*, **26**, 409–462, 1973.

Control of Enzyme Levels in Animal Tissues. Schimke, R. T. and Doyle, D. *Annual Reviews of Biochemistry*, **39**, 929–976, 1970.

SECTION 15

Tissue Interrelationships

CHAPTER 49

Tissue Interrelationships

All tissues remove material from the blood stream and contribute material to it. The nature and amount of the contribution depends on the particular metabolic state of the animal and on the enzyme make-up of the particular tissue. All tissues are able to take up glucose with, in many cases, the rate being limited by the availability of insulin (see p. 422). However, the end product of the glucose utilization differs depending on the tissue and the balance of its enzymic activities. Figure 15.1 compares the activities of some of the enzymes concerned with the utilization of glucose-6-phosphate in the liver, the kidney and the adipose tissue. These relative activities result in the net flow of carbon along the

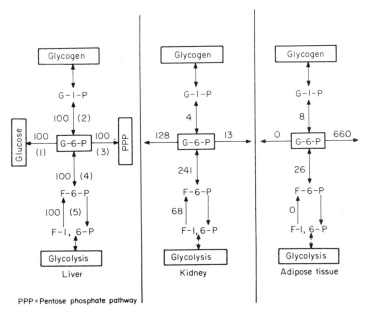

PPP = Pentose phosphate pathway

Figure 15.1 Carbohydrate related enzyme activities in liver, kidney and adipose tissue in rat. The activities were compared on a percentage basis, taking the values found in liver arbitrarily as 100 per cent. (1) Glucose-6-phosphatase; (2) phosphoglucomutase; (3) glucose-6-phosphate dehydrogenase; (4) phosphohexose isomerase; (5) fructose-1,6-diphosphatase. From Weber, *Advances in Enzyme Regulation*, Vol. 1, Pergamon, London, 1963

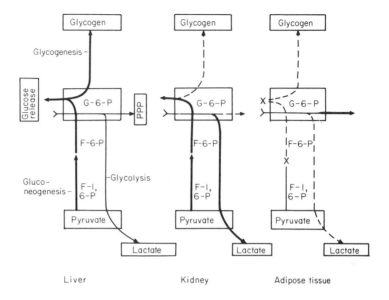

PPP = Pentose phosphate pathway

Figure 15.2 Comparison of carbohydrate metabolism in liver, kidney cortex and adipose tissue. The relative activities of metabolic pathways are indicated by the thickness of the arrows. The interrupted lines indicate pathways of very low activity. The × sign means complete absence of an enzyme activity. From Weber, *Advances in Enzyme Regulation*, Vol. 1, Pergamon, London, 1963

pathways shown in Figure 15.2 which are in accordance with the known function of these tissues; thus, for example, the adipose tissue preferentially synthesizes fatty acids, whereas the liver preferentially synthesizes glycogen. Tissue individuality is also conferred by the possession of special enzymes; for example, the liver mitochondria have the special acetyl glutamate activated carbamyl phosphate synthetase that initiates urea synthesis and a high concentration of pyruvate carboxylase for the initiation of gluconeogenesis. Isoenzyme balance within a tissue will also influence the direction of carbon flow, for example ensuring that heart muscle is always a net consumer of blood lactate and skeletal muscle is always a net contributor to blood lactate.

All tissues under conditions of excess food supply are capable of laying down some reserves of carbohydrate, fat and protein although these are slight except in the case of the liver, the adipose tissues and the muscle. There is an important difference between the carbohydrate and fat stores on the one hand and the protein stores on the other. Whilst glycogen and triglycerides are storage materials *per se*, the protein stores are all molecules which have specific biological functions and therefore the consumption of these molecules must result in a diminution of the activity that they mediate. This is shown, for example, in the muscle weakness that occurs with the wasting of prolonged starvation. The availability of the stored material depends on the enzymes converting it to the form suitable for transport in the blood. The key enzyme as far as the carbohydrate stores are concerned is glucose-6-phosphatase which is present in the liver and the kidney but not in the muscle. Thus glucose-6-phosphate arising from stored liver glycogen or glucose-6-phosphate arising in liver or kidney by gluconeogenesis can be hydrolysed to maintain blood glucose. On the other hand, the substantial stores of muscle glycogen cannot give rise to blood glucose because muscle does not possess glucose-6-phosphatase.

The enzyme which makes the triglyceride store of the adipose tissue available is the hormone sensitive lipase (see p. 257). The activity of this enzyme determines the level of free fatty acids (attached to albumin) in the blood and also the level of circulating glycerol. There is no special mechanism available to tap the stores of proteins but it is known that all cells contain proteinases and all cells have a pool of free amino acids. The pool of amino acids and the cellular proteins are constantly turning over and blood amino acids are constantly entering the cellular pool. It may be that in times of dietary protein shortage with a consequent lowering of the level of blood amino acids there is a natural movement of these compounds from the cellular pool into the blood.

Cycles Between Tissues

As far as we know all possible metabolic pathways in a human are taking place all the time although of course their rates vary tremendously according to the circumstances. At all times (fed or unfed) the following cycles between tissues are taking place:

(A) The lactate (Cori) cycle. The glycolytic tissues (white muscle, smooth muscle and erythrocytes) and other tissues according to the circumstances contribute lactate to the blood. The smooth muscle contributes its lactate directly into the portal circulation but the majority of the lactate arising from the contracting skeletal muscle (particularly after exercise) must first traverse the heart, the lungs and then the heart again (which abstracts some for oxidation) before reaching the liver via the hepatic artery. Roughly 10 per cent (500 ml per min) of the heart's output passes through the hepatic artery. A further 20 per cent of the heart's output passes through the kidneys. Thus about one-third of the blood lactate at any one time can be used for gluconeogenesis. In the liver some 40 per cent of the lactate received is converted to glucose and in the kidney about 10 per cent. In the resting subject, whether starved or fed, this amounts to a daily synthesis of about 40 g of glucose. The synthesized glucose is then available for use by the glycolytic tissues, thus completing the cycle.

(B) The fatty acid, triglyceride cycle. Fatty acids and glycerol are liberated from the adipose tissue (at rates dependent on the activity of the hormone sensitive lipase) and pass into the blood stream. The relative contributions of the omental fat (whose products pass into the portal circulation) and the other fat depots (draining into the systemic circulation) are not clear. The fatty acids are attached to albumin (which in effect solubilizes them) whereas the glycerol is in

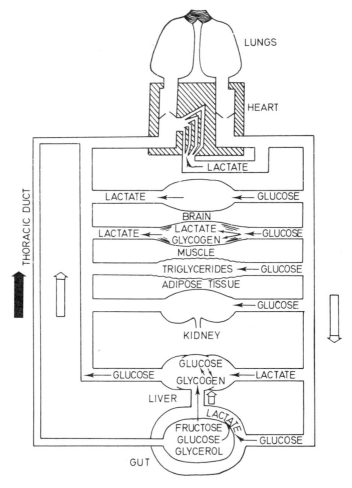

Figure 15.3 Net uptake or loss of blood metabolites related to carbohydrate metabolism by tissues when carbohydrate is in excess. Contribution of lungs ignored

free solution. Both supply streams enter the liver where the glycerol is phosphorylated to enter the pool of α-glycerol phosphate with the major options of being used for triglyceride synthesis or for gluconeogenesis. The fatty acids have the major options of being converted to triglycerides or to ketone bodies. The distribution to the competing pathways depends on the control systems already discussed (see Section 14) and is determined by the overall needs of the body. Thus the lipid-related substances in the venous outflow from the liver will be triglycerides (as very-low-density lipoproteins), ketone bodies, unused fatty acids and glycerol in proportions determined by the liver as appropriate to metabolic needs. The ketone bodies in the metabolic mix have virtually one fate only, that of oxidation to CO_2 and H_2O. Apart from the use by the liver, the glycerol may

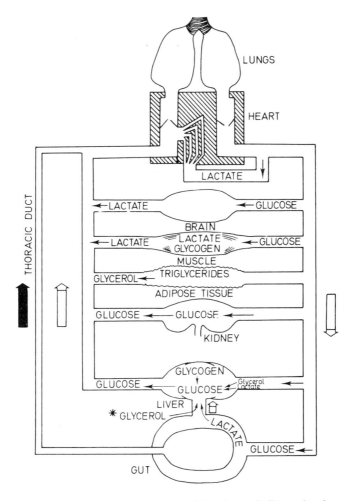

Figure 15.4 Net uptake or loss of blood metabolites related to carbohydrate metabolism by tissues when carbohydrates are lacking. Contribution of lungs ignored. ∗ This glycerol is derived from the omental fatty tissue

be used by the kidney for gluconeogenesis. Triglycerides (reinforced by any that may be entering via the thoracic duct as chylomicra) can make their constituent fatty acids available to any tissue that possesses an active clearing factor lipase (see p. 249). Thus a certain fraction of the triglycerides will have their fatty acids taken up by the adipose tissue for resynthesis to triglycerides if glucose is available to supply α-glycerophosphate. This then completes the cyclic process.

(C) The ammonia, amino acid, amide cycle. All tissues donate amino acids to the blood and remove amino acids from it. The overall balance results in an amino acid concentration which is remarkably constant at about 4 mM amino nitrogen. Tissues also contribute amide nitrogen (glutamine) and ammonia to the

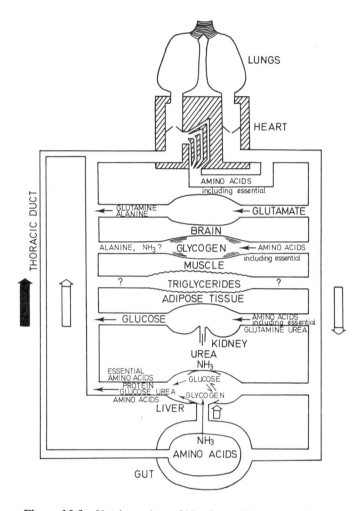

Figure 15.5 Uptake or loss of blood metabolites related to protein metabolism by tissues when protein is in excess. Contribution of lungs ignored

blood in amounts characteristic of the individual tissue. In the special case of the liver, ammonia is removed from the blood. The amide group of glutamine and the amino group of alanine are particularly mobile as ammonia precursors and their residual carbon skeletons enter the gluconeogenic pathway very readily. At all times the liver is the recipient of ammonia and amino acids in the portal circulation as a result of the combined effect of gut bacteria and digestive enzymes. Similarly on the systemic side, the liver receives amino acids and glutamine but very little ammonia. The venous outflow from the liver is almost completely free from ammonia but always contains urea, amino acids and glutamine. After traversing the heart and lungs, the amino acids are available to complete the

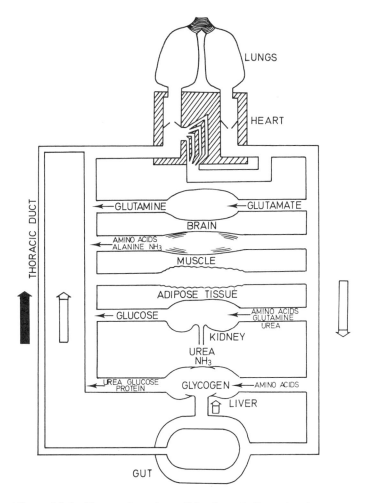

Figure 15.6 Net uptake or loss of blood metabolites related to protein metabolism by tissues when protein is lacking. Contribution of lungs ignored

cycle by being taken up by the tissues. Glutamate is much favoured by the brain because of its special metabolism (see p. 327) and glutamine is particularly required by the kidney as an ammonia source for urinary neutralization (see p. 319). The urea is a dead-end product whose only fate is to be excreted by the kidney.

Substrate Flow Under Different Metabolic Conditions

The direction of net flow of carbon either into the tissues or out from the tissues is decided by the metabolic pressure of carbon entering the circulation

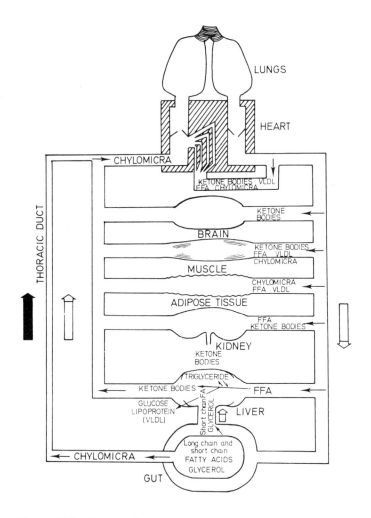

Figure 15.7 Net uptake or loss of blood metabolites related to lipid metabolism by tissues when lipid is in excess. Contribution of lungs ignored. Chylomicra next to a tissue implies that the capillary walls contain active clearing factor lipase

from the gut. If it is high, metabolism moves in the direction of deposition of material in the tissues; if it is low, carbon moves out from the tissues. Figures 15.3 to 15.8 illustrate the individual contributions of the main tissues to this directional flow in cases where carbohydrate, protein or fat are in excess or when they are unavailable. Total starvation is therefore expressed by combining the figures showing deficiencies in each of the three foodstuffs (Figures 15.4, 15.6 and 15.8).

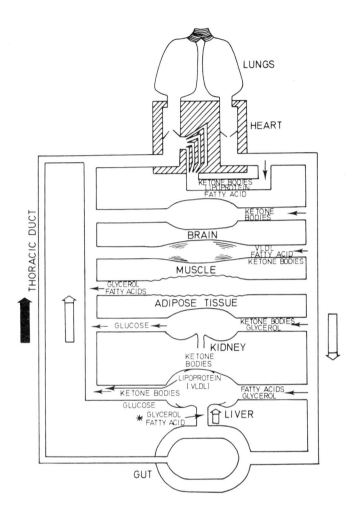

Figure 15.8 Net uptake or loss of blood metabolites related to lipid metabolism by tissues when lipid is lacking. Contribution of lungs ignored. *From omental fat tissue. Lipoprotein or VLDL next to a tissue implies that the capillary wall contains clearing factor lipase

SECTION 16

Selected Topics in Human Metabolism

CHAPTER 50

Obesity and Starvation

In the long run an individual's weight is determined by the balance between the quantity of foodstuff eaten and quantity excreted as CO_2, H_2O and urea (together with minor excretion products). Strictly speaking, it is the amount of food absorbed by the gut rather than the amount eaten that is relevant, but the difference between these two values is quite small. For example, of the 200 g or so of faeces produced daily on a mixed diet, 50 g is dry material whereas the dry weight of the food taken in for a mixed diet yielding 2500 kcal is about 420 g. Of the dry faecal material only about 20 g will be food residues that were potentially absorbable. The non-digestible cellulose and other fibrous material of the diet is mainly carbohydrate in nature but it is important to realize when using food tables of nutritive values that the roughage will not be included in the carbohydrate analysis of the foodstuff. For example 100 g of cabbage contains 92 g water, 1·5 g protein, 5 g of carbohydrate and about 100 mg of various minerals mainly calcium. The 1·4 g deficit is the non-digestible material. Thus of the dry weight of the green vegetable material we eat, about 20 per cent does not contribute to the diet. With potatoes the value of unabsorbable material is somewhat less.

Of the food consumed it is the dry material content that is important; the accompanying water is excreted or retained to help maintain the normal water composition of the body. In the average man, water amounts to about 60 to 65 per cent of the body mass. The non-fatty tissues contain on average about 70 to 73 per cent of water* whilst the adipose tissue contains about 20 per cent of water. Broadly speaking the total amount of the adult's protein and carbohydrate is constant. The two variables that determine the body weight are the content of water and the content of fat. Food surplus to current requirements (including the maintenance of glycogen stores as part of the requirements) has one fate only, that of deposition as triglycerides. Long-term loss or gain in weight therefore directly reflects the size of the fat depots. Short-term changes in weight may simply reflect variations in the water content of the body, for example a 1 per cent change in body water is roughly equivalent to 0·7 kg change in body weight. Such changes occur spontaneously in response to changes in hormone balance (e.g. menstrual cycle) or can be induced artificially by administering a diuretic.

Calorie excess and obesity is associated with prosperous 'western' cultures where it is coupled with a quite unnecessarily high consumption of protein.

* Excluding the skeleton (44% water).

Calorie deficiency and protein lack are characteristics of the underdeveloped and emergent societies. Within these societies obesity tends to reflect the individual's prosperity. The realization that adiposity is not a good prognosis for longevity has prompted developed societies to examine the make-up of their diets with the apparent hope that they will be able to overeat and yet not become fat. As has already been pointed out, the human is a chemical engine fully subject to the laws of physics and chemistry and therefore it is not possible to dispose of calories that are surplus to the work output of the body.

Excess calories are deposited as fat for the simple reason that there is nowhere else for them to go. High-protein-containing diets are often recommended as nonfattening on the grounds that metabolic work has to be done to produce the excreted urea from the protein. This is true but it should be remembered that the food tables of calorific values deduct the energy value of the urea and the value of 4 kcal/g for protein expresses the utilizable calories. Also in real life foodstuffs do not come as protein or fat or carbohydrate. Table 16.1 gives weights and composition of individual foods yielding 2000 kcal. Steak, eggs and cheese, which often figure in diets as high-protein foods, in fact contribute most of their calories from their fat content. It is interesting to note that if it were possible to consume and digest the huge pile of lettuce necessary to obtain 2000 kcal daily you would be obtaining a substantially higher protein intake than if you had chosen steak. White fish is almost the only natural food where it is possible to achieve anything approaching a calorie sufficiency that is very largely from protein and the thought of approximately 3 kilos of cod daily would daunt most individuals.

If food is taken largely in the form of fat some of the calorific intake is excreted. Such a diet (275 g of butter would produce 2000 cal) raises the ketone bodies of the blood to a level where appreciable amounts are excreted in the urine. In practical terms reduction in the body adipose tissue can be brought about only by a reduction of the calorie intake below that used in the daily activities.

The deposition of fats in the adipose tissue is controlled by the clearing factor lipase, which in a state of calorie excess is high in the capillaries of the adipose

Table 16.1 Amounts of foodstuffs equivalent to 200 kcal. Protein and carbohydrate ≡ 4 kcal/g, fat ≡ 9 kcal/g

Food	Fresh weight to give 2000 kcal (g)	Protein content (g)	Fat content (g)	Carbohydrate content (g)
Steak	630	88	185	0
Eggs	1270	155	147	11
Cheese	490	122	168	0
Potatoes	2750	55	0	444
Bread	800	68	9	410
Lettuce	18,400	200	0	290
Fish (cod)	2900	465	15	0

tissue. Thus blood triglycerides are hydrolyzed and fatty acid enters the adipose tissue to be stored as triglyceride. If the calorie excess is from carbohydrate, the triglycerides in the blood are largely those manufactured by the liver from the carbohydrate. If the calorie excess is from dietary fat the triglycerides will be largely from the intestine in the form of chylomicra. Clearing factor lipase is also found at high activity in calorie excess in the skeletal and heart muscles and in the kidney and the lung. In time of calorie deficiency where the blood triglycerides originate from the liver (using fatty acids supplied from the adipose tissue) the clearing factor lipase of the adipose tissue falls to very low levels whereas the activity in the muscle increases, thus positively directing the fatty acids towards the muscle for oxidation.

Because of the appreciation of the adverse effects of obesity there have been several medically controlled studies of prolonged starvation in overweight subjects. In these subjects measurements have been made of the arteriovenous differences across many of the body's organs and thus the amounts of substrate used by these organs can be calculated. If a normal man is fasted briefly he uses each day about 300 g of carbohydrate, 100 g of fat and 75 g of protein from body reserves of 150–200 g, 15,000 g and 7500 g respectively. The carbohydrate deficiency must be eaten or made within the body by gluconeogenesis. It can be calculated that the daily substrate flow in such a man is as shown in Figure 16.1. Such use of protein for gluconeogenesis cannot continue since all the protein of the body would be consumed within comparatively few weeks. After three days of fasting the blood insulin falls to about 20 per cent of the fed value and the blood glucose to about 75 per cent of normal. These levels are then maintained more or less indefinitely. By this time blood-free fatty acids have doubled and thereafter increase more slowly to reach values some three times (1·9 mM) the normal. Ketone bodies which are of a comparable level with the free fatty acids by three days continue to increase for about three weeks to some 4 to 5 times the concentration of the free fatty acids. During this period nitrogen excretion drops, being halved by three days, and thereafter falling to reach a value around 20 per cent of the initial by about four weeks. Of the 4 g or so excreted at this time the majority is as ammonia rather than urea.

Whilst before fasting, glucose and gluconeogenic substrates in the blood together were roughly 10 mM (1·5 g/l) and fatty acids and ketone bodies were roughly 0·8 mM (0·2 g/l), at the end of the fast glucose and precursors were roughly 8 mM (1·1 g/l) and fatty acids and ketone bodies were about 9 mM (1·2 g/l). Because of the increase in the level of the ketone bodies the renal threshold is exceeded and about 100 mmoles are lost daily (roughly 11 g) or about 12 per cent of the daily production of ketone bodies. It is to neutralize the excreted ketone bodies that so much of the nitrogen excretion is in the form of ammonia. To ensure that as much carbohydrate is preserved as possible it is important to protect the lactate emanating from the glycolytic tissues from further oxidation and also to ensure that glycerol produced from triglyceride hydrolysis enters the pathway of gluconeogenesis as completely as possible. To this end pyruvate dehydrogenase is inhibited by the mechanisms already described (see p. 435). The high levels of

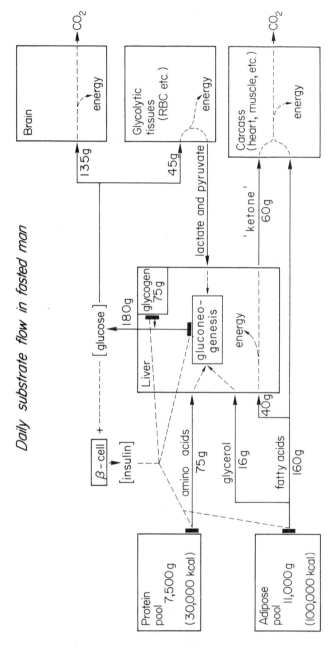

Figure 16.1 Estimated substrate flow in man fasted overnight. The β-cell detects the concentration of glucose, which when low is associated with a decrease in insulin release. Since insulin is rapidly metabolized ($t_{\frac{1}{2}} = 10$ minutes), insulin levels fall. Insulin serves to check amino acid release, free fatty acid release, and probably also gluconeogenesis as well as glycogen breakdown, both in liver. These concentrations of insulin are far lower than those which occur in the fed state and which are associated with entry of glucose into muscle and adipose tissue. Thus glucose, with the low concentration of insulin present in the fasted state, is metabolized solely by brain to CO_2 and by the glycolytic tissues to lactate and pyruvate. From G. F. Cahill jnr. in *The Human Adrenal Cortex* (ed. N. P. Christy), Harper and Row, 1971

NADH + H$^+$ and of ATP brought about by the rapid oxidation of fatty acids by the liver help to drive the gluconeogenic pathway. Because of the changed level of substrates the brain makes much greater use of ketone bodies so that roughly half its energy comes from this source and half from the metabolism of glucose. Between them, the brain and the kidney use most of the ketone bodies produced by the liver but the heart and skeletal (red) muscles greatly reduce their demands on them turning rather to fatty acids for their energy supply.

The deficit in glucose during prolonged starvation (because of its continued oxidation by the brain) is replenished with the minimum gluconeogenesis from protein. Of the 80 or so grams of glucose that are formed from gluconeogenic precursors daily (in roughly equal amounts by the liver and the kidney) half is the product of recycling lactate and pyruvate, a further quarter is derived from glycerol and the remainder from protein. This greatly reduced gluconeogenesis (50 per cent) from all sources during prolonged starvation is roughly equal to the amount of glucose that is synthesized from protein alone during a one-day fast. The changes in the enzymes of gluconeogenesis during starvation in man have not been measured but in the rat there are large increases in them as well as in the enzymes of protein catabolism.

The activity of an enzyme pathway gives no information about the flow of carbon down that pathway unless the supply of substrate is also defined. With the decreased supply of substrate occurring in starvation, it may be necessary to increase the activity of the gluconeogenic pathways to ensure that the limited substrate is directed efficiently and swiftly down those pathways.

Diabetes

This condition has been compared to that of starvation but there are several differences between the controlled breakdown of tissue during starvation and the uncontrolled tissue wasting of diabetes. Ketone bodies exert a feedback inhibition of lipolysis so that there is some relationship between fatty acid release and ketone body utilization. This control appears absent in the diabetic. In starvation gluconeogenesis and protein catabolism are decreased whereas in diabetes they are increased. In starvation the blood glucose falls but in diabetes it rises. However, in both cases, at least in the experimental rat, the enzymes of gluconeogenesis and protein catabolism show marked increases.

CHAPTER 51

Alcohol and Sucrose Consumption

As pointed out previously, food eaten is a mixture of complex materials and it is very unusual for a food to be a single chemical. There are, however, two articles of diet, sucrose and alcohol, where the substances are taken with little else than added water. These two substances have rather special metabolic effects and are moreover consumed in very substantial amounts in the advanced civilizations. Britain's consumption of sucrose is somewhat greater than 2·5 million tons annually and if all this is eaten in one form or the other it amounts to an average intake of 130 g daily or about 500 kcal per person. A substantial fraction of this is simply added to liquids that are drunk. For example, the addition of two teaspoonfuls of sugar to a cup of tea or coffee means a consumption of at least 10 g of sucrose or about 40 kcal. Six cups of such beverages is not at all uncommon. The consumption of 100 g of boiled sweets would contribute about 350 kcal. It is thus easily possible that about one-quarter of the daily calories can arise from sucrose. The ethanol consumption of Great Britain is not easy to ascertain but values from the European countries and from the United States suggest that on average 5 to 10 per cent of the daily calories are derived from ethanol in one form or another. Assuming adults only are alcohol consumers, 10 per cent seems a reasonable value for the average contribution to the dietary calories. Both sucrose and ethanol are valuable sources of dietary calories; however, it is likely that both are consumed for pleasurable reasons rather than to satisfy the need for calories. In both cases the excessive consumption leads to undesirable metabolic consequences. After sucrose has been eaten the liver receives a mixture which is mainly glucose since some of the fructose resulting from the hydrolysis of the sucrose by gut sucrase is converted to glucose in its passage through the gut wall. This glucose enters the metabolic pool of glucose and that in excess of requirements will end as fat. The enzyme fructokinase that forms fructose-1-phosphate from the fructose is of such a high activity that it can deplete the liver's store of ATP in carrying out the phosphorylation. The result of this is to greatly increase the quantity of AMP that is available for catabolism to uric acid. The whole reaction sequence may be sufficiently fast to raise blood uric acid and precipitate an attack of gout in the susceptible individual. Fructose-1-phosphate is split by aldolase to give dihydroxyacetone phosphate and glyceraldehyde. Both of these substances are reducible by $NADH + H^+$ in reactions catalysed by α-glycerophosphate dehydrogenase or by alcohol dehydrogenase to give α-glycerophosphate and glycerol respectively. The glycerol can in turn give rise to further α-glycerophosphate. It is generally believed that the rate of triglyceride synthesis by the liver is dependent on the concentration of α-glycerophosphate and thus the situation described will favour liver triglyceride synthesis. The glucose-6-phosphate formed from the glucose component of sucrose hydrolysis is not so easily available for triose and α-glycerophosphate formation since their rate of formation from glucose is restricted by the low activity of phosphofructokinase.

Alcohol is a natural product of microbial fermentation in the human gut which yields a portal blood concentration of about 0·5 mM. Thus even teetotallers consume about 3 g of ethanol daily. The first step in the metabolism of alcohol is the formation of acetaldehyde catalysed by the enzyme alcohol dehydrogenase in the cytosol

$$CH_3CH_2OH + NAD^+ \rightleftharpoons CH_3CHO + NADH + H^+$$

The enzyme is found very largely in the liver with a small amount in the stomach and very tiny amounts in other tissues. The enzyme activity of the liver is sufficient to oxidize 3·5 kg of ethanol daily but this activity is limited by the rate at which the NADH + H$^+$ can be oxidized. Perfusion studies have shown that the actual rate of oxidation of ethanol is only one-tenth of the liver's enzymic potential capacity. Theoretically the consumption of 1·5 litres of wine (8 per cent ethanol) spread over a 12-hour period should be within the liver's capacity for oxidation. This would amount to some 40 per cent of the daily calorie requirement. However, the regular intake of ethanol much in excess of 30–40 per cent of daily calorie requirements usually results in the development of a fatty liver. Acetaldehyde is oxidized further in the liver by the rather non-specific group of aldehyde dehydrogenases:

$$CH_3CHO + NAD^+ + H_2O \longrightarrow CH_3COOH + NADH + H^+$$

This reaction is virtually irreversible. Some four-fifths of the activity is in the cytosol and is about equal to the rate at which alcohol is oxidized by the liver. In spite of this, a substantial amount of the acetaldehyde (about 50 per cent) passes into the blood if the alcohol concentration in the liver exceeds 8 mM. (The statutory limit of blood alcohol for drivers is 80 mg per cent or 18 mM.) The acetaldehyde can be oxidized in other tissues but the aldehyde dehydrogenases are much less active than the liver enzymes. The next most active tissue is the kidney (one-seventh of the specific activity of the liver) and most of the other tissues have activities about one-thirtieth of the liver (heart, brain, intestine). Skeletal muscle has an even lower but measurable activity. The acetic acid produced from the oxidation of acetaldehyde gives rise to acetyl-CoA in the reaction catalysed by acetyl thiokinase. It is of interest that the female liver and kidneys (at least in the rat) have almost twice the aldehyde dehydrogenase activity of the male. Aldehyde dehydrogenase is inhibited by disulfiram (Antabuse); thus if alcohol is taken after the drug, the blood level of acetaldehyde rises. This is thought to give rise to the nausea, hypotension and vomiting that occurs when alcohol is taken after antabuse.

Interaction of Alcohol Metabolism with other Metabolic Pathways

Gluconeogenesis from lactate starts by oxidizing the lactate to pyruvate. This reaction requires NAD^+. The oxidation of ethanol also requires NAD^+ and, because of the very high activity of alcohol dehydrogenase, it competes with lactate dehydrogenase. A very high $NADH + H^+/NAD^+$ ratio is established in the liver cell and this, besides preventing gluconeogenesis from lactate, drives the reduction of dihydroxyacetone phosphate to α-glycerophosphate. A high level of α-glycerophosphate is conductive to triglyceride synthesis since it is the availability of that substance that is rate limiting. The inhibition of gluconeogenesis by ethanol oxidation may result in a hypoglycemia which in susceptible subjects can lead to fainting. This is particularly noticeable after exercise where the need to replenish blood glucose from muscle lactate becomes acute. It is not clear whether the fatty liver of alcoholism is connected with the stimulation of triglyceride synthesis but since in general the calories of alcohol are superfluous to the daily requirements it is likely that the acetyl-CoA produced in its oxidation will be synthesized to fatty acids and will contribute to the liver's triglyceride output and ultimately to an increase in the amount of adipose tissue.

Comparison of Sucrose and Ethanol as Dietary Components

Both the substances tend to be taken in the diet as 'extras' and not simply to satisfy hunger. In both cases there are clear metabolic pathways which allow the extra calories to be easily laid down as fat. In the case of sucrose (in experimental animals) there is an increase in the enzymes of glycolysis and of fatty acid synthesis and there appears to be no limit to calorie intake that can be used by the body. In the case of ethanol there is no clear evidence that the enzymes change to meet the metabolic load and there is a limitation on the calories of alcohol that can be metabolized. Sucrose metabolism does not inhibit other metabolic pathways and does not alter the cells redox state whereas alcohol metabolism does both of these. However, both substances taken at a reasonable level as part of the calorie requirements provide a source of quickly accessible energy.

References

Some Endocrine and Metabolic Aspects of Obesity. Rabinowitz, D. *Annual Review of Medicine,* **21**, 241–258, 1970.

Liver and Kidney Metabolism during Prolonged Starvation. Owen, O. E., Felig, P., Morgan, A. P., Wahren, J. and Cahill, G. F. *Journal of Clinical Investigation,* **48**, 574–583, 1969.

Index

A-band, 95
A-Protein, of lactose synthase, 380
A site, of ribosome, 178
Absorption, of food, 159
  of lipids, 163
Acetaldehyde, in blood after ethanol ingestion, 471
Acetanilide, 226
Acetic anhydride, 42
Acetoacetate, effect on blood pH, 317
Acetoacetic acid, decarboxylation, 17
Acetoacetyl-CoA, formation, 233
Acetone, formation, 233
p-Acetyl aminophenol, 226
Acetylchloride, 43
Acetylcholine, 75, 340
  structure, 340
  synthesis, 340
Acetylcholinesterase, 75, 340
Acetyl-CoA, activator of pyruvate carboxylase, 436
  carboxylation, 253
  effect on pyruvic dehydrogenase, 435
  exit from mitochondria, 254
  formation of citrate, 125
  formation of pyruvate, 123
  hydrolysis, 15
Acetyl-CoA carboxylase, 253
  activation by insulin, 258
N-acetyl glutamic acid, activation of carbamyl-phosphate synthetase, 221
N-acetyllactosamine synthetases, 381
  identity with A protein, 381
Acetyl salicylic acid, detoxication, 228
Achlorhydria, and iron absorption, 295
Acid anhydride, 110
Acids, 53
  Brønsted definition, 54
  dissociation constant, 54
Acid–base balance, 316
Acid phosphatase, in prostatic cancer, 291
  in seminal plasma, 366

Aconitase, 125
cis-Aconitic acid, 125
Actin, 97
  and myosin ATP-ase, 98
Actinomycin D, action, 181
  structure, 182
Action potential, 99, 335
Activated complex, 46
Active site, 74
Actomyosin, 98
Acyl-CoA dehydrogenase, 231
Addition reactions, 16
Adenine nucleotide transport, inhibitors of, 146
Adenosine phosphates, structures, 43
S-Adenosyl homocysteine, formation, 238
S-Adenosyl methionine, formation, 238
Adenylate deaminase, NH$_3$ formation, 219
Adenylate kinase, amplifier for nucleotide changes, 434
Adenyl cyclase, 401
  adipose tissue, 257
  in pineal gland, 348
Adenylic acid deaminase, 330
Adenylosuccinate lyase, 190, 191
Adenylosuccinate synthetase, 191
Adipose tissue, brown, 258
  carbon inputs, 249
  effects of hormones, 257
  location, 249
  metabolism, 249
  protein content, etc., 89
  triglyceride content, 249
  water content, 465
  weight, 249
ADP, and PFK, 108
  phosphorylation, 51
  respiratory control by, 140
Adrenal cortex, 404
Adrenaline, on adipose tissue, 257
  phosphopyruvate carboxykinase synthesis, 447
  synthesis from tyrosine, 411

Adrenal medulla, 410
Adrenocorticotrophic hormone, controller of glucocorticoid synthesis, 405
  on adipose tissue, 257
Adsorption chromatography, 59
Alanine, absolute configuration, 13
  in blood, 219
L-Alanine, structure, 26
Albinism, 385
Albumins, fatty acid transport, 257
Alcaptonuria, 385
Alcohol, consumption, 470
  effect on retina, 359
  in breath, 308
  production in gut, 470
Alcohol dehydrogenase, 471
  intestinal, 167
  role in formation of Δ^{11}-cis-retinal, 357
  specificity, 74
Alcohol metabolism, interaction with other metabolic pathways, 472
Aldactone, structure, 408
Aldehyde dehydrogenase, 471
  tissue activities, 471
Aldolase, 108
  muscular dystrophy, 116, 291
  perinatal increase, 375
Aldolase reductase, in seminal vesicles, 366
  in lens, 354
Aldosterone, antidiuretic properties, 408
  functions, 408
  slow acting hormone, 409
  sodium retention, 320
Aldosterone antagonists, as diuretics, 320
Alkaline phosphatase, in bone calcification, 272
Allicin, in expired air, 308
Allopurinol, gout treatment, 193
Allosteric control, 83
Allosteric sites, 80
Alveolar epithelial cells of lungs, 307
Alveolar macrophages of lungs, 307
Amethopterin, 188, 189
Amidophosphoribosyl transferase, 190
Amino acid absorption, 165
Amino **acids**, activating enzymes, 178
  activation, 45, 178
  analogues, 184
  catabolism, 214
  chemistry, 25
  essential, 213
  glucogenic, 215
  ketogenic, 215
  optical activity, 13
  reabsorption in kidney, 303
  sequence in proteins, 63
  structures, 26
  supply, 213
Amino acid oxidase, D and L, 74
Amino acyl site of ribosome, 178
Amino acyl-t-RNA, formation, 45
Amino acyl-t-RNA synthetases, 178
γ-Aminobutyric acid, brain, 328
  cycle, 329
α-Amino-β-oxoadipic acid, structure, 278
δ-Amino laevulinic acid, in haem synthesis, 277
  structure, 278
Amino peptidases, 64, 160
p-Aminophenol, 227
Aminopterin, 188, 189
Ammonia, amino acid, amide cycle, 457
  buffering, 219
  concentrations in tissues, 218
  metabolism, 218
  transport in blood, 218
Ammonium chloride ingestion, effect on blood pH, 317
Ammonium sulphate precipitation, 59
AMP and PFK, 108
  activation of phosphofructokinase, 434
  formation, 189
c-AMP, adipose tissue, 257
  second messenger, 403
  superior cervical ganglion, 345
Amylase, 160
  in pancreatitis, 291
Amylo-1,6-glucosidase, 106
Amytal, electron transport inhibition, 138
Anderson's disease, 208
Angiotensin I, 320
Angiotensin II, 320
  controlling aldosterone release, 409
  controlling aldosterone synthesis, 409
Angiotensinase, 320
Anserine, formula, 150
Antabuse, inhibitor of aldehyde dehydrogenase, 471
Antibodies, 60, 297
  production, 299
Antidiuretic hormone controlling urine volume, 321
Antigen binding sites, 298
Antigen sensitized lymphocytes, 301
Antimycin, inhibitor of electron transport, 145
Antimycin A, electron transport inhibition, 138
Anti-Rhesus immunoglobulins, 302
Arachidonic acid, structure, 343
Arginase, 75, 221

Arginine, hydrolysis, 221
L-Arginine, structure, 27
Argininosuccinase, 222
Argininosuccinate, 222
Argininosuccinate synthetase, 221
  perinatal increase, 377
Argininosuccinic acid, formation, 221
Arsenate, as uncoupler, 146
Arsenite, 144
  reaction with lipoate, 124
Ascorbic acid, formation, 212
  in collagen biosynthesis, 267
Aspartate transcarbamylase, 73, 186
L-Aspartic acid, structure, 27
Aspirin, absorption, 58
  detoxication, 228
  inhibition of **prostaglandin** synthesis, 342
  structure, 58
Atherosclerosis, 241
Atoms, structure of, 9
ATP, and creatine phosphate, 107
  and PFK, 108
  energy supply, 42
  formation from GTP, 127
  formation from PEP, 113
  glucose phosphorylation, 107
  $\Delta G^{\circ\prime}$ hydrolysis, 44
  in chromaffin granules, 410
  inhibition of phosphofructokinase, 434
  synthesis from GTP, 141
ATP/ADP ratio, energetic implications, 434
ATP-ase, Na^+/K^+-dependent, 286
  and red cell glycolysis, 287
ATP-citrate lyase, 255
Atractyloside, inhibitor of adenine nucleotide transport, 146
Atropine, 345
Avidin, 200
Azaserine, 191

B-Protein, lactose synthase, 380
Barbiturates, inhibitors of electron transport, 145
Base, Brønsted definition, 54
Benzene, structure, 14
Benzoic acid, 227
Benzoyl-CoA, 227
Benzylamine, 227
Bicarbonate, pancreatic, 160
Bile acids, formation, 241
Bile pigments, 279
Bile salts, 160
Bilirubin, structure, 279

Biliverdin, structure, 279
Bioenergetics, 38
Biotin, 204
  acetyl-CoA carboxylation, 253
  role, 200
  structure, 200
Black widow spider, 340
Blood, 276
  coagulation, 291
  composition, 314
  flow to tissues, 89
  glucose levels, 199
  protein content, etc., 89
Blood glucose, effect of glucocorticoids, 408
Blood group antigens, 302
Body, tissue weights, etc., 89
Bohr effect, 282
Bone, calcium phosphate deposition, 272
  collagen content, 265
  mucopolysaccharide content, 272
  protein content, etc., 89
Bongkrekic acid inhibitor of adenine nucleotide transport, 147
Brain, ammonia metabolism, 327
  changes in substrates with age, 448
  cholesterol content, 327
  DNA-phosphorus content, 327
  fuel for respiration, 325
  general metabolism, 325
  glucose consumption in starvation, 326
  ketone body consumption in starvation, 326
  ketone body oxidation, 234
  glutamate metabolism, 327
  lipids, 331
  phospholipid content, 327
  phospholipid turnover, 332
  postmortem ammonia production, 330
  protein content, etc., 89
  protein turnover, 331
  RQ in prolonged starvation, 326
  use of ketone bodies in infancy, 326
  weight, 327
Branching enzyme, 202
  Anderson's disease, 208
Buffer solutions, 53

Cabbage, composition, 465
Calcium, as uncoupler, 146
  binding to troponin, 97
  muscle contraction, 99
Calcium phosphate, chromatography, 59
Calories, conversion to Joules, 51
Carbamylcholine, 75

Carbamylphosphate, 221
  synthesis from glutamine, 185
Carbamylphosphate synthetase, 186, 221
Carbamino bound CO_2, 282
Carbanions, 17
Carbodiimide, peptide formation, 6
Carbohydrate, absorption, 162
  chemistry, 20
  content of foodstuffs, 466
  inhibition by phlorizin, 162
  role of Na^+ ions, 162
Carbohydrate metabolism, activity in various tissues, 454
  diseases, 206
  of foetus, 374
Carbon, properties of its compounds, 17
Carbon dioxide, carriage by haemoglobin, 281
  fixation, 200, 203
  formation in pentose shunt, 209
  uncoupler, 146
Carbon flow, in tissues, 453
Carbonic anhydrase, acid–base balance, 316
  red cells, 283
  tissue buffering, 316
  tissue localization, 316
Carbonic anhydrase inhibitors, diuretics, 320
Carbonium ion, 16, 84
Carbon tetrachloride, fatty liver, 235
  structure, 10
Carboxymethyl cellulose, 59
Carboxypeptidase, 64, 160
Cardiac glycosides, 285
Cardiac muscle, 151
Carnitine, fatty acid oxidation, 230
Carnitine acetyl transferase, 255
Carnitine acyl transferase, 230
Carnitine palmitoyl transferase, activity in muscle, 149
Carnosine, formula, 150
Carrier molecules, control of oxaloacetate utilization, 436
  control of pyruvate utilization, 435
Catalysis, 45
Catalytic subunits, 73
Cataracts, 354
Catecholamines, effect on heart muscle, 413
  effect on liver, 413
  effect on smooth muscle, 413
  effect on white adipose tissue, 413
  inactivation, 415
  insulin secretion, 413
  metabolic actions, 413
Catechol-O-methyl transferase, 343
  in catecholamine inactivation, 415

Cathepsin, uterine changes during pregnancy, 371
CDP-choline, formation, 237
Cell mediated immunity, 301
Cell surfaces, differentiation, 392
Cells, protodifferentiated, 391
Cellulose, structure, 24
Cephalin, structure, 30
Chemiosmotic theory, oxidative phosphorylation, 143
Chenodeoxycholic acid, 243
Chloramphenicol, action, 182
  structure, 182
Chloride shift, red cells, 283
Cholesterol, atherosclerosis, 241
  bile acid formation, 241
  formation, 239
  in membranes, 32
  structure, 32
Cholesterol synthesis, control, 241
Cholic acid, 243
Choline, deficiency, 235
  phosphorylation, 237
  seminal plasma, 366
  uptake into nerve endings, 340
Choline acetylase, 341
  concentration in human brain, 344
Chondroitin sulphate, structure, 23, 270
Chorionic gonadotrophin, 369
Chromaffin granules, 410
Chromatography, 59
Chromogranins, 410
Chylomicra, 249
  composition, 165
  formation, 164
Chymotrypsin, specificity, 65
α-Chymotrypsin, structure, 71
Chymotrypsinogen, 160
Citrate, acetyl-CoA carboxylase, 253
  concentration increase during fat oxidation, 435
  exit from mitochondria, 255
  formation, 125
  inhibitor of phosphofructokinase, 434
Citrate cleavage enzyme, 255
  increase at weaning, 377
  perinatal increase, 375
  malate/pyruvate cycle, 256
Citrate synthetase, 125
  activity in muscle, 149
  malate/pyruvate cycle, 256
Citric acid, seminal plasma, 364
Citric acid cycle, 119
  inhibitors, 151

rate of operation, 129
subcellular location, 129
summary of functions, 130
Citrulline, formation, 221
Clearing factor lipase, 249, 425
  variation with calorie intake, 466
Cocarboxylase, 123
Codon, 176
Coenzyme A, as allosteric effector, 200
  α-oxoglutarate dehydrogenase, 126
  structure, 124
Coenzyme Q/CoQ2H, $E^{\circ\prime}$, 49
Coenzymes, 109
Collagen, 264
  aging, 267
  amino acid composition, 266
  biosynthesis, 266
  cordae tendinae content, 265
  formation, 265
  structure, 67
  synthesis, 264
  tendon content, 265
  uterine changes during pregnancy, 371
Complement, 299
Complementary surfaces, 73
Condensation reactions, 16
Condensing enzyme, 125
Cone–rod transition time, 358, 359
Conformational change, 73, 83, 86, 87, 98
Conjugation, detoxication, 227
Connective tissue, changes in senescence, 395
Control, carrier molecules, 433
  enzyme activity by small molecules, 433
  enzyme amount, 442
  glycolysis in liver, 435
  glycolysis in muscle, 434
  phosphorylation and dephosphorylation of enzymes, 472
  substrate flow by small molecules, 433
Cooperative interactions, 73, 81
Coproporphyrin, formation, 278
Coproporphyrin II, structure, 278
Cori cycle, 199, 455
Cornea, collagen content, 265
Corticosteroids, adipose tissue, 257
  inactivation, 409
  physiological function, 405
  synthesis from cholesterol, 404
Cortisol, effect on enzymes, 408
Coupling factors, 142
Covalent bonds, 75
Creatine, biosynthesis, 102
  structure, 102
Creatine phosphate, $\Delta G^{\circ\prime}$ hydrolysis, 44

structure, 102
Creatine phosphokinase, 102
  in muscular dystrophy, 291
Creatinine, formation, 224
α-Crystallin, 354
β-Crystallin, 354
γ-Crystallin, 354
CTP synthetase, 186
Cycles between tissues, 455
Cycloheximide, action, 183
  structure, 183
Cysteine, 62
  S–S bridges, 25
L-Cysteine, structure, 26
L-Cystine, 27
Cystinuria, 385
Cytochrome P_{450}, hydroxylation, 225
  mixed function oxidase, 404
Cytochromes, $E^{\circ\prime}$, 49
  respiratory, chain, 135

Debranching enzyme, 106
  Forbes' disease, 207
Dehydrosphingosine, 238
Delayed hypersensitivity, 301
Denaturation, of proteins, 76
Deoxycholic acid, 243
Deoxycytidine monophosphate deaminase, 186
Deoxyribonucleotides, formation, 188
Dermis, 264
Desmosine, formula, 268
Detoxication, 225
Dextran, gel filtration, 59
Diabetes, difference from starvation, 469
  fatty liver, 235
Diabetic cataract, 354
Diazooxo norleucine, 191
Dibutyryl cAMP, 346
Dicoumarol, as uncoupler, 145
  inhibitor of blood clotting, 293
Diet, composition, 161
Dietary fat, composition, 163
Diethylaminoethylcellulose, 59
Differentiation, and cell surfaces, 392
Digestion, of food, 159
  of lipids, 163
Digestive enzymes, activation of, 160
Diglyceride, triglyceride formation, 252
Diglyceride transacylase, 252
Dihydroorotase, 186
Dihydroorotic acid dehydrogenase, 186
Dihydroxyacetone phosphate, condensation with glyceraldehyde-3-phosphate, 16
  formation, 108

Dihydroxyacetone phosphate—*continued*
  from fructose-6-phosphate, 204
  reduction, 252
3,4-Dihydroxymandelic aldehyde, structure, 341
2,4-Dinitrophenol, uncoupler, 141
Dipalmityl lecithin, lung surfactant, 309
Dipeptidase, 160
1,3-Diphosphoglyceric acid, formation, 110
  $\Delta G^{o\prime}$ hydrolysis, 44
2,3-Diphosphoglyceric acid, 112
Dipoles, 15
Disaccharides, 22
Distal tubule, lack of carbonic anhydrase, 317
Disulphiram, 471
Diurnal variations in enzymes, 448
Diuretics, 320
DNA, base pairing, 169
  biosynthesis, 168
  double helix, 36
  hydrogen bonding, 169
  repair, 171
  replication, 170
  sense strand, 177
  structure, 168
DNA-dependent RNA polymerase, effect of tri-iodothyronine, 420
  messenger RNA synthesis, 176
DNA polymerase, function, 170
Donnan equilibrium, 334
Dopadecarboxylase, catecholamine synthesis, 411
Dopamine, concentration in human brain, 344
Dopamine-β-hydroxylase, catecholamine synthesis, 411
  in neurones, 341
  location in storage granules, 410
Drug metabolism, intestinal conjugation reactions, 167
  intestinal hydrolysis reactions, 167
Dulcitol, 354

Eicosa-5,8,11,14-tetraynoic acid, inhibition of prostaglandin synthesis, 342
Elastin, 267
  amino acid composition, 267
  copper deficiency, 268
  synthesis, 264
  uterine changes during pregnancy, 371
Electrode potential, 49
Electron delocalization, 15
  heterocyclic compounds, 36
Electron transport, $\Delta G^{o\prime}$, 51
  inhibitors, 145

Electrophoresis, 60
Elongation factor, 180
Embryo, changes in enzyme activity, 373
  substrates utilized, 373
Endocrine tissues, 400
Endocytosis, of thyroglobulin, 417
Endopeptidases, 65
Endothelial cells, of lungs, 307
Energy consumption, renal amino-acid reabsorption, 316
Energy metabolism, changes at birth, 375
Energy production, regulation, 432
Energy reserves, in tissues, 454
Enolase, 112
Enoyl hydrase, 232
Entropy, 38
  and free energy, 39
  of water and protein conformation, 67, 72
Enzyme activity, control by small molecules, 433
  diurnal variations, 448
  effect of pH, 77
  factors changing, 445
Enzyme changes, during development, 447
  in foetal muscle, 448
  in kidney, 448
Enzyme quantity, control of, 442
  role of diet, 445
  role of hormones, 445
Enzymes, catalytic role, 45
  control by phosphorylation, 437
  effect of pH, 77
  effect of temperature, 75
  inhibition, 81
  kinetics, 77
  mechanism of action, 83
  organ specific, 390
  specificity, 74
Enzyme–substrate complex, 79
Epidermis, 264
Equilibria, chemical, 39
Equilibrium constant, and $\Delta G^{o\prime}$, 41
Equilibrium potential, 334
Error catastrophe theory, 395
Erythrose-4-phosphate, formation, 209
Eserine, 341
Essential amino acids, biochemical role, 444
  tissue distribution of transaminases, 447
Essential fatty acids, biochemical role, 444
  deficiency and atherosclerosis, 241
Esterase, 74
Ethane, structure, 11
Ethanol, contribution to daily calories, 470
  formation, 116

livers potential for oxidation, 471
 oxidation, 74
Ethanolamine, phosphorylation, 237
Ethionine, 184
Ethylene, structure, 13
Ethylene diamine tetraacetic acid, 99
Excitation–contraction coupling, 99
Exercise, energy expenditure, 113
Exocytosis, 412
 insulin secretion, 422
 noradrenaline secretion, 342
Exopeptidases, 64
Eye, 353

FAD, fatty acid oxidation, 231
 succinic dehydrogenase, 133
Faecal fat, amount, 164
Faeces, composition, 465
Farnesyl pyrophosphate, 239
Fat, content of foodstuffs, 466
 metabolism, 230
 sink for excess calories, 465
Fat cells, ultrastructure, 250
Fatty acids, absorption of short chain, 163
 activation, 163
 active transport, 163
 desaturation, 253
 elongation, 253
 mobilization, 257
 oxidation, 230
 oxidation, energy yield, 233
 oxidation in brown adipose tissue, 259
 saturated, unsaturated, 30
 solubility, 29
 structure, 29
 synthesis, 253
 water solubility, 163
Fatty acid synthetase, 253
Fatty acid/triglyceride cycle, 455
Fatty acyl-CoA synthetase, 230
Fatty livers, 235
Favism, 280
Fenestrated collar, sarcoplasmic reticulum, 99
Ferritin, 295
Fertility control, 378
Fibrin, 292
Fibrinogen, 292
Fibrinolysin, 293
Fibroblasts, 264
Fluoroacetate, 144
Fluoride, inhibition of enolase, 113
1-Fluoro-2,4-dinitrobenzene, 63
p-Fluorophenylalanine, 184
FMN, structure, 128

Foetus, 373
 energy metabolism, 374
Folic acid, 188
Follicle-stimulating hormone, 368
 releasing hormone, 368, 369
Foodstuffs, calorific values, 161
 equivalent to 2000 kcals, 466
Forbes' disease, 207
Formaldehyde, inhibition of oxidative phosphorylation in retina, 359
Free energy, entropy, 39
Free energy change, effect of concentration, 40
Free fatty acids, blood level in starvation, 467
 foetal blood, 374, 375
Fructokinase, 204
 activity in liver, 470
 fructosuria, 206
Fructose, metabolism, 470, 204
 lens, 354
 seminal plasma, 367
D-Fructose, structure, 22
Fructose-1,6-diphosphate, activator of liver pyruvate kinase, 435
 formation, 16, 108
Fructose-1,6-diphosphatase, activity in muscle, 149
 activity in tissues, 453
 gluconeogenesis, 201
 inhibition by AMP, 436
 perinatal increase, 375
 role in white muscle, 150
Fructose-1-phosphate, cleavage, 204
 formation, 204
 $\Delta G^{\circ\prime}$ hydrolysis, 44
Fructose-6-phosphate, formation, 108
 PFK, 108
 sedoheptulose phosphate, 209
Fructosuria, 206
Fumarase, 128
 specificity, 75
Fumaric acid, 14
 conversion to malic acid, 16, 128
 formation, 127
Functional groups of organic compounds, 18

Galactokinase, 205
 perinatal increase, 375
Galactosaemia, 206
Galactose, metabolism, 204
D-Galactose, structure, 22
Galactose-1-phosphate uridyl transferase, 205
Galactosaemia, 206, 354
Gall stones, 241
Ganglioside, structure, 388
Gas gangrene, 238

Gel filtration, 59
Genetic code, 176
  degeneracy, 176
Genetic defects, 383
Glial cells, 332
β-Globulin, impaired synthesis, 235
Glucagon, 425
  adipose tissue, 257
  phosphopyruvate carboxykinase synthesis, 447
Glucocorticoids, 404
  function, 407
Glucokinase, 199
  action, 107
  increase at weaning, 377
  K_m, 107
Gluconeogenesis, 199
  control by acidity in kidney, 319
  kidney ammonia production, 319
  prolonged starvation, 469
Glucose, degradation and energy transformation, 132
  in foetal blood, 374
  levels in blood, 199
  metabolism of mammary gland, 380
  synthesis from lactate, 199
  uptake by adipose tissue, 251
D-Glucose, structure, 20
Glucose-1,6-diphosphate, 106
Glucose-6-phosphatase, 201
  activity in tissues, 453
  blood glucose level, 455
  perinatal increase, 375
  von Gierke's disease, 207
Glucose-1-phosphate, equilibrium with glucose-6-phosphate, 41
  formation from glycogen, 106
  $\Delta G^{\circ\prime}$ hydrolysis, 44
Glucose-6-phosphate, equilibrium with glucose-1-phosphate, 41
  $\Delta G^{\circ\prime}$ of formation, 44
  $\Delta G^{\circ\prime}$ hydrolysis, 44
  isomerization, 106
  routes of formation, 437
  routes of utilization in liver, 437
Glucose-6-phosphate dehydrogenase, 209
  activity in tissues, 453
  red cell deficiency, 280
Glucose transport in kidney, 313
Glucuronic acid, formation, 211
Glutamate–oxaloacetate transaminase in myocardial infarction, 291
Glutamic acid, in blood, 219
L-Glutamic acid, structure, 27

Glutamic dehydrogenase, in glutamate oxidation, 215
  removal of NH_3, 218
Glutaminase, control by glutamate, 319
  in urine neutralization, 318
  NH_3 formation, 219
Glutamine, in blood, 219
Glutamine synthetase, brain, 327
  removal of NH_3, 218
γ-Glutamyl cycle, amino acid transport, 315
γ-Glutamyl transpeptidase, 316
Glutathione, kidney amino acid reabsorption, 316
  lens, 354
  red cells, 280
  structure, 280
Glutathione peroxidase, different activities in males and females, 364
Glutathione reductase, 354
Glyceraldehyde, absolute configurations, 12
  from fructose-1-phosphate, 204
Glyceraldehyde-3-phosphate, condensation with DHAP, 16
  formation, 108
  oxidation, 110
Glyceraldehyde-3-phosphate dehydrogenase, mechanism of action, 111
Glycerol kinase, absence in adipose tissue, 252
  obese mice, 257
  perinatal increase, 377
Glycerol oxidation, energy yield, 233
Glycerol-1-phosphate, $\Delta G^{\circ\prime}$ hydrolysis, 44
L-α-Glycerophosphate, formation, 116
  formation from DHAP, 252
  triglyceride synthesis, 252
L-α-Glycerophosphate dehydrogenase, 252
  activities in muscle, 149
  distribution in brain, 325
  white muscle, 150
Glycerophosphatides, solubility, 30
  structure, 30
Glyceryl phosphoryl choline, in seminal plasma, 366
Glycine, 26
  conjugation, 227
  haem synthesis, 277
  titration curve, 57
Glycocyamine, 102
Glycogen, breakdown, 105
  foetus, 374
  liver and muscle store, 199
  storage diseases, 206
  structure, 23, 104

Glycogen synthetase, 202
 congenital absence, 208
 control of, 438
 non-phosphorylated a form, 438
 phosphorylated b form, 438
Glycogen synthetase phosphatase, control of activity, 439
Glycolysis, 104
 control in liver, 435
 control in muscle, 434
 energy balance, 114
 energy yield, 117
 universality, 116
Glycoproteins, cell surface, 392
GMP, synthetase, 191
Gonadotrophic releasing hormones, 369
Gout, 192
Ground substance, 269
Growth hormone, adipose tissue, 257
GTP, formation from succinyl-CoA, 127
 structure, 127
Guanase, 192
Gut, protein content, etc., 89

H-Antigen, structure, 302
Haem, in myoglobin, 71
 synthesis, 277
Haemoglobin, oxygen dissociation curve, 282
 species, 281
 structure, 72
Haemoglobin S, biochemical difference from Haemoglobin A, 386
 properties, 386
Haemoglobins, fingerprints of, 388
Haemopoiesis, 277
Haemosiderin, 296
Halophenols, as uncouplers, 145
Heart, lactate consumption, 117
 oxidation of ketone bodies, 233
 protein content, etc., 89
Heart muscle, ultrastructure, 120
Heat production, brown adipose tissue, 258
α-Helix, 70, 71
 structure, 66
Hemicholinium, 341
Henderson–Hasselbalch equation, 56
Heparin, inhibitor of blood clotting, 293
 lung content, 308
 synthesis by mast cells, 269
Hers' disease, 208
Heterocyclic compounds, chemistry, 34
Heterotropic effectors, 73, 83, 87, 98,
 effect on enzyme activity, 80
Hexamethonium, 345

Hexobarbital oxidase, different activities in males and females, 364
Hexokinase, action, 107
 activity in muscle, 149
 activity with fructose, 204
 K_m, 107
 lactose formation, 381
Hexosaminidase A, lack in Tay–Sachs disease, 389
Hippuric acid, 227
 liver function test, 227
Histaminase, in pregnancy, 291
Histamine, formation from histidine, 269
 in mast cells, 269
 structure, 269
L-Histidine, structure, 27
Histocompatibility antigens, 302
HMG-CoA Synthetase, perinatal increase, 377
Hodgkin's equation, 334
Homogentisic acid, excretion in alcaptonuria, 385
Homogentisic acid oxidase, lack in alcaptonuria, 385
Homotropic effect, 81, 86
Hormones, 401
 activation of adenyl cyclase, 401
 general description, 401
Hormone sensitive lipase, blood fatty acid levels, 455
 control of, 442
Hyaluronic acid, amount in skin, 269
 in arthritic subject, 270
 structure, 24, 269
Hyaluronidase, in sperm, 367
Hydrocarbons, solubility, 53
Hydrochloric acid, gastric, 160
Hydrogen bonds, 52, 67, 70, 75
Hydrogen electrode, 49
Hydrogen/H$^+$, $E^{\circ\prime}$, 49
Hydrophilic groups, 53
Hydrophobic bonds, 53, 71, 75
Hydrophobic groups, 53
β-Hydroxyacyl-CoA dehydrogenase, 231
 perinatal increase, 377
Hydroxyapatite, in bone, 272
D-(−)-β-Hydroxybutyrate, 233
β-Hydroxybutyrate, effect on blood pH, 317
β-Hydroxybutyrate dehydrogenase, 233
L-(+)-β-Hydroxybutyryl-CoA, 234
Hydroxyethyl thiamine pyrophosphate, 123
Hydroxyindole-O-methyltransferase, 347
Hydroxylations, detoxication, 225
β-Hydroxy-β-methylglutaryl-CoA, cholesterol formation, 239

β-Hydroxy-β-methylglutaryl-CoA—*continued*
  formation, 233
17-α-Hydroxyprogesterone acetate, as contraceptive, 378
L-Hydroxyproline, 67
  structure, 28
5-Hydroxytryptamine, concentration in human brain, 344
5-Hydroxytryptamine N-acetyltransferase, 347
5-Hydroxytryptophan hydroxylase, 347
Hyperthyroidism, 419
Hypoglyccmia, after ethanol ingestion, 472
Hypophosphatasia, 272
Hypothyroidism, 420
Hypoxanthine, 192
Hypoxanthine–guanine phosphoribosyl transferase, 191
Human, weight, etc. of tissues, 89

Ice, structure, 53
Imiprimine, inhibitor of noradrenaline uptake, 343
Immunoglobulins, heavy chains, 297
  light chains, 297
  structure, 297
Immunological tests, protein homogeneity, 60
IMP, formation, 189
IMP-cyclohydrolase, 190
IMP-hydrogenase, 191
Inborn errors of metabolism, 193, 206
Induced fit, 86
Inert gases, electronic structure, 9
Infectious hepatitis, 60
Inhibitors, competitive, 82
  enzyme, 82
  non-competitive, 82
Initiation complex, 178
Initiation factors, 178
Initiator codon, 178
Insulin, 421
  activators of secretion, 422
  coma, 423
  deficiency, 425
  effects on adipose tissue, 251, 257, 424
  effect on cell permeability, 423
  effect on clearing factor lipase, 425
  effect on enzymes, 424
  effect on glucose uptake, 423
  effect on liver enzymes, 424
  effect on metabolic pathways, 424
  glucose carrier mechanism, 423
  inactivation, 425
  inhibitors of secretion, 422
  metabolic effects, 422

NADPH production, 255
  primary structure, 64
Insulin secretion, effect of catecholamines, 413
Interferon, 301
Internal environment, magnitude of changes in, 431
Intestinal secretions, contributions to gut protein, 159
Intestine, 158
  calcium absorption, 166
  chloride contents, 166
  drug metabolism, 166
  epithelial replacement, 159
  sodium contents, 166
  water contents, 166
  weight of epithelial cells, 159
Invertase, 160
Iodide peroxidase, 416
Iodoacetate, inhibition of glyceraldehyde phosphate dehydrogenase, 111
Ion transport, 336
  ATP-ase, 286, 339
  brain, 333
  mechanism, 287
  red cells, 283
  sarcoplasmic reticulum, 99
Iproniazid, 341
  inhibitor of MAO, 415
Iron, absorption, 295
  metabolism, 294
Isocitric acid, 125
Isocitric dehydrogenase, 125
Isodesmósine, formula, 268
Isoelectric point, 59
Isoelectric precipitation, 59
Isoenzymes, 60
L-Isoleucine, structure, 26
Isomerase, 108
Isomerization, geometrical, 14
Δ^3-Isopentenylpyrophosphate, 239
Isozymes, 60

Keratin, amino acid composition, 264
  synthesis, 264
Keratinization, 264
Ketone bodies, blood level in starvation, 467
  excretion on high fat diet, 466
  foetal blood, 374, 375
  formation, 233
  insulin release, 234
  metabolic significance, 234
  use by tissues in prolonged starvation, 469
β-Ketothiolase, 231
Kidneys, 313

collagen content, 265
mucopolysaccharide content, 272
protein content, etc., 89
Kinase, 191
Kinetics, enzyme, 77
K_m, 78

α-Lactalbumin, stimulation of lactose production, 380
Lactase, 160
Lactate, oxidation to pyruvate, 119
  production by muscle, 117
  use by heart, 117
Lactate cycle, 455
Lactate/pyruvate, $E°'$, 49
Lactation, hormonal control, 381
Lactic acid, disposal, 117
  formation, 114
Lactic dehydrogenase, 113, 119
  in glucose synthesis, 200
  in myocardial infarction, 291
  isoenzyme ratio in, 149
  isoenzymes in muscle, 148
  isoenzymes in plasma, 60
Lactose, biosynthesis, 380
  formation, 381
  structure, 22
Lactose synthetase, 380
Latent heat, fusion and evaporation, 52
Lathyrism, 268
Lecithin, structure, 30
Lecithinase, 160, 238
Lecithin–cholesterol acyl transferase, 251
Lens, 353
  glucose usage, 353
L-Leucine, structure, 26
Leucocytes, O_2 consumption, 289
Life, origins, 6
Life span of humans, 393
Lipase, 160
  hormone sensitive, 257
  triglyceride, 257
Lipid bilayers, structure, 30
Lipid metabolism of foetus, 374
  of lung, 309
Lipids, chemistry, 28
Lipoic acid, α-oxoglutarate dehydrogenase, 126
  structure, 123
Lipo-fuschin granules, in aging tissue, 396
Lipolysis, adipose tissue, 257
Lipoprotein lipase, 230, 249, 425
  of lung, 309
  in pregnancy, 251
  in starvation, 251

Lipoproteins, high density, 251
  low density, 251
  very low density, 230, 249
Lipoyl acetyl transferase, 73, 125
Lipoyl dehydrogenase, 73, 125
Lithocholic acid, 243
Liver, collagen content, 265
  glucostatic function, 199
  glutamate oxidation, 217
  metabolism, 197
  mucopolysaccharide content, 272
  protein content, etc., 89
  ultrastructure, 244
Liver function test, 228
Liver phosphorylase, control of, 441
Lungs, 307
  collagen content, 265
  dimensions, 301
  lipid metabolism, 309
  oxygen consumption, 307
  protein content, etc., 89
  scavenging functions, 307
  mucopolysaccharide content, 272
Lung surfactant, 309
Luteinizing hormone, 368
Luteinizing hormone releasing hormone, 368, 369
Lymphocytes, B, 301
  T, 301
Lymphoid tissue, protein content, etc., 89
Lymphotoxins, 301
Lysine, binding to lipoate, 123
L-Lysine, structure, 27
Lysolecithin, 238
Lysosomes, 417, 419
  Pompe's disease, 207
Lysozyme, covalent structure, 62
  3D structure, 83
  in alveolar macrophages, 308
  mechanism of action, 83

M-bridges, 95
Malate, $\Delta G°'$ oxidation to oxaloacetate, 50
Malate/oxaloacetate, $E°'$, 49
Maleic acid, 14
Malic acid, formation, 128
  formation from fumaric acid, 16
  oxidation, 128
Malic dehydrogenase, 128
  adipose tissue, 255
  in myocardial infarction, 291
  malate/pyruvate cycle, 256
Malic enzyme, adipose tissue, 255
  control by thyroxine, 448
  malate/pyruvate cycle, 256

Malonate, 144
Malonic acid, 82
Malonyl-CoA, formation, 253
Maltase, 160
   Pompe's disease, 207
Mammary gland, enzyme changes during lactation, 379
   glutamate oxidation, 217
   metabolism, 379
Marrow, protein content, etc., 89
Mast cells, 308
   of lungs, 307
McArdle's disease, 208
Melatonin, 346
Membrane permeability, 431
Membrane potentials in brain, 333
Membranes, fluid mosaic structure, 32
   proteins, 32
   unsaturated fatty acids, 32
H-Meromyosin, 95
L-Meromyosin, 95
Messenger RNA, 175
Metabolic control, 431
Metaraminol, structure, 342
Methaemoglobin, 280
Methane, structure, 10, 11
Methanol, metabolism by retina, 359
Methionine, initiator for protein synthesis, 180
L-Methionine, structure, 27
Methionyl-t-RNA, 178
Methotrexate, 188, 189
Methyl group, donor, 237
Methylmaleimide, 99
Methylmalonyl-CoA from propionate, 203
Methylmalonyl-CoA isomerase, 204
Mevalonic acid, 239
Michaelis constant, 78
Microtubules, in axons, 344
Milk fat, biosynthesis, 379
Milk protein, biosynthesis, 379
Mineralocorticoids, 404
Mitochondria, 119
   amino acid transport in, 133
   dicarboxylic acid transport, 133
   fatty acid oxidation, 231
   fatty acid transport, 134
   muscle, 95
   nucleotide transport, 134
   phosphate transport, 133
   redox system, 49
   respiratory chain, 135
   transaminases, 215
   tricarboxylic acid transport, 133
Mitochondrial function, inhibitors, 143

Mitochondrial permeability, 132
Molecular sieving, 59
Monoamine oxidase, 341
   in catecholamine inactivation, 415
Monosaccharides, structures, 20
Mucopolysaccharides, definition, 269
   linkage with protein, 270
   synthesis, 264, 269
      by chondrocytes, 269
      by fibroblasts, 269
Multienzyme complex, 123
Multienzyme system, 73, 253
Multiple myeloma, immunoglobulin production, 299
Muscle, contraction, 93, 100
   cross innervation, 151
   effects of denervation, 151
   enzyme activities of red, 149
   enzyme activities of white, 149
   lactate dehydrogenase isoenzymes, 148
   metabolism, 148
   red, 148
   relaxation, 93
   white, 148
Muscular dystrophy, 115
Myocardial infarction, 60
Myofibres, 95
Myofibrils, 95
Myoglobin, muscle content, 148
   structure, 70
Myokinase, amplifier for nucleotide changes, 434
Myosin, ATP-ase, 95, 98
   in thick filaments, 95
   proteolysis, 95
   red muscles, 151
   structure, 95
   white muscles, 151

NAD^+, fatty acid oxidation, 231
   reduction by pyruvic dehydrogenase, 122
   structure, 109
$NADH/NAD^+$, $E^{\circ\prime}$, 49
$NADP^+$, 125
   reduction for fat synthesis, 255
   reduction in pentose shunt, 209
   hydroxylation, 242
   reoxidation, 113
$NADP^+$-malate dehydrogenase, increase at weaning, 377
Nernst equation, 333
Nervous system, 324
Neuromuscular junction, 341
Neurone, motor, 99

Neurotransmitters, 340
Nicotinic acid, brain, 331
Night blindness, 359
Nitrate, inhibitor of iodide uptake, 417
Nitrobenzene, 227
Nitrogen balance, 214
Nitrogen excretion, 218
  in starvation, 467
Nitrophenols, uncouplers, 145
Non-haem iron, mixed function oxidase, 404
Non-polar groups, 53
Noradrenaline, adipose tissue, 257
  brown adipose tissue, 258
  concentration in human brain, 344
  neuronal, 341
  storage particles, 341
  synthesis from tyrosine, 411
19-Nortestosterone, as contraceptive, 378
Nucleic acids during embryonal development, 373
Nucleoside diphosphatase, 186
Nucleoside diphosphate kinase, 127, 186
Nucleoside monophosphate kinase, 186
5′ Nucleotidase, 192

Obesity, 258, 465
Oestrogen, action, 371
  synthesis by placenta, 371
Oligomers, 86, 71
Oligomycin, coupling factors, 142
  inhibitor of transphosphorylation, 145
Oocyte, development, 371
Opsin, 357
Optical rotation, 12
Orbitals, atomic, 9
  molecular, 10
Organic mercurials as diuretics, 320
Ornithine, formation, 221
Ornithine transaminase, activities in males and females, 364
Ornithine transcarbamylase, 221
Orotic acid decarboxylase, 186
Orotidylic acid pyrophosphorylase, 186
Ouabain, inhibition of active transport, 285
Ovarian development, 369
Oxaloacetate, formation of citrate, 125
Oxaloacetic acid, formation, 128
  from propionate, 203
  inhibition of succinic dehydrogenase, 129
  inhibitor of TCA cycle, 144
Oxidase, mixed function, 225, 254, 404
Oxidation, 47
  detoxication, 227
β-Oxidation of fatty acids, 230

Oxidative phosphorylation, 135, 139
  chemiosmotic theory, 143
  energetics, 49
  mechanism, 142
3-Oxoacid-CoA transferase, 233
α-Oxoglutarate, 125
α-Oxoglutarate dehydrogenase, 126
5-Oxoproline, 316
Oxygen, carriage by haemoglobin, 281
  consumption of human tissues, 89
  uptake and phosphorylation, 140
Oxygen/H_2O, $E^{\circ\prime}$, 49
Oxygenase, mixed function, 225, 254, 404
Oxyhaemoglobin, 281

P site, attachment of m-RNA, 178
Pancreatic hormones, 421
Pantotheine, structure, 124
Parapyruvate, 144
Pentose phosphate pathway, 208
  adrenal cortex, 405
  decrease in cataracts, 355
  fat synthesis, 255
Pentosuria, 212
Pepsin, specificity, 65
Pepsinogen, 160
Peptide bond, 25, 62
  formation, 178
  geometry, 65
Peptidyl site, attachment of m-RNA, 178
Peptidyl transferase, 180
Perinatal biochemistry, 373
Perinatal enzyme changes, 375
Periodic table, of elements, 9
pH, 54
  drug absorption, 58
  protein conformation, 58
  regulation by kidney, 316
Phagocytosis, 288
Phentolamine, 345
L-phenylalanine, structure, 28
Phenylalanine hydroxylase, deficiency in phenylketonuria, 383
Phenylethanolamine N-methyl transferase, catecholamine synthesis, 411
Phenylketonuria, 383
Phenylpyruvic acid, excretion in phenylketonuria, 383
Phosphatase, 160
  dephosphorylating muscle phosphorylase a, 441
Phosphatases, dephosphorylating enzymes, 437
Phosphate and PFK, 108
Phosphatidic acid, formation, 240

Phosphatidic acid—*continued*
  triglyceride formation, 252
Phosphatidic acid phosphatase, 252
Phosphatidyl choline, formation, 240
  structure, 30
Phosphatidyl ethanolamine, formation, 240
Phosphatidyl inositol, formation, 240
Phosphatidyl serine, formation, 238, 240
  structure, 30
3′-Phosphoadenosine-5′-phosphosulphate,
  structure, 228
Phosphodiesterase, decay of hyperpolarization,
  346
  degrading c-AMP, 403
Phosphoenolpyruvate, $\Delta G^{\circ\prime}$ hydrolysis, 44
Phosphoenolpyruvate carboxykinase, gluconeogenesis, 201
  oxaloacetate formation, 201
  white muscle, 151
Phosphoenolpyruvic acid, formation, 112
Phosphofructokinase, 204
  action, 108
  activity in muscle, 149
  control, 108, 434
Phosphoglucomutase, 106
  activity in tissues, 453
  lactose formation, 381
  muscular dystrophy, 116
6-Phosphogluconate dehydrogenase, 210
6-Phosphogluconic acid, formation, 209
2-Phosphoglyceric acid, formation, 112
3-Phosphoglyceric acid, formation, 112
  $\Delta G^{\circ\prime}$ of formation, 44
Phosphoglyceric acid kinase, 111
Phosphoglycerol transacylase, 252
Phosphoglyceromutase, 112
Phosphohexoisomerase, 108
  activity in tissues, 453
Phosphoinositide, structure, 30
Phospholipids, biosynthesis, 236
  breakdown by phospholipase-A, 163
  breakdown by β-phospholipase, 163
  solubility, 30
  structure, 30
  turnover, 236
  turnover in brain, 331
Phosphoprotein phosphatase, decay of hyperpolarization, 346
Phosphopyruvate carboxykinase, perinatal increase, 377
  switch on at birth, 447
Phosphoribose pyrophosphokinase, 190
Phosphoribosyl aminoimidazolecarboxamide formyltransferase, 190

Phosphoribosyl aminoimidazole carboxylase, 190
Phosphoribosyl aminoimidazole succinocarboxamide synthetase, 190
Phosphoribosyl aminoimidazole synthetase, 190
Phosphoribosyl formylglycineamidine synthetase, 190
Phosphoribosyl glycineamide formyltransferase, 190
Phosphoribosyl glycineamide synthetase, 190
Phosphoribosyl pyrophosphate, formation, 187
Phosphoribosyl pyrophosphate synthetase, 187
Phosphorylase action, 105
  activity in muscle, 149
  control of, 439
  dephosphorylated b form, 439
  Hers' disease, 208
  McArdle's disease, 208
  muscular dystrophy, 116
  phosphorylated a form, 439
Phosphorylase b kinase, control of, 440
  dephosphorylated b form, 440
  phosphorylated a form, 440
Phosphorylase b kinase b, activation by Ca^{2+}, 440
Phosphorylation, substrate level, 112, 127
Phosphorylation sites, 141
Phosphorylcholine, formation, 237
  in seminal plasma, 366
Phosphoryl group transfer potential, 43
Phosphoserine phosphatase, in development, 391
Photosynthesis, 47
Picramic acid, 227
Picric acid, 227
Pineal gland, 346
pK, 56
Plasma, 290
  mucopolysaccharide content, 272
  potassium concentration, 284
  sodium concentration, 284
Plasma acidity, adjustment by kidney, 317
Plasma alkalinity, adjustment by kidney, 317
Plasma cells, antibody production, 299
Plasma proteins, 290
  in disease, 291
Platelets, in clot retraction, 293
Pleated sheet, 67
  in α-chymotrypsin, 71
P/O ratio, values of, 140
Polar groups, 53
Polyacrilamide, gel electrophoresis, 60
Polypeptide, 60
Polypeptides, formation from amino acids, 25

Polysomes, 177
Polyuridylic acid, synthetic m-RNA, 176
Pompe's disease, 207
Porphobilinogen, haem synthesis, 277
    structure, 278
Porphyrias, 278
Porphyrin, cytochromes, 137
Prekeratin, 264
Primaquine, red cell lysis, 280
    sensitivity, 280
Profibrinolysin, activation in lung, 308
Progesterone, action, 371
    synthesis by placenta, 369
Proinsulin, 421
    conversion to insulin, 421
    structure, 421
Prolactin-release inhibiting hormone, 382
Proline, 67
L-Proline, structure, 28
Propionyl-CoA carboxylase, 204
Prostaglandins, biochemical effects, 366
    noradrenaline function, 342
    seminal plasma, 366
Prostaglandin, E_2, structure, 343
Prosthetic group, 127
Proteinases, selective attack of enzyme classes, 444
Proteinase, sperm, 367
Protein conformation, effect of pH, 58
Protein kinase, adipose tissue, 257
    c-AMP, 402
    control of, 439, 440
    nerve hyperpolarization, 346
Protein kinases, phosphorylating enzymes, 437
Protein–mucopolysaccharide complex, synthesis, 270
Proteins, amino acid sequence, 63
    ammonium sulphate precipitation, 59
    3D conformation, 65
    content of foodstuffs, 466
    content of human tissues, 89
    covalent structure, 62
    daily requirement, 214
    daily synthesis in tissues, 181
    homogeneity, 59
    hydrolytic cleavage, 64
    isolation, 59
    loss in starvation, 213
    structure, 60
    turnover, 181
Protein synthesis, 168
    basic concepts, 168
    inhibitors, 181
Protein turnover, 443

brain, 331
Prothrombin, 292
Protomers, 71, 86
Protoporphyrin IX, structure, 278
Purine bases, in DNA, 168
Purine biosynthesis, control, 192
    inhibitors, 189
Purine nucleoside phosphorylase, 192
Purines, biosynthesis, 189
    catabolism, 192
Puromycin, action, 182
    structure, 182
Pyridoxal, brain, 331
Pyridoxal phosphate, 214
    transamination, 36, 215
Pyridoxamine phosphate, 214
Pyridoxine, 214
Pyrimidine bases, biosynthesis, 185
    in DNA, 168
Pyrimidine biosynthesis, inhibitors, 188
Pyrophosphatase, 222, 237
Pyruvate, control of utilization in muscle, 435
Pyruvate dehydrogenase, control of, 441
Pyruvate kinase, increase at weaning, 377
    perinatal increase, 375
Pyruvic acid, conversion to acetyl-CoA, 123
Pyruvic carboxylase, adipose tissue, 255
    allosteric, control, 200
    control by acetyl-CoA, 436
    glucose formation, 200
    location, 200
    malate/pyruvate cycle, 256
Pyruvic dehydrogenase, 123
    activation by insulin, 258
    control by acetyl-CoA, 435
    malate/pyruvate cycle, 256
    structure, 73
Pyruvic kinase, 113
    forms in liver, 435

Racemases, 74
Raney, nickel, conversion of cysteine to alanine, 178
Reaction mechanisms, 15
Red cells, 277
    carbohydrate metabolism, 279
    cold storage, 283
    drug sensitivity, 279
    potassium concentration, 284
    sodium concentration, 284
Redox reactions, 47
Reduction, 47
    detoxication, 227
Regulatory subunits, 73, 98

Relaxation, muscle, 100
Renin, 320
Reproduction, 357
Reserpine, effect on noradrenaline storage, 341
Resonance stabilization, 15
Respiration, 47
  coupling to phosphorylation, 140
  energy yield, 117
Respiratory chain, 135
  components, 135
  energy transformations, 139
  flavins, 135
  functioning, 139
  sequence of components, 137
Respiratory chain phosphorylation, 139
Resting potential, in neurones, 333
Retina, 356
all trans-Retinal, 356
Δ^{11}-cis-Retinal, 357
Retinal in visual processes, 357
Retinal isomerase, 357
Rhesus factor, 302
Rhodopsin, 357
Ribonucleotide reductase, 186, 191
D-Ribose, structure, 22
Ribose-5-phosphate, formation, 209
Ribulose-5-phosphate, formation, 209
Rigor mortis, 100
RNA, function, 172
  ribosomal, 175
  structure, 169, 172
m-RNA, 175
  codons, 176
  combination with ribosomes, 176
  life time, 175–6
  long-lived, 391
t-RNA, 173
  acylation, 45
  anticodon region, 175
  configuration, 173
  in developmental processes, 394
Rod threshold, 358, 359

Salicylic acid, conjugations, 228
Salt absorption, 165
Salting out, 59
Sarcolemma, 93
Sarcomere, 95
Sarcoplasmic reticulum, 95
  calcium pump, 99
  structure, 99
Secretion, hydrochloric acid, 162
Sedoheptulose-7-phosphate, formation, 209
Self-assembly, 71

Seminal fluid, 364
Senescence, 393
Sephadex, 59
L-Serine, structure, 26
Serum complement system, 299
Sex, biochemistry of, 363
  determination of, 363
Sex steroids, 404
Sexual maturity, enzyme changes at, 448
Sickle cell anaemia, 385
Silk, structure, 67
Skeletal muscle, collagen content, 265
  contraction, 93
  lactate production, 117
  mucopolysaccharide content, 272
  protein content, etc., 89
  relaxation, 93
  structure, 93
  ultrastructure, 121
Skin, 263
  collagen content, 265
  fatty acids, 271
  general biochemistry, 270
  glucose content, 270
  glucose utilization, 271
  glycolysis, 270
  mucopolysaccharide content, 272
  oxygen consumption, 270
  protein content, etc., 89
  synthesis of lipids, 271
  synthesis of steroids, 271
Smooth muscle, 151
  structure, 154
Snake venom, 238
Sodium citrate ingestion, effect on blood pH, 317
Sodium glutamate ingestion, effect on blood pH, 317
Sodium pump, properties, 336
Sodium transport, kidney, 313
Solubility, carbon compounds, 18
Sorbitol dehydrogenase, in seminal vesicles, 367
  in lens, 354
Specific immune responses, 297
Sperm, biochemical changes during maturation, 364
  capacitation, 368
  metabolism, 364
  motility, 367
  structure, 364
Spermatogenesis, hormonal control, 368
Spermine, seminal plasma, 366
  structure, 367
Sphingolipids, solubility, 30
  structure, 30

Sphingomyelin, 239
　formation, 240
　structure, 30
Sphingosine, 238
Spironolactone, structure, 408
Spleen, protein content, etc., 89
Squalene, cholesterol formation, 239
　skin lipids, 271
Standard electrode potential, 49
　values, 49
Standard electrode potential change, and $\Delta G^{\circ\prime}$, 51
Standard free energy change, 40
　and $\Delta E^{\circ\prime}$, 51
　equilibrium constant, 41
Starch, gel electrophoresis, 60
Starvation, 465
　biochemical effects, 467
　changes in blood substrates, 467
　enzyme changes in, 446
　substrate use in, 446
Steroid hormone contraceptives, 378
Steroid hormones, biosynthesis, 405
Steroid metabolism, skin, 271
Stomach contents, HCl, 161
　pH, 161
Streptomycin, action, 183
　structure, 183
Structural tissues, 263
Subfraction 1, myosin, 97
Subfraction 2, myosin, 96
Substitution reactions, 15
Substrate, 74
Substrate flow, and metabolic state, 459
Substrate level phosphorylation, 112, 127
Succinate/fumarate, $E^{\circ\prime}$, 49
Succinic acid, conformations, 13
　oxidation, 127
Succinic dehydrogenase, 127
　inhibition, 82, 129
Succinic thiokinase, 127
Succinyl-CoA, activation of acetoacetate, 233
　formation, 126
　haem synthesis, 277
　methylmalonyl CoA, 203
Sucrase, 160
Sucrose, consumption, 470
　consumption in Britain, 470
　contribution to daily calories, 470
　structure, 22
Sulphonamides, acetylation, 229
Superior cervical ganglion, 345
Symmetry, 86
Synaptic vesicles, 340

Tangier disease, 257
Tay–Sachs disease, 389
Temperature, effect on enz
Tension, muscle, 101
Terminal cisternae, sarcoplasmic
Terminal differentiation, 390
Terminator codons, 180
Testosterone, functions, 369
Tetracyclines, action, 183
　structure, 183
Tetrahydrofolic acid, 188, 189
　methyl transfer, 238
Tetrodotoxin, 341
Theophylline, 345
Thermodynamics, 2nd Law, 38
Thiamine, brain content, 330
Thiamine pyrophosphate, 123, 209
　α-oxoglutarate dehydrogenase, 126
Thick filaments, muscle, 95
Thin filaments, muscle, 97
Thiocyanate, inhibitor of iodide uptake, 417
Thiokinase, 230, 251
Thiophorase, 233
Thioredoxin, 188
Thioredoxin reductase, 188
Thiouracil, inhibitor of iodide peroxidase, 417
　structure, 417
Thiourea, inhibitor of iodide peroxidase, 417
　structure, 417
L-Threonine, structure, 26
Thrombin, 292
Thromboplastins, 292
Thymidylate synthetase, 186
Thymus dependent cells, and delayed hypersensitivity, 301
Thymus independent lymphocytes, and antibody production, 301
Thyroglobulin, 416
　hydrolysis by lysosomal proteases, 417
Thyroid, 416
Thyroid hormones, effects, 420
　inactivation, 420
Thyroid stimulating hormone, 417
Thyroxine, 416
　as uncoupler, 145
Tissues, blood flow, 89
　individuality, 454
　interrelationships, 453
　O_2 consumption, 89
　protein content, 89
　weight, 89
Titration curve, glycine, 57
　weak acid vs. strong base, 56
Tolbutamide, structure, 422

Transaldolase, 209
Transamination, 214
Transaminidase, different activities in males and females, 364
Transferrin, 295
Transglucosylase, action, 106
Transhydrogenase, 126
   energy dependent, 256
   respiratory chain, 135
Transketolase, 209
Transverse tubules, skeletal muscle, 95, 99
Triad, sarcoplasmic reticulum, 99
Tricarboxylic acid cycle, final common pathway, 130
   inhibitors of, 144
   rate of operation, 129
   subcellular location, 129
   summary of functions, 130
Triglyceride, 240
   synthesis, 251
   uptake by adipose tissue, 249
Triglyceride lipase, activity in muscle, 149
Triglyceride oxidation, energy yield, 233
Triglycerides, breakdown by lipase, 163
   resynthesis by intestine, 164
   solubility, 29
   structure, 29
Triiodothyronine, 416
Triosephosphate dehydrogenase, 110
Triosephosphate isomerase, 109
Triphosphatase, 186
Tropocollagen, polymerization, 265
   structure, 67
Tropomyosin, 97
Troponin, 97, 100
Trypsin, specificity, 65
Trypsinogen, 160
L-Tryptophan, structure, 28
Tryptophan hydroxylase, 347
D-Tubocurare, 341
L-Tyrosine, structure, 28
Tyrosine hydroxylase, 343
   catecholamine synthesis, 411
   lack in albinism, 385

Ubiquinone, formula, 137
UDP-galactose, 205
UDP-galactose-4-epimerase, 205
   lactose formation, 381
UDP-galactose pyrophosphorylase, galactosaemia, 206
UDP-glucose, 205
   glycogen formation, 202
   oxidation, 211
   structure, 202
UDP-glucose concentration, controller of glycogen synthetase, 439
UDP-glucose pyrophosphorylase, lactose formation, 381
UDP-glucuronic acid, conjugating agent, 229
Ultracentrifuge, 59
Uncoupling agents, 145
Urea, dissociating agent, 72
   formation, 220
   $\Delta G°'$ hydrolysis, 45
   reabsorption and urine flow rate, 316
Urea cycle, 220
   absence in brain, 223
   location of enzymes, 222
Urease, 46, 75
Uric acid, formation, 192
Urine, composition, 314
UTP, structure, 202

L-Valine, structure, 26
van der Waals bonds, 53, 67, 70
Vasopressin, controlling urine volume, 321
all trans Vitamin A, 357
Vitamin A, bone metabolism, 272
   cartilage metabolism, 272
   deficiency, 357
   visual processes, 357
Vitamin B_1, 123, 209
Vitamin B_6, 214
Vitamin B_{12}, 204
Vitamin C, formation, 212
Vitamin D, formation from 7-dehydrocholesterol, 271
Vitamins, and enzyme activity, 443
   biochemical role, 444
   brain metabolism 330
Von Gierke's disease, 207

Water, body, 465
   dissociation constant, 54
   entropy, 53
   entropy and protein conformation, 67
   movement across membranes, 432
   protein structure, 72
   solvent, 53
   structure and properties, 52
Water absorption, 165
Water/O_2, $E°'$, 49
Weaning, associated enzyme changes, 447
White blood cells, 288

X-ray crystallography, 67

Xanthine, 192
Xanthine oxidase, 192
  inhibition, 193
Xylitol, formation, 212

Xylulose, formation, 212
Xylulose-5-phosphate, formation, 209

Z-line, 95